T0201177

The Physics and Mathematics
of Electromagnetic Wave Propagation
in Cellular Wireless Communication

The Physics and Mathematics of Electromagnetic Wave Propagation in Cellular Wireless Communication

Tapan K. Sarkar

Magdalena Salazar Palma

Mohammad Najib Abdallah

With Contributions from:

Arijit De

Walid Mohamed Galal Diab

Miguel Angel Lagunas

Eric L. Mokole

Hongsik Moon

Ana I. Perez-Neira

The right of Tapan K. Sarkar, Magdalena Salazar Palma and Mohammad Najib Abdallah to be identified as the authors of this work has been asserted in accordance with law.

Registered Offices
John Wiley & Sons, Inc., 111 River Street, Hoboken, NJ 07030, USA

Editorial Office
111 River Street, Hoboken, NJ 07030, USA

For details of our global editorial offices, customer services, and more information about Wiley products visit us at www.wiley.com.

Wiley also publishes its books in a variety of electronic formats and by print-on-demand. Some content that appears in standard print versions of this book may not be available in other formats.

Library of Congress Cataloging-in-Publication Data
Names: Sarkar, Tapan (Tapan K.), author. | Salazar Palma, Magdalena, author. |
    Abdallah, Mohammad Najib, 1983– author.
Title: The physics and mathematics of electromagnetic wave propagation in cellular wireless
    communication / Tapan K. Sarkar, Magdalena Salazar Palma, Mohammad Najib Abdallah ;
    with contributions from Arijit De, Walid Mohamed Galal Diab, Miguel Angel Lagunas,
    Eric L. Mokole, Hongsik Moon, Ana I. Perez-Neira.
Description: Hoboken, NJ, USA : Wiley, 2018. | Includes bibliographical references and index. |
Identifiers: LCCN 2017054091 (print) | LCCN 2018000589 (ebook) |
    ISBN 9781119393139 (pdf) | ISBN 9781119393122 (epub) | ISBN 9781119393115 (cloth)
Subjects: LCSH: Cell phone systems–Antennas–Mathematical models. |
    Radio wave propagation–Mathematical models.
Classification: LCC TK6565.A6 (ebook) | LCC TK6565.A6 S25 2018 (print) |
    DDC 621.3845/6–dc23
LC record available at https://lccn.loc.gov/2017054091

Cover design by Wiley
Cover image: © derrrek/Gettyimages

Set in 10/12pt Warnock by SPi Global, Pondicherry, India

Printed in the United States of America

V077696_062018

Contents

Preface

Wireless communication is an important area of research these days. However, the promise of wireless communication has not matured as expected. This is because some of the important principles of electromagnetics were not adhered to during system design over the years. Therefore, one of the objectives of this book is to describe and document some of the subtle electromagnetic principles that are often overlooked in designing a cellular wireless system. These involve both physics and mathematics of the concepts used in deploying antennas for transmission and reception of electromagnetic signals and selecting the proper methodology out of a plethora of scenarios. The various scenarios are but not limited to: is it better to use an electrically small antenna, a resonant antenna or multiple antennas in a wireless system? However, the fact of the matter as demonstrated in the book is that a single antenna is sufficient if it is properly designed and integrated into the system as was done in the old days of the transistor radios where one could hear broadcasts from the other side of the world using a single small antenna operating at 1 MHz, where an array gain is difficult to achieve!

The second objective of this book is to illustrate that the main function of an antenna is to capture the electromagnetic waves that are propagating through space and prepare them as a signal fed to the input of the first stage of the radio frequency (RF) amplifier. The reality is that if the signal of interest is not captured and available for processing at the input of the first stage of the RF amplifier, then application of various signal processing techniques cannot recreate that signal. Hence the modern introduction of various statistical concepts into this deterministic problem of electromagnetic wave transmission/reception is examined from a real system deployment point of view. In this respect the responses of various sensors in the frequency and the time domain are observed. It is important to note that the impulse response of an antenna is different in the transmit mode than in the receive mode. Understanding of this fundamental principle can lead one to transmit ultrawideband signals through space using a pair of antennas without any distortion. Experimental results are provided to demonstrate how a distortion free tens of gigahertz bandwidth signal can be transmitted and received to justify this claim. This technique can be achieved by recasting the Friis's transmission formula (after Danish-American radio engineer Harald Trap Friis) to an alternate form which clearly illustrates that if the physics of the transmit and receive antennas are factored in the channel modelling then the path loss can be made independent of frequency. The other important point to note is that in deploying an antenna in a real system one should focus on the radiation efficiency of the antenna and not on the maximum power transfer theorem which has resulted in the misuse of the S-parameters. Also two antennas which possess a century bandwidth (i.e., a 100:1 bandwidth) are also discussed.

The next topic that is addressed in the book is the illustration of the shortcomings of a MIMO system from both theoretical and practical aspects in the sense that it is difficult if not impossible to achieve simultaneously several orthogonal modes of transmission with good radiation efficiency. In this context, a new deterministic methodology based on the principle of reciprocity is presented to illustrate how a signal can be directed to a desired user and simultaneously be made to have nulls along the directions of the undesired ones without an explicit characterization of the operational environment. This is accomplished using an embarrassingly simple matrix inversion technique. Since this principle also holds over a band of frequencies, then the characterization of the system at the uplink frequency can be used to implement this methodology in the downlink or vice versa.

Another objective of the book is to point out that all measurements related to propagation path loss in electromagnetic wave transmission over ground illustrate that the path loss from the base station in a cellular environment is approximately 30 dB per decade of distance within the cell of a few Km in radius and the loss is 40 dB per decade outside this cell. This is true independent of the nature of the ground whether it be urban, suburban, rural or over water. Also the path loss in the cellular band appears to be independent of frequency. Therefore in order to propagate a signal from 1 m to 1 kilometer the total path loss, based on the 30 dB per decade of distance, is 90 dB. And compared to this free space path loss over Earth, the attenuation introduced by buildings, trees and so on has a second order effect as it is shown to be of the order of 30–40 dB. Even though this loss due to buildings, trees and the like is quite large, when compared to the free space path loss of approximately 90 dB over a 1 km, it is negligible! Also, the concept of slow fading appears to be due to interference of the direct wave from the transmitting antenna along with the ground wave propagation over earth and also emanating from it and generally occurs when majority of the cell area is located in a near field environment of the base station antenna. These concepts have been illustrated from a physics based view point developed over a hundred years ago by German theoretical physicist Arnold Johannes Wilhelm Sommerfeld and have been validated using experimental data where possible. Finally, it is shown how to reduce the propagation loss by deploying the transmitting antenna closer to the ground with a slight vertical tilt – a rotation about the horizontal axis – a very non-intuitive solution. Deployment of base station antennas high above the ground indeed provides a height-gain in the far field, but in the near field there is actually a height loss. Also, the higher the antenna is over the ground the far field starts further away from the transmitter.

Finally we introduce the concept of simultaneous transfer of information and power. The requirements for these two issues are contradictory in the sense that transmission of information is a function of the bandwidth of the system whereas the power transfer is related to the resonance of the system which is invariably of extremely narrow bandwidth. To this end, the various concepts of channel capacities are presented including those of an American mathematician and electrical engineer Claude Elwood Shannon, a Hungarian-British electrical engineer and physicist Dennis Gabor, and an American electrical engineer William G. Tuller. It is rather important to note that each one of these methodologies is suitable for a different operational environment. For example, the Shannon capacity is useful when one is dealing with transmission in the presence of thermal noise and Shannon's discovery made satellite communication possible. The Gabor channel capacity on the other hand is useful when a system is operating in the presence of interfering signals which is not white background noise. And finally the Tuller capacity is useful in a realistic near field noisy environment where the concept of power flow through the Poynting vector is a complex quantity. Since the Tuller capacity is defined in terms of the smallest discernable voltage levels that the first stage of the RF amplifier can handle and is not related to power, the Tuller formula can be and has been used in the design of a practical system. Tuller himself designed and constructed the first private ground to air communication system and it worked in the first trial and provided a transmission rate which was close to the theoretical design. It is also important to point out that in the development of the various properties of channel capacity it makes sense to talk about the rate of transmission only when one is using coding at the RF stage. To Shannon a transmitter was an encoder and not an RF amplifier and similarly the receiver was a decoder! Currently only two systems use coding at RF. One is satellite communication where the satellite is quite far away from the Earth and the other is in Global Positioning System (GPS) where the code is often gigabits long. In some radar systems, often a Barker code (R. H. Barker, "Group Synchronizing of Binary Digital Systems". *Communication Theory*. London: Butterworth, pp. 273–287, 1953) is used during transmission. It is also illustrated how the effect of matching using both conventional and non Foster type devices have an impact on the channel capacity of a system.

The book contains four chapters. In Chapter 1, the principle of electromagnetics is developed through the Maxwellian principles where it is illustrated that the superposition of power does not apply in electrical engineering. It is either superposition of the voltages or the currents (or electric and magnetic fields). The other concept is that the energy flow in a wire, when we turn on a switch to complete the electrical circuit, does not take place through the flow of electrons. For an alternating current (AC) system the electrons never actually leave the switch but simply move back and forth when an alternating voltage is applied to excite the circuit and cause an AC current flow. The energy flow is external to the wire where the electric and the magnetic fields reside and they travel at the speed of light in the given dielectric medium carrying the energy from the source to the load. Also, the transmitting and receiving responses of simple antennas both in time and frequency domains are presented to illustrate the various subtleties in their properties. Maxwell also developed and introduced the first statistical law into physics and formulated the concept of ensemble averaging. In this context, the concepts of information and channel capacity are related to the Poynting's theorem of electromagnetic energy transmission. This introduces the principle of conservation of energy into the domain of signal analysis which is missing in the context of information theory. The concepts of the various channel capacities are also introduced in this chapter.

In Chapter 2, the properties of an antenna in the frequency domain is described. These refer to the commonly used wire antennas. One of the major topic discussed is the difference between the near field and the far field of an antenna. Understanding of this basic principle is paramount to a good system design. Even though wireless communication has been an important area of research these days, one obvious conclusion one can reach is that the promise of wireless communication has not matured as expected. This is because some of the important principles of electromagnetics were not adhered to during system design over the years. The first of the promises has to do with the introduction of space division multiple access (SDMA) which really never matured. This section will illustrate why and how it is possible to do SDMA and why it has not happened to-date. This has to do with the definition of the radiation pattern of an antenna and that is only defined in the far field of the antenna as SDMA can only be carried out using antenna radiation patterns. This chapter will explain where does the far field of an antenna starts when the antenna is operating in free space and over a ground plane. In addition, it is illustrated that in designing an antenna the emphasis should be on maximizing the radiation efficiency and not put emphasis on the maximum power transfer principles. Under the input energy constraint, the radiation of electrically small versus resonant sized antennas is analyzed under different terminating conditions. In this context, both classical and non-Foster matching systems are described. Next the performance of antennas in free space and over an earth is discussed and it is shown that sometimes presence of obstacles in the direct line-of-sight path may actually enhance the signal levels. Also, the principle of antenna diversity and the use of multiple antennas over a single antenna is examined. This brings us to the topic of a multiple-input-multiple-output (MIMO) system and its performance in comparison to a single-input-single-output (SISO) is discussed. Finally, an embarrassingly simple solution based on the principle of reciprocity is presented to illustrate the competitiveness of this simple system in deployment both in terms of radiation efficiency and cost over a MIMO system.

Chapter 3 deals with the characterization of propagation path loss in a cellular wireless environment. The presentation starts with a summary of the various experimental results all of which demonstrate that inside a cell the radio wave propagation path loss is 30 dB per decade of distance and outside the cell it is 40 dB per decade. This is true irrespective of the nature of the ground whether it be rural, urban, suburban or over water. The path loss is also independent of the operating frequency in the cellular band, height of the base station antennas and so on. Measurement data also illustrate the effect of buildings, trees and the like to the propagation path loss is of a second order effect and that the major portion of the path loss is due to the propagation in space over ground. A theoretical macro model based on the classical Sommerfeld formulation can duplicate the various experimental data carried out by Y. Okumura and coworkers in 1968. This comparison can be made using a theoretical model based on the Sommerfeld formulation without any massaging in the details of the environment for transmission and reception. Thus, the experimental data generated by Y. Okumura and co-workers can be duplicated using the Sommerfeld theory. It is important to point out that there are also many statistical models but they do not conform to the results of the experimental data available. And based on the analysis using the macro model developed after Sommerfeld's classic century old analytical formulation, one can also explain the origin of slow fading which is due to the interference between the direct wave from the base station antenna and the reflection of the direct wave from the ground and occurs only in the near field of the transmitting antenna. The so called height gain occurs in the far field of a base station antenna deployment which is generally outside the cell of interest and in the near field within the cell there is actually a height loss, if the antenna is deployed high above the ground. It will also be illustrated using both theory and experiment that the signal strength within a cell can significantly be improved by lowering the height of the base station antenna towards the ground. Based on the evidences available both from theory and experiment, a novel method will be presented on how to deploy base-station antennas by lowering them towards the ground and then slightly tilting them towards the sky, which will provide improvement of the signal loss in the near field over current base station antenna deployments.

Chapter 4, the final chapter deals with ultrawideband antennas and the mechanisms of broadband transmission of both power and information. Broadband antennas are very useful in many applications as they operate over a wide range of frequencies. To this effect two century bandwidth antennas will be presented and their performances described. Then the salient feature of time domain responses of antennas will be outlined. If these subtleties in time domain antenna theory are followed it is possible to transmit gigahertz bandwidth signals over large distances without any distortion. As such, the phase responses of the antennas as a function of frequency are of great interest for wideband applications. Configurations and schematic of two century bandwidth antennas are presented. The radiation and reception properties of various conventional ultrawideband (UWB) antennas in the time domain are shown. Experimental results are provided to verify how to transmit and receive a tens of gigahertz bandwidth waveform without any distortion when propagating through space. It is illustrated how to generate a time limited ultrawideband pulse fitting the Federal Communication Commission (FCC) mask in the frequency domain and describe a transmit/receive system which can deal with such type of pulses without any distortion. Finally, simultaneous transmission of power and information is also illustrated and shown how their performances can be optimized over a finite band.

This book is intended for engineers, researchers and educators who are or planning to work in the field of wireless communications. The prerequisite to follow the materials of the book is a basic undergraduate course in the area of dynamic electromagnetic theory. Every attempt has been made to guarantee the accuracy of the contents of the book. We would however appreciate readers bringing to our attention any errors that may have appeared in the final version. Errors and/or any comments may be emailed to one of the authors, at tksarkar@syr.edu

Acknowledgments

Thanks are due to Ms. Rebecca Noble (Syracuse University) for her expert typing of the manuscript. Grateful acknowledgement is also made to Dr. John S. Asvestas for suggesting ways to improve the readability of the book.

Syracuse, New York
September 2017

Tapan K. Sarkar (tksarkar@syr.edu)
Magdalena Salazar Palma (salazar@tsc.uc3m.es)
Mohammad Najib Abdallah (mnabdall@syr.edu)

1

The Mystery of Wave Propagation and Radiation from an Antenna

Summary

An antenna is a structure that is made of material bodies that may consist of either conducting or dielectric materials or may be a combination of both. Such a structure should be matched to the source of the electromagnetic energy so that it can radiate or receive the electromagnetic field in an efficient manner. The interesting phenomenon is that an antenna displays selectivity properties not only in the frequency domain but also in the space domain. In the frequency domain an antenna is capable of displaying an external resonance phenomenon where at a particular frequency the current density induced on it can be sufficiently significant to cause radiation of electromagnetic fields from that structure. An antenna also possesses a spatial impulse response that is a function of both the azimuth and elevation angles. Thus, an antenna displays spatial selectivity as it generates a radiation pattern that can selectively transmit or receive electromagnetic energy along certain spatial directions in the far field as in the near field even a highly directive antenna has essentially an omnidirectional pattern with no selectivity. That is the reason researchers have been talking about space division multiple access (SDMA) where one directs a beam along the direction of the desired user but places a null along the direction of the undesired user. This has not materialized as we shall see in the next chapter as most of the base station antennas operate in the near field of an antenna. As a receiver of electromagnetic field, an antenna also acts as a spatial sampler of the electromagnetic fields propagating through space. The voltage induced in the antenna is related to the polarization and the strength of the incident electromagnetic fields. The objective of this chapter is to illustrate how an electromagnetic wave propagates and how an antenna extracts the energy from such a wave. In addition, it will be outlined why the antenna was working properly for the last few decades where one could receive electromagnetic energy from the various parts

The Physics and Mathematics of Electromagnetic Wave Propagation in Cellular Wireless Communication, First Edition. Tapan K. Sarkar, Magdalena Salazar Palma, and Mohammad Najib Abdallah.
© 2018 John Wiley & Sons, Inc. Published 2018 by John Wiley & Sons, Inc.

of the world (with the classical transistor radios) without any problems but now various deleterious effects have propped up which are requiring deployment of multiple antennas, which as we shall see does not make any sense! Is it an aberration in basic understanding of electromagnetic theory or is it related to new physics that has just recently been discovered in MIMO system and the like? Another goal is to demonstrate that the principle of superposition applies when using the reciprocity theorem but does not hold for the principle of correlation which represents power. In general, power cannot be simply added or subtracted in the context of electrical engineering. It is also illustrated that the impulse response of an antenna when it is transmitting, is different from its response when the same structure operates in the receive mode. This is in direct contrast to antenna properties in the frequency domain as the transmit radiation pattern is the same as the receive antenna pattern. An antenna provides the matching necessary between the various electrical components associated with the transmitter and receiver and the free space where the electromagnetic wave is propagating. From a functional perspective an antenna is thus analog to a loudspeaker, which matches the acoustic generation/receiving devices to the open space. However, in acoustics, loudspeakers and microphones are bandlimited devices and so their impulse responses are well behaved. On the other hand, an antenna is a high pass device and therefore the transmit and the receive impulse responses are not the same; in fact, the former is the time derivative of the latter. An antenna is like our lips, whose instantaneous change of shapes provides the necessary match between the vocal cord and the outside environment as the frequency of the voice changes. By proper shaping of the antenna structure one can focus the radiated energy on certain specific directions in space. This spatial directivity occurs only at certain specific frequencies, providing selectivity in frequency. The interesting point is that it is difficult to separate these two spatial and temporal properties of the antenna, even though in the literature they are treated separately. The tools that deal with the dual-coupled space-time analysis are called *Maxwell's equations*. We first present the background of Maxwell's equations and illustrate how to solve for them analytically. Then we utilize them in the subsequent sections and chapters to illustrate how to obtain the impulse responses of antennas both as transmitting and receiving elements and demonstrate their relevance in the saga of smart antennas. We conclude the section with a note on the channel capacity which evolved from the concept of entropy and the introduction of statistical laws (the concept of ensemble averaging) into physics by Maxwell himself. The three popular forms of the channel capacity due to Shannon, Gabor and Tuller are described and it is noted that for practical applications the Tuller form is not only relevant for practical use and can make direct connection with the electromagnetic physics but is also easy to implement as Tuller built the first "private line" communication link between the aircraft traffic controller and the aircraft under their surveillance and it worked.

1.1   Historical Overview of Maxwell's Equations

In the year 1864, James Clerk Maxwell (1831–1879) read his "Dynamical Theory of the Electromagnetic Field" [1] at the Royal Society (London). He observed theoretically that electromagnetic disturbance travels in free space with the velocity of light [1–7]. He then conjectured that light is a transverse electromagnetic wave by using dimensional analysis [7] as he did not have the boundary conditions to solve the wave equation except in source free regions. In his original theory Maxwell introduced 20 equations involving 20 variables. These equations together expressed mathematically virtually all that was known about electricity and magnetism. Through these equations Maxwell essentially summarized the work of Hans C. Oersted (1777–1851), Karl F. Gauss (1777–1855), André M. Ampère (1775–1836), Michael Faraday (1791–1867), and others, and added his own radical concept of displacement *current* to complete the theory.

Maxwell assigned strong physical significance to the magnetic vector and electric scalar potentials A and ψ, respectively (**bold** variables denote vectors; *italic* denotes that they are function of both time and space, whereas roman variables are a function of space only), both of which played dominant roles in his formulation. He did not put any emphasis on the sources of these electromagnetic potentials, namely the currents and the charges. He also assumed a hypothetical mechanical medium called *ether* to justify the existence of displacement currents in free space. This assumption produced a strong opposition to Maxwell's theory from many scientists of his time. It is well known that Maxwell's equations, as we know them now, do not contain any potential variables; neither does his electromagnetic theory require any assumption of an artificial medium to sustain his displacement current in free space. The original interpretation given to the displacement current by Maxwell is no longer used; however, we retain the term in honor of Maxwell. Although modern Maxwell's equations appear in modified form, the equations introduced by Maxwell in 1864 formed the foundation of electromagnetic theory, which together with his radical concept of displacement current is popularly referred to as *Maxwell's electromagnetic theory* [1–7]. Maxwell's original equations were modified and later expressed in the form we now know as Maxwell's equations independently by Heinrich Hertz (1857–1894) [8, 9] and Oliver Heaviside (1850–1925) [10]. Their work discarded the requirement of a medium for the existence of displacement current in free space, and they also eliminated the vector and scalar potentials from the fundamental equations. Their derivations were based on the impressed sources, namely the current and the charge. Thus, Hertz and Heaviside, independently, expressed Maxwell's equations involving only the four field vectors E, H, B, and D: the electric field intensity, the magnetic field intensity, the magnetic flux density, and the electric flux density or displacement, respectively. Although priority is given to

Heaviside for the vector form of Maxwell's equations, it is important to note that Hertz's 1884 paper [2] provided the Cartesian form of Maxwell's equations, which also appeared in his later paper of 1890 [3]. Thus, the coordinate forms of the four equations that we use nowadays were first obtained by Hertz [2, 7] in a scalar form in 1885 and then by Heaviside in 1888 in a vector form [9, 10].

It is appropriate to mention here that the importance of Hertz's theoretical work [2] and its significance appear not to have been fully recognized [5]. In this 1884 paper [2] Hertz started from the older action-at-a-distance theories of electromagnetism and proceeded to obtain Maxwell's equations in an alternative way that avoided the mechanical models that Maxwell used originally and formed the basis for all his future contributions to electromagnetism, both theoretical and experimental. In contrast to the 1884 paper where he derived them from first principles, in his 1890 paper [3] Hertz postulated Maxwell's equations rather than deriving them alternatively. The equations were written in component form rather than in the vector form as was done by Heaviside [10]. This new approaches of Hertz and Heaviside brought unparalleled clarity to Maxwell's theory. The four equations in vector notation containing the four electromagnetic field vectors are now commonly known as Maxwell's equations. However, Einstein referred to them as *Maxwell–Hertz–Heaviside equations* [6, 7].

Although the idea of electromagnetic waves was hidden in the set of 20 equations proposed by Maxwell, he had in fact said virtually nothing about electromagnetic waves other than light, nor did he propose any idea to generate such waves electromagnetically. It has been stated [6, Ch. 2, p. 24]: "*There is even some reason to think that he [Maxwell] regarded the electrical production of such waves an impossibility.*" There is no indication left behind by him that he believed such was even possible. Maxwell did not live to see his prediction confirmed experimentally and his electromagnetic theory fully accepted. The former was confirmed by Hertz's brilliant experiments, his theory received universal acceptance, and his original equations in a modified form became the language of electromagnetic waves and electromagnetics, due mainly to the efforts of Hertz and Heaviside [7].

Hertz discovered electromagnetic waves around the year 1888 [8]; the results of his epoch-making experiments and his related theoretical work (based on the sources of the electromagnetic waves rather than on the potentials) confirmed Maxwell's prediction and helped the general acceptance of Maxwell's electromagnetic theory. However, it is not commonly appreciated that "*Maxwell's theory that Hertz's brilliant experiments confirmed was not quite the same as the one Maxwell left at his death in the year 1879*" [6]. It is interesting to note how the relevance of electromagnetic waves to Maxwell and his theory prior to Hertz's experiments and findings are described in [6]: "*Thus Maxwell missed what is now regarded as the most exciting implication of his theory, and one with enormous practical consequences. That relatively long electromagnetic waves or perhaps light itself, could be generated in the laboratory with ordinary electrical apparatus was unsuspected through most of the 1870's."*

Maxwell's predictions and theory were thus confirmed by a set of brilliant experiments conceived and performed by Hertz, who generated, radiated (transmitted), and received (detected) electromagnetic waves of frequencies lower than light. His initial experiment started in 1887, and the decisive paper on the finite velocity of electromagnetic waves in air was published in 1888 [3]. After the 1888 results, Hertz continued his work at higher frequencies, and his later papers proved conclusively the optical properties (reflection, polarization, etc.) of electromagnetic waves and thereby provided unimpeachable confirmation of Maxwell's theory and predictions. English translation of Hertz's original publications [8] on experimental and theoretical investigation of electric waves is still a decisive source of the history of electromagnetic waves and Maxwell's theory. Hertz's experimental setup and his epoch-making findings are described in [9].

Maxwell's ideas and equations were expanded, modified, and made understandable after his death mainly by the efforts of Heinrich Hertz, George Francis Fitzgerald (1851–1901), Oliver Lodge (1851–1940), and Oliver Heaviside. The last three have been christened as "the Maxwellians" by Heaviside [7, 11].

Next we review the four equations that we use today due to Hertz and Heaviside, which resulted from the reformulation of Maxwell's original theory. Here in all the expressions we use SI units (Système International d'unités or International System of Units).

1.2 Review of Maxwell–Hertz–Heaviside Equations

The four Maxwell's equations are among the oldest sets of equations in mathematical physics, having withstood the erosion and corrosion of time. Even with the advent of relativity, there was no change in their form. We briefly review the derivation of the four equations and illustrate how to solve them analytically [12]. The four equations consist of Faraday's law, generalized Ampère's law, generalized Gauss's law of electrostatics, and Gauss's law of magnetostatics, respectively, along with the equation of continuity.

1.2.1 Faraday's Law

Michael Faraday (1791–1867) observed that when a bar magnet was moved near a loop composed of a metallic wire, there appeared to be a voltage induced between the terminals of the wire loop. In this way, Faraday showed that a magnetic field produced by the bar magnet under some special circumstances can indeed generate an electric field to cause the induced voltage in the loop of wire and there is a connection between the electric and magnetic fields. This physical principle was then put in the following mathematical form:

$$V = -\oint_L \boldsymbol{E} \cdot \mathbf{d}\ell = -\frac{\partial \Phi_m}{\partial t} = -\frac{\partial}{\partial t} \iint_S \boldsymbol{B} \cdot \mathbf{d}s \tag{1.1}$$

where: V = voltage induced in the wire loop of length L,
     $\mathbf{d}\ell$ = differential length vector along the axis of the wire loop,
     $\boldsymbol{E}$ = electric field along the wire loop,
     Φ_m = magnetic flux linkage with the loop of surface area S,
     $\boldsymbol{B}$ = magnetic flux density,
     S = surface over which the magnetic flux is integrated (this surface is bounded by the contour of the wire loop),
     L = total length of the loop of wire,
     $\bullet$ = scalar dot product between two vectors,
     $\mathbf{d}s$ = differential surface vector normal to the surface.

This is the integral form of Faraday's law, which implies that this relationship is valid over a region. It states that the line integral of the electric field is equivalent to the rate of change of the magnetic flux passing through an open surface S, the contour of which is the path of the line integral. In this chapter, the variables in italic, for example $\boldsymbol{B}$, indicate that they are functions of four variables, x, y, z, t. This consists of three space variables (x, y, z) and a time variable, t. When the vector variable is written as $\mathbf{B}$, it is a function of the three spatial variables (x, y, z) only. This nomenclature between the variables denoted by italic as opposed to roman is used to distinguish their functional dependence on spatial-temporal variables or spatial variables, respectively.

To extend this relationship to a point located in a space, we now establish the differential form of Faraday's law by invoking Stokes' theorem for the electric field. Stokes' theorem relates the line integral of a vector over a closed contour to a surface integral of the curl of the vector, which is defined as the rate of spatial change of the vector along a direction perpendicular to its orientation (which provides a rotary motion, and hence the term curl was first introduced by Maxwell), so that

$$\oint_L \boldsymbol{E} \cdot \mathbf{d}\ell = \iint_S (\nabla \times \boldsymbol{E}) \cdot \mathbf{d}s \tag{1.2}$$

where the curl of a vector in the Cartesian coordinates is defined by

$$
\nabla \times E(x, y, z, t) = determinant\ of\ \begin{vmatrix} \hat{x} & \hat{y} & \hat{z} \\ \dfrac{\partial}{\partial x} & \dfrac{\partial}{\partial y} & \dfrac{\partial}{\partial z} \\ E_x & E_y & E_z \end{vmatrix}
$$

$$
= \hat{x}\left[\frac{\partial E_z}{\partial y} - \frac{\partial E_y}{\partial z}\right] + \hat{y}\left[\frac{\partial E_x}{\partial z} - \frac{\partial E_z}{\partial x}\right] + \hat{z}\left[\frac{\partial E_y}{\partial x} - \frac{\partial E_x}{\partial y}\right]
$$

(1.3)

Here $\hat{x}, \hat{y}$, and $\hat{z}$ represent the unit vectors along the respective coordinate axes, and E_x, E_y, and E_z represent the x, y, and z components of the electric field intensity along the respective coordinate directions. The surface S is limited by the contour L. ∇ stands for the operator $[\hat{x}(\partial/\partial x) + \hat{y}(\partial/\partial y) + \hat{z}(\partial/\partial z)]$. Using (1.2), (1.1) can be expressed as

$$
\oint_L E \cdot d\ell = \iint_S (\nabla \times E) \cdot ds = -\frac{\partial}{\partial t}\iint_S B \cdot ds
$$

(1.4)

If we assume that the surface S does not change with time and in the limit making it shrink to a point, we get Faraday's law at a point in space and time as

$$
\nabla \times E(x, y, z, t) = \frac{1}{\varepsilon}\nabla \times D(x, y, z, t)
$$

$$
= -\frac{\partial B(x, y, z, t)}{\partial t} = -\mu\frac{\partial H(x, y, z, t)}{\partial t}
$$

(1.5)

where the constitutive relationships (here ε and μ are assumed to be constant of space and time) between the flux densities and the field intensities are given by

$$
B = \mu H = \mu_0 \mu_r H
$$
(1.6a)

$$
D = \varepsilon E = \varepsilon_0 \varepsilon_r E
$$
(1.6b)

D is the electric flux density and H is the magnetic field intensity. Here, ε_0 and μ_0 are the permittivity and permeability of vacuum, respectively, and ε_r and μ_r are the relative permittivity and permeability of the medium through which the wave is propagating.

Equation (1.5) is the point form or the differential form of Faraday's law or the first of the four Maxwell's equations. It states that at a point the negative rate of the temporal variation of the magnetic flux density is related to the spatial change of the electric field along a direction perpendicular to the orientation of the electric field (termed the curl of a vector) at that same point.

1.2.2 Generalized Ampère's Law

André M. Ampère observed that when a current carrying wire is brought near a magnetic needle, the magnetic needle is deflected in a very specific way determined by the direction of the flow of the current with respect to the magnetic needle. In this way Ampère established the complementary connection with the magnetic field generated by an electric current created by an electric field that is the result of applying a voltage difference between the two ends of the wire. Ampère first illustrated how to generate a magnetic field using the electric field or current. Ampère's law can be stated mathematically as

$$I = \oint_L H \cdot d\ell \tag{1.7}$$

where I is the total current encircled by the contour. We call this the *generalized Ampère's law* because we are now using the total current, which includes the displacement current due to Maxwell and the conduction current. The conduction current flows in conductors whereas the displacement currents flow in dielectrics or in material bodies. In principle, Ampère's law is connected strictly with the conduction current. Since we use the term *total current*, we use the prefix *generalized* as it is a sum of both the conduction and displacement currents. Therefore, the line integral of *H*, the magnetic field intensity along any closed contour L, is equal to the total current flowing through that contour.

To obtain a point form of Ampère's law, we employ Stokes' theorem to the magnetic field intensity and integrate the current density *J* over a surface to obtain

$$I = \iint_S J \cdot ds = \oint_L H \cdot d\ell = \iint_S (\nabla \times H) \cdot ds = \frac{1}{\mu} \iint_S (\nabla \times B) \cdot ds \tag{1.8}$$

This is the integral form of Ampère's law, and by shrinking S to a point, one obtains a relationship between the electric current density and the magnetic field intensity at the same point, resulting in

$$J(x,y,z,t) = \nabla \times H(x,y,z,t) \tag{1.9}$$

Physically, it states that the spatial derivative of the magnetic field intensity along a direction perpendicular to the orientation of the magnetic field intensity is related to the electric current density at that point. Now the electric current density *J* may consist of different components. This may include the conduction current (current flowing through a conductor) density J_c and displacement current density (current flowing through air, as from a transmitter to a receiver without any physical connection, or current flowing through the dielectric between the plates of a capacitor or in any material bodies) J_d, in addition to an externally applied impressed current density J_i. So in this case we have

$$J = J_i + J_c + J_d = J_i + \sigma E + \frac{\partial D}{\partial t} = \nabla \times H \tag{1.10}$$

where D is the electric flux density or electric displacement and σ is the electrical conductivity of the medium. The conduction current density is given by *Ohm's law*, which states that at a point the conduction current density is related to the electric field intensity by

$$J_c = \sigma E \tag{1.11}$$

The displacement current density introduced by Maxwell is defined by

$$J_d = \frac{\partial D}{\partial t} \tag{1.12}$$

We are neglecting the convection current density, which is due to the diffusion of the charge density at that point. We consider the impressed current density only as the source of all the electromagnetic fields.

1.2.3 Gauss's Law of Electrostatics

Karl Friedrich Gauss established the following relation between the total charge enclosed by a surface and the electric flux density or displacement D passing through that surface through the following relationship:

$$\oiint_S D \cdot \mathrm{d}s = Q \tag{1.13}$$

where integration of the electric displacement is carried over a closed surface and is equal to the total charge Q enclosed by that surface S.

We now employ the divergence theorem. This is a relation between the flux of a vector function through a closed surface S and the integral of the divergence of the same vector over the volume V enclosed by S. The divergence of a vector is the rate of change of the vector along its orientation. It is given by

$$\oiint_S D \cdot \mathrm{d}s = \iiint_V \nabla \cdot D \; \mathrm{d}v \tag{1.14}$$

Here $\mathrm{d}v$ represents the differential volume, whereas $\mathbf{d}s$ defines the surface element with a unique well-defined normal that points away to the exterior of the volume. In Cartesian coordinates the divergence of a vector, which represents the rate of spatial variation of the vector along its orientation, is given by

$$\nabla \cdot \boldsymbol{D} = \left[\hat{x} \frac{\partial}{\partial x} + \hat{y} \frac{\partial}{\partial y} + \hat{z} \frac{\partial}{\partial z} \right] \cdot \left[\hat{x} D_x + \hat{y} D_y + \hat{z} D_z \right]$$

$$= \frac{\partial D_x(x, y, z, t)}{\partial x} + \frac{\partial D_y(x, y, z, t)}{\partial y} + \frac{\partial D_z(x, y, z, t)}{\partial z} \tag{1.15}$$

So the divergence ($\nabla \cdot$) of a vector represents the spatial rate of change of the vector along its direction, and hence it is a scalar quantity, whereas the curl defined mathematically by ($\nabla \times$) of a vector is related to the rate of spatial change of the vector perpendicular to its orientation, which is a vector quantity and so possesses both a magnitude and a direction. All of the three definitions of *grad*, *Div* and *curl* were first introduced by Maxwell.

By applying the divergence theorem to the vector $\boldsymbol{D}$, we get

$$\oint_S \boldsymbol{D} \cdot d\boldsymbol{s} = \iiint_V \nabla \cdot \boldsymbol{D} \, dv = Q = \iiint_V q_v \, dv \tag{1.16}$$

Here q_v is the volume charge density and V is the volume enclosed by the surface S. Therefore, if we shrink the volume in (1.16) to a point, we obtain

$$\nabla \cdot \boldsymbol{D} = \frac{\partial D_x(x, y, z, t)}{\partial x} + \frac{\partial D_y(x, y, z, t)}{\partial y} + \frac{\partial D_z(x, y, z, t)}{\partial z} = q_v(x, y, z, t) \tag{1.17}$$

This implies that the rate change of the electric flux density along its orientation is influenced by the presence of a free charge density only at that point.

1.2.4 Gauss's Law of Magnetostatics

Gauss's law of magnetostatics is similar to the law of electrostatics defined in Section 1.2.3. If one uses the closed surface integral for the magnetic flux density $\boldsymbol{B}$, its integral over a closed surface is equal to zero, as no free magnetic charges occur in nature. Typically, magnetic charges appear as pole pairs. Therefore, we have

$$\oiint \boldsymbol{B} \cdot d\boldsymbol{s} = 0 \tag{1.18}$$

From the application of the divergence theorem to (1.18), one obtains

$$\iiint_V \nabla \cdot \boldsymbol{B} \, dv = 0 \tag{1.19}$$

which results in

$$\nabla \cdot \boldsymbol{B} = 0 \tag{1.20}$$

Equivalently in Cartesian coordinates, this becomes

$$\frac{\partial B_x(x, y, z, t)}{\partial x} + \frac{\partial B_y(x, y, z, t)}{\partial y} + \frac{\partial B_z(x, y, z, t)}{\partial z} = 0 \tag{1.21}$$

This completes the presentation of the four equations, which are popularly referred to as Maxwell's equations, which really were developed by Hertz in scalar form and cast by Heaviside into the vector form that we use today. These four equations relate all the spatial-temporal relationships between the electric and magnetic fields. In addition, we often add the equation of continuity, which is presented next.

1.2.5 Equation of Continuity

Often, the equation of continuity is used in addition to equations (1.18)–(1.21) to relate the impressed current density J_i to the free charge density q_v at that point. The equation of continuity states that the total current is related to the negative of the time derivative of the total charge by the following relationship

$$I = \frac{-\partial Q}{\partial t} \tag{1.22}$$

By applying the divergence theorem to the current density, we obtain

$$I = \oiint_S J \cdot ds = \iiint_V (\nabla \cdot J) dv = \frac{-\partial}{\partial t} \iiint_V q_v \, dv \tag{1.23}$$

Now shrinking the volume V to a point results in

$$\nabla \cdot J = \frac{-\partial q_v}{\partial t} \tag{1.24}$$

In Cartesian coordinates this becomes

$$\frac{\partial J_x(x, y, z, t)}{\partial x} + \frac{\partial J_y(x, y, z, t)}{\partial y} + \frac{\partial J_z(x, y, z, t)}{\partial z} = -\frac{\partial q_v(x, y, z, t)}{\partial t} \tag{1.25}$$

This states that there will be a spatial change of the current density along the direction of its flow if there is a temporal change in the charge density at that point. Next we obtain the solution of Maxwell's equations.

1.3   Development of Wave Equations

To obtain the electromagnetic wave equation, which every propagating wave must satisfy, we summarize the laws of Maxwell's equations developed in the last section. This is necessary to visualize the fundamental properties related to wave propagation so that it does not lead to any erroneous conclusions! In free space where there are no available sources the Maxwell's equations can be written as

$$\nabla \times E = -\frac{\partial B}{\partial t}; \quad \mu_0 \varepsilon_0 \frac{\partial E}{\partial t} = \nabla \times B; \quad \nabla \cdot E = 0; \quad \nabla \cdot B = 0 \qquad (1.26)$$

Now taking the curl of the first two curl equations one obtains

$$\nabla \times \nabla \times E = -\frac{\partial(\nabla \times B)}{\partial t} = -\mu_0 \varepsilon_0 \frac{\partial^2 E}{\partial t^2}; \qquad (1.27)$$

$$\nabla \times \nabla \times B = \mu_0 \varepsilon_0 \frac{\partial(\nabla \times E)}{\partial t} = -\mu_0 \varepsilon_0 \frac{\partial^2 B}{\partial t^2} \qquad (1.28)$$

Next we use the Laplacian of a vector Π as

$$\nabla \times \nabla \times \Pi = \nabla(\nabla \cdot \Pi) - \nabla^2 \Pi \qquad (1.29)$$

Since the divergence of either E or B in free space is zero (i.e., the first term in the vector identity drops out), one can then obtain

$$\nabla \times \nabla \times E = -\nabla^2 E = -\mu_0 \varepsilon_0 \frac{\partial^2 E}{\partial t^2} = -\frac{1}{c_0^2} \frac{\partial^2 E}{\partial t^2}; \quad \text{where} \quad \frac{1}{c_0^2} = \mu_0 \varepsilon_0; \quad (1.30)$$

Similarly,

$$-\nabla^2 B = -\frac{1}{c_0^2} \frac{\partial^2 B}{\partial t^2} \qquad (1.31)$$

Here c_0 is the velocity of light in free space. The spatial and the temporal derivatives for E and B constitute the wave equations in free space for an electromagnetic wave and the speed of light in free space is $c_0 = 2.99 \times 10^8$ m/sec.

In one dimension, the wave equation is reduced to

$$\frac{\partial^2 B}{\partial x^2} = \frac{1}{c_0^2} \frac{\partial^2 B}{\partial t^2} \qquad (1.32)$$

A general solution to the electromagnetic wave equation is a linear superposition of waves of the form $B(x,t)$ where this function can have any of the two special form as

$$B(x, t) = f(x - c_0 t) = F(t - x/c_0).$$ (1.33)

The function f denotes a fixed pattern in x which travels towards the positive x-direction with a speed c_0. This is illustrated in Figure 1.1 where the wave-shape is propagating. The other function F states equivalently the same thing! So that if an observer is located at a point on this function of f, then the observer's movement will occur at the phase velocity of the waveform.

An electromagnetic wave can be imagined to compose of a propagating transverse wave of oscillating electric (E) and magnetic fields (B). As shown in Figure 1.2 the electromagnetic wave is propagating from left to right (along the x-axis). The electric field E is along a vertical plane (y-axis) and the magnetic field B is in a horizontal plane (z-axis). The electric and the magnetic fields in an electromagnetic propagating wave are always in temporal phase but spatially displaced by 90°. The direction of the propagation of the wave is orthogonal to the directions of both the electric and the magnetic fields. This is displayed in Figure 1.2.

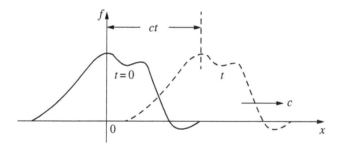

Figure 1.1 A general solution of the wave equation.

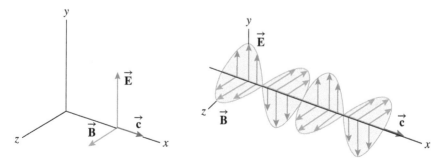

Figure 1.2 Propagation of an electromagnetic wave in free space.

A particular solution of (1.30) for the electric field is of the form $E = E_m \sin(kx - \omega t)$ and is illustrated in Figure 1.2. The magnetic field is of a similar form $B = B_m \sin(kx - \omega t)$ where the subscripts m represents a magnitude. So that in this case, the ratio $E_m/B_m = c_0$ represents the velocity of light in vacuum and $E_m/H_m = \eta = 377\Omega$ represents the characteristic impedance of free space. The magnetic field B is perpendicular to the electric field E in the orientation and where the vector product $E \times B$ is along the direction of the propagation of the wave. As illustrated in Figure 1.2 when a wave is propagating in free space as in a wireless communication scenario, the wave shape moves with time and space and hence the location of neither the minimum nor the maximum in both space and time are stationary. In other words the wave pattern changes as a function of time and space as shown in Figure 1.3 so that the location of the maxima and the minima are not fixed. This is the property of an alternating current wave and is not in any way related to fading.

The component solution (1.33) represents a propagating electromagnetic wave in free space. It tells us that both the maximum and the minimum of the wave moves in time and space. Therefore if its value is zero at a particular instant of space and time it may not be zero at the next spatio-temporal instance. So there is no fading associated with a travelling wave as its property for propagation is that it changes not only its amplitude continuously but also its position of minima and maxima. So there is no stationary point at which the field is always zero. Hence, it is difficult to conceive then how can one attribute the property of fading to such a signal!

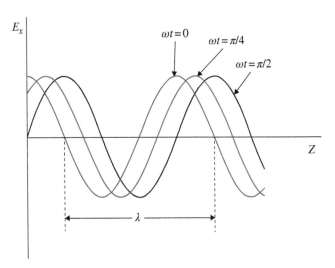

Figure 1.3 Temporal and spatial variation of a propagating electromagnetic wave in free space.

Since the wave equation contains only c_0^2, changing the sign of c_0 makes no difference in the final result. In fact, the *most general* solution of the one-dimensional wave equation is the sum of two arbitrary functions, both of which has to be twice differentiable with space and time. This results in the solution

$$B(x, t) = f(x - ct) + \Im(x + ct) \tag{1.34}$$

holds in general. The first term represents a wave travelling toward positive x, and the second term constitutes an arbitrary wave travelling toward negative x direction. The general solution is the superposition of two such waves both existing at the same time. Although the function f can be and often is a monochromatic sine wave, it does not have to be sinusoidal, or even periodic. In practice, f cannot have infinite periodicity because any real electromagnetic wave must always have a finite extent in time and space. As a result, and based on the theory of Fourier decomposition, a real wave must consist of the superposition of an infinite set of sinusoidal frequencies.

When a forward going wave (given by the first term in (1.34)) interacts with a backward going wave or equivalently a reflected wave (given by the second term in (1.34)) then one obtains a standing wave where the position of the maxima and the minima in amplitude does not change as a function of position even though its amplitude changes as a function of time. Hence, it is a wave that oscillates in time but has a stationary spatial dependence. In that case one may encounter locations of zero field strength for all times but for that to occur one has to operate in an environment which has multiple reflections. Reflections from buildings, trees and the like which are located in an open environment, it is most probable that a standing wave does not occur in such circumstances as seen in the Chapter 2 as one is operating in the near field of the transmitting antenna where there are no pattern nulls and the rays are not defined.

Another point to be made here is the following: when one turns on a switch in the power line how does the energy travel on the wires? The contribution of Maxwell which is often missed in this context is that the energy does not travel through the electrons in the wires but through the E and B fields which reside outside the wire and they essentially travel at the speed of light in the media in which they are located. The electrons in the conducting wire travel typically at a velocity given by $v = I/(n A Q)$, where v is the velocity, I is the magnitude of the current flow, n is the number of atoms in a cubic meter of the conductor, A is the cross section area of the wire and Q is the charge of the electron. So for a current flow of 1 A in a copper wire of radius 1 mm, $n = 8.5 \times 10^{28}$ m^3 and the charge on the electron being -1.6×10^{-19} Coulombs, the velocity of the flow of electrons becomes $v = [8.5 \times 10^{28} \times (\pi \times 10^{-6}) \times (-1.6 \times 10^{-19})]^{-1} = -0.000023$ m/sec. When a DC voltage is applied, the electron velocity will increase in proportion to the strength of the electric field. AC voltages cause no net movement of the electrons as they oscillate back and forth in response to the alternating electric field (over a distance of a few micrometers). For a 60 Hz alternating current, this means that within half a cycle the electrons drift less than 0.2 μm in a copper conductor. In other words, electrons flowing across the contact point in a switch will never actually leave the switch. In contrast to this slow velocity, individual velocity of the electron at room temperature in the absence of an electric field is ~1570 km/sec.

So the energy transmission in an electrical engineering context is due to the propagating electric and the magnetic fields at the velocity of light and is in no way related to the flow of electrons. The electric and the magnetic fields actually exist and propagate outside the structure. This follows directly from Maxwell's theory and it is this philosophy that revolutionized twentieth century science.

1.4   Methodologies for the Solution of the Wave Equations

The wave equation is a differential equation containing both space and time variables and its unique solution can be obtained from the specific boundary conditions for the fields which should be given for the nature of problem at hand. The solution is quite complex when the excitations are arbitrary functions of time. However, when the fields are AC, that is when the time variation of the fields is harmonic of time, the mathematical analysis can be simplified by using complex quantities and invoking Euler's identity which is given by

$$e^{j\alpha} = \cos\alpha + j\sin\alpha \tag{1.35}$$

where $j = \sqrt{-1}$. This relates real sinusoidal functions with the complex exponential function. Any AC quantity can be represented by a complex quantity. A scalar quantity is then interpreted according to

$$v = \sqrt{2}|V|\cos(\omega t + \alpha) = \sqrt{2}\Re e\left(Ve^{j\omega t}\right) \tag{1.36}$$

where v is called the *instantaneous quantity* and $V = |V|e^{j\alpha}$ is called the complex quantity. The notation $\Re e(\bullet)$ stands for "the real part of", that is the part not associated with the imaginary part j. It is important to note that the convention $v = \sqrt{2}\,\Im m(Ve^{j\omega t})$ can also be used, where the notation $\Im m(\bullet)$ stands for "the imaginary part of". The factor $\sqrt{2}$ can be omitted if it is desired that $|V|$ be the peak value of v instead of the rms (root mean square) value.

The other names for V are *phasor quantity* and vector quantity, the last name causing confusion with space vectors. Awful past practices that refuse to go away! In our notation v represents a voltage and is a real number which is a function of time, and hence V is a complex voltage in frequency domain.

As an example consider the waveform that is being used in power frequencies. If we say the voltage is 110 V at 50 Hz, then v represents a sinusoidal voltage whose amplitude is 110 and the waveform is similar to the one shown in Figure 1.3. Here we are restricting ourselves to the voltage and hence to the propagating electric field. To observe this waveform we need to use an oscilloscope where the sinusoidal waveform will be displayed illustrating that this voltage waveform changes as a function of time at the location where we are observing the time varying voltage waveform. This in no way implies that the waveform is displaying fading characteristics as its amplitude changes as a function of time. It is a time domain representation and the waveform should change as a function of time in an alternating current waveform. Therefore, for meaningful measurements that illustrate fading, the measurements should always be carried out in the frequency domain displaying the phasor quantities, particularly its magnitude which is generally a rms (root mean square) value and the associated phase. This magnitude and phase characteristics can be measured by a vector voltmeter or by a vector network analyzer and assuming that the waveform being watched is not modulated. And if this rms value changes with time then it is meaningful to claim that this wave has an amplitude that is time dependent — in other words there may be fading. The phasor voltage V can be measured by a vector voltmeter which will display a reading of 110 and the needle of the voltmeter will remain stationary and will not change with time, even though the waveform is alternating. Now if V is changing with time then the vector voltmeter needle position will change with time and this implies that that there is some sort of variation in the amplitude of the waveform. Then this situation can be characterized by a waveform going through a fade. What is termed slow fading in wireless communication is the interference between the various vector components of the fields (both direct and the reflections from the Earth) and not necessarily multipath components of the signal. As discussed in Chapter 2, multipaths are ray representation of the propagation of the signals and this analysis can only be done in the far field of the antenna. This will be addressed in Chapter 2. Therefore the misuse of the term fading in wireless communication comes because of its misinterpretation and misconception of the fundamentals of electrical engineering principles and thus making a wrong association with a phenomenon that occurs in long wave radio communication where the change of the signal amplitude occurs due to its reflection from the ionosphere and the temporal variation of the electrical properties of the ionosphere which takes place over a time period of seconds or even hours and nothing happens at the milliseconds scale!

In addition, the term fast fading is a complete erroneous characterization of the physics of electromagnetics as it is trying to relate this to the properties of a transient travelling wave! By characterizing a waveform in the time domain one is simply misinterpreting and erroneously interpreting natural phenomenon of the properties of a propagating wave in a nonscientific way.

Finally, there is another phenomenon which is wrongly associated with fast fading and that is Doppler. And again it is due to not comprehending the fundamental physics of the complete problem! The Doppler frequency is the shift in the carrier frequency when the source or the receiver is moving with a finite velocity. If the source/ receiver is moving with a velocity ϑm/sec then the carrier frequency of the source will display a Doppler shift in frequency of the signal and is given by

$$f_d = \frac{2\vartheta}{\lambda} \text{Hz} \tag{1.37}$$

where λ is the wavelength of the original frequency. The shift in frequency is either positive or negative depending on whether the source/receiver is moving towards or away from each other. So if the frequency source is moving at a relative velocity of 360 kM/hour (= 100 m/sec) which is beyond any real movement in a wireless communication scenario, but may exist more in one's dreams or in an euphoric nonconscious state, then the Doppler shift of a 1 GHz carrier frequency will be $f_d = \frac{2 \times 100}{0.3} = 666\text{Hz}$. So the Doppler shift is equal to 666 Hz in a carrier frequency of 10^9 Hz when one is moving at 360 km/hour. If one takes the best crystal oscillator available in the market, its frequency stability can be at best 1 part in a million implying that the carrier frequency of 1 GHz may vary within $10^9 \pm 10^3$ Hz. So, if the crystal oscillator of 1 GHz is moving at 360 km/hour, the Doppler shift in its frequency is only 666 Hz. Thus, the fact of the matter is that a variation of less than 1 kHz in frequency is simply impossible to visualize in a 1 GHz carrier. Hence fast fading due to Doppler is at best a mythology!

Another point to be made here is that one can look at the expression of the propagation of a wave either from the time domain or in the frequency domain using phasors. It is not possible to combine them in any way! However one will find phrases like this in some modern text books on wireless communication: "*In response to a transmitted sinusoid cos (2πft), we can express the electric far field at time t as* $E\left[f, t, (r, \theta, \varphi)\right] = \dfrac{\alpha_s\left(\theta, \varphi, f\right)\cos\left\{2\pi f\left(t - r/c\right)\right\}}{r}$. *Here* (r, θ, φ) *represents the point u in space at which the electric field is being measured, where r is the distance between the transmit antenna to u and where* (θ, φ) *represents the vertical and horizontal angles from the antenna to u, respectively. The constant c is the speed of light, and* $\alpha_s(r, \theta, \varphi)$ *is the radiation pattern of the sending Antenna at frequency f in the direction* (θ, φ); *it also contains a scaling factor to account for antenna losses. The phase of the field varies with* (fr/c) *corresponding to the delay caused by the radiation travelling at the speed of light.*"

Such a representation has no meaning in electrical engineering and one has to be careful in what one reads in many textbooks on wireless communication these days!! In summary, the voltage across an inductor is $L\dfrac{di(t)}{dt}$ in the temporal domain and in the phasor domain it is $j\omega L I(\omega)$. However, one cannot write it as $j\omega L i(t)$- this is actually a meaningless expression in electrical engineering

1.5   General Solution of Maxwell's Equations

Instead of solving the four coupled differential Maxwell's equations directly dealing with the electric and magnetic fields, we introduce two additional variables A and ψ. Here A is the magnetic vector potential and ψ is the scalar electric potential. The introduction of these two auxiliary variables facilitates the solution of the four coupled differential equations.

We start with the generalized Gauss's law of magnetostatics, which states that

$$\nabla \cdot \mathbf{B}(x, y, z, t) = 0 \tag{1.38}$$

Since the divergence of the curl of any vector A is always zero, that is,

$$\nabla \cdot \nabla \times \mathbf{A}(x, y, z, t) = 0 \tag{1.39}$$

one can always write

$$\mathbf{B}(x, y, z, t) = \nabla \times \mathbf{A}(x, y, z, t) \tag{1.40}$$

which states that the magnetic flux density can be obtained from the curl of the magnetic vector potential A. So if we can solve for A, we obtain B by a simple differentiation. It is important to note that at this point A is still an unknown quantity. In Cartesian coordinates this relationship

$$\mathbf{B}(x, y, z, t) = \hat{x}B_x(x, y, z, t) + \hat{y}B_y(x, y, z, t) + \hat{z}B_z(x, y, z, t)$$

$$= determinant\ of \begin{bmatrix} \hat{x} & \hat{y} & \hat{z} \\ \dfrac{\partial}{\partial x} & \dfrac{\partial}{\partial y} & \dfrac{\partial}{\partial z} \\ A_x & A_y & A_z \end{bmatrix}$$

which is known as the *Lorenz gauge condition* [13]. It is important to note that this is not the only constraint that is possible between the two newly introduced variables A and ψ in our solution procedure. This choice which we have made is only a particular assumption, and other choices will yield different forms of the solution of the Maxwell–Hertz –Heaviside equations. Interestingly, Maxwell in his treatise [1] chose the Coulomb gauge [7], which is generally used for the solution of static problems and not for dynamic time varying problems.

Next, we observe that by using (1.52) in (1.51), one obtains

$$(\nabla\cdot\nabla)A - \mu\varepsilon\frac{\partial^2 A}{\partial t^2} = -\mu\, J_i \tag{1.53}$$

In summary, the solution of Maxwell's equations starts with the solution of equation (1.53) first, for A, given the impressed current J_i. Then the scalar potential ψ is solved for by using (1.52). Once A and ψ are obtained, the electric and magnetic field intensities are derived from

$$H = \frac{1}{\mu}B = \frac{1}{\mu}\nabla\times A \tag{1.54}$$

$$E = -\frac{\partial A}{\partial t} - \nabla\psi \tag{1.55}$$

This completes the solution in the time domain, even though we have not yet provided an explicit form of the solution. We now derive the explicit form of the solution in the frequency domain and from that obtain the time domain representation. We assume the temporal variation of all the fields to be time harmonic in nature, so that

$$E(x, y, z, t) = E(x, y, z)e^{j\omega t} \tag{1.56}$$

$$B(x, y, z, t) = B(x, y, z)e^{j\omega t} \tag{1.57}$$

where $\omega = 2\pi f$ and f is the frequency (whose unit is Hertz abbreviated as Hz) of the electromagnetic fields. By assuming a time variation of the form $e^{j\omega t}$, we now have an explicit form for the time differentiations, resulting in

$$\frac{\partial}{\partial t}\Big[A(x, y, z, t)\Big] = \frac{\partial}{\partial t}\Big[A(x, y, z)e^{j\omega t}\Big] = j\omega A(x, y, z)e^{j\omega t} \tag{1.58}$$

Therefore, (1.52) and (1.53) are simplified in the frequency domain after eliminating the common time variations of $e^{j\omega t}$ from both sides to form

$$H(x, y, z) = \frac{1}{\mu}B(x, y, z) = \frac{1}{\mu}\nabla\times A(x, y, z) \tag{1.59}$$

$$E(x, y, z) = -j\omega A(x, y, z) - \nabla\psi(x, y, z) \tag{1.60}$$

Furthermore, in the frequency domain (1.52) transforms into

$$\nabla \cdot A + j\omega\mu\varepsilon\,\psi = 0$$

or equivalently,

$$\psi = -\frac{\nabla \cdot A}{j\omega\mu\varepsilon} \tag{1.61}$$

In the frequency domain, (1.53) transforms into

$$\nabla^2 A + \omega^2\mu\varepsilon A = -\mu J_i \tag{1.62}$$

The solution for A in (1.62) can now be written explicitly in an analytical form as illustrated in [14, 15] as

$$A(x, y, z) = \frac{\mu}{4\pi} \int_V \frac{J_i(x', y', z')\,e^{-jkR}}{R}\,dv' \tag{1.63}$$

where

$$\mathbf{r} = \hat{x}x + \hat{y}y + \hat{z}z \tag{1.64}$$

$$\mathbf{r}' = \hat{x}x' + \hat{y}y' + \hat{z}z' \tag{1.65}$$

$$R = |\mathbf{r} - \mathbf{r}'| = \sqrt{(x-x')^2 + (y-y')^2 + (z-z')^2} \tag{1.66}$$

$$k = \frac{2\pi}{\lambda} = \frac{2\pi f}{c} = \sqrt{\omega^2\mu\varepsilon} = \sqrt{(2\pi)^2 f^2 \mu\varepsilon} \tag{1.67}$$

$$c = \text{velocity of light in the medium} = \frac{1}{\sqrt{\mu\varepsilon}} \tag{1.68}$$

$$\lambda = \text{wavelength in the medium} \tag{1.69}$$

In summary, first the magnetic vector potential A is solved for in the frequency domain given the impressed currents $J_i(\mathbf{r})$ through

$$A(\mathbf{r}) = A(x, y, z) = \frac{\mu}{4\pi} \iiint_V \frac{J_i(\mathbf{r}')\,e^{-jk|\mathbf{r}-\mathbf{r}'|}}{|\mathbf{r}-\mathbf{r}'|}\,d\mathbf{r}' \tag{1.70}$$

then the scalar electric potential ψ is obtained from (1.61). Next, the electric field intensity E is computed from (1.60) and the magnetic field intensity H from (1.59).

In the time domain the equivalent solution for the magnetic vector potential A is then given by the time-retarded potentials:

$$A(\mathbf{r}, t) = A(x, y, z, t) = \frac{\mu}{4\pi} \iiint_V \frac{J_i\left(\mathbf{r}', t - \frac{|\mathbf{r} - \mathbf{r}'|}{c}\right)}{|\mathbf{r} - \mathbf{r}'|} d\mathbf{r}' \qquad (1.71)$$

It is interesting to note that the time and space variables are now coupled and they are not separable. That is why in the time domain the spatial and temporal responses of an antenna are intimately connected and one needs to look at the complete solution. From the magnetic vector potential we obtain the scalar potential ψ by using (1.52). From the two vector and scalar potentials the electric field intensity E is obtained through (1.55) and the magnetic field intensity H using (1.54).

We now use these expressions to calculate the impulse response of some typical antennas in both the transmit and receive modes of operations. The reason the impulse response of an antenna is different in the transmit mode than in the receive mode is because the reciprocity principle in the time domain contains an integral over time forming a convolution as it is a simple product in the transformed frequency domain. Thus the mathematical form of the reciprocity theorem in the time domain is quite different from its counterpart in the frequency domain. For the former a time integral is involved, whereas for the latter no such integral is involved as it is a simple product. This relationship comes directly from the Fourier transform theory where a convolution in the time domain is translated into a product in the frequency domain. Because of the simple product in the frequency domain reciprocity theorem, the antenna radiation pattern in the transmit mode is equal to the antenna pattern in the receive mode, except for a scale factor. This is discussed next.

1.6 Power (Correlation) Versus Reciprocity (Convolution)

In electrical engineering there are two principles that are quite important in understanding the principles of electrical engineering aka electromagnetic theory. The two principles are correlation and convolution. It is important to note that the principle of superposition applies to convolution and not to correlation. Convolution which is related to the computation of the response when a system is excited by an input of arbitrary shape given the response to an impulsive input. One can obtain the output responses due to various inputs to a system by applying the principle of superstition to the various inputs namely summing up their individual contributions. So the output $y(t)$ from a system with the impulse response $h(t)$ due to an applied input $x(t)$ is given by the following integral representing the convolution of $x(t)$ with $h(t)$ symbolically written as $x \, \textbf{\textcircled{\tiny$\bullet$}} \, h$ as

$$y(t) = h \, \textbf{\textcircled{\tiny$\bullet$}} \, x = \int h(t-\tau)\, x(\tau)\, d\tau. \tag{1.72}$$

Next, we discuss the context of reciprocity in electromagnetics. In electromagnetics the reciprocity relationship in general starts with the Lorentz theorem. To establish the Lorentz reciprocity theorem, assume that one has a current density $\mathbf{J}_1$ in a volume V bounded by a closed surface S which produces an electric field $\mathbf{E}_1$ and a magnetic field $\mathbf{H}_1$, where all three are periodic functions of time with angular frequency ω, and in particular they have a time-dependence $\exp(-j\omega t)$. Suppose that we similarly have a second current source $\mathbf{J}_2$ at the same frequency ω which (by itself) produces fields $\mathbf{E}_2$ and $\mathbf{H}_2$. These fields satisfy Maxwell's equations and therefore

$$\nabla \times \mathbf{E}_1 = -j\omega\mu\mathbf{H}_1 \quad \nabla \times \mathbf{H}_1 = j\omega\varepsilon\mathbf{E}_1 + \mathbf{J}_1 \tag{1.73}$$

$$\nabla \times \mathbf{E}_2 = -j\omega\mu\mathbf{H}_2 \quad \nabla \times \mathbf{H}_2 = j\omega\varepsilon\mathbf{E}_2 + \mathbf{J}_2 \tag{1.74}$$

By expanding $\nabla \cdot (\mathbf{E}_1 \times \mathbf{H}_2 - \mathbf{E}_2 \times \mathbf{H}_1)$ and simultaneous use of Maxwell's equations lead to (as explained in [16])

$$\begin{aligned}\nabla \cdot \left(\mathbf{E}_1 \times \mathbf{H}_2 - \mathbf{E}_2 \times \mathbf{H}_1\right) &= \left(\nabla \times \mathbf{E}_1\right)\cdot \mathbf{H}_2 - \left(\nabla \times \mathbf{H}_2\right)\cdot \mathbf{E}_1 - \left(\nabla \times \mathbf{E}_2\right)\cdot \mathbf{H}_1 \\ &\quad + \left(\nabla \times \mathbf{H}_1\right)\cdot \mathbf{E}_2 \\ &= -\mathbf{J}_2 \cdot \mathbf{E}_1 + \mathbf{J}_1 \cdot \mathbf{E}_2 \end{aligned} \tag{1.75}$$

Integrating now both sides over the volume V and using the divergence theorem yields

$$\begin{aligned}\int_V \nabla \cdot \left(\mathbf{E}_1 \times \mathbf{H}_2 - \mathbf{E}_2 \times \mathbf{H}_1\right)dV &= \oint_S \left(\mathbf{E}_1 \times \mathbf{H}_2 - \mathbf{E}_2 \times \mathbf{H}_1\right)\cdot \boldsymbol{n}\, dS \\ &= \int_V \left(-\mathbf{J}_2 \cdot \mathbf{E}_1 + \mathbf{J}_1 \cdot \mathbf{E}_2\right)dV \end{aligned} \tag{1.76}$$

where $\boldsymbol{n}$ is the unit outward normal to S. This is the Lorentz reciprocity theorem for an isotropic medium. This mathematical expression is sometimes also termed as *reaction* in the computational electromagnetics literature [16].

A few special cases arise where the surface integral vanishes [16]. For example when S is a perfectly conducting surface then it is zero. Also the surface integral vanishes when S is chosen as a spherical surface at infinity for which $\boldsymbol{n} = \boldsymbol{a}_r$, where $\boldsymbol{a}_r$ is the vector along the radial direction of a spherical coordinate system. Actually, for any surface S which encloses all the sources for the fields the surface integral will be zero. This can be seen if S becomes the surface of a sphere of infinite radius. Hence, the surface integral taken over any closed surface S surrounding all the sources vanishes. When the surface integral vanishes one obtains

$$\int_V \mathbf{E}_2 \cdot \mathbf{J}_1 \, dV = \int_V \mathbf{E}_1 \cdot \mathbf{J}_2 \, dV \tag{1.77}$$

If $\mathbf{J}_1$ and $\mathbf{J}_2$ are infinitesimal current elements, then

$$\mathbf{E}_1 \cdot \mathbf{J}_2 = \mathbf{E}_2 \cdot \mathbf{J}_1 \tag{1.78}$$

which states that the electric field $\mathbf{E}_1$ produced by $\mathbf{J}_1$ has a component along $\mathbf{J}_2$ that is equal to the component along $\mathbf{J}_1$ of the field radiated by $\mathbf{J}_2$ when both $\mathbf{J}_1$ and $\mathbf{J}_2$ have unit magnitude. This is the form that is essentially used in circuit analysis called the Rayleigh-Carson reciprocity theorem except that $\mathbf{E}$ and $\mathbf{J}$ are replaced by the voltage $\mathbf{V}$ and the current $\mathbf{I}$ resulting in

$$V_1 I_2 = V_2 I_1 \tag{1.79}$$

Therefore, the principle of superposition can be applied to both the voltage $\mathbf{V}$ and the current $\mathbf{I}$ when applying the principle of reciprocity. Next we consider the principle of correlation, which represents power, in general.

Circuit theory describes the excitation of a two terminal element in terms of the voltage $\mathbf{V}$ applied between the terminals and the current $\mathbf{I}$ into and out of the respective terminals. The power supplied through the terminal is VI^* (where the superscript * implies a complex conjugate). So in an AC circuit the complex power flow needs to have both $\mathbf{V}$ and $\mathbf{I}$ to compute the power [14–16]. However, for purely resistive circuits the power can be given solely by either the voltage or the current as they are scalar multiple of each other related to the resistance of the component on which the voltage and the currents are determined. So the correlation between the voltage and the current represents power which is VI^*. In this case, superposition does not apply and this holds true even for resistive circuits. As an example consider the two simple circuits of Case A and Case B of Figure 1.4 where two different voltages of 6V and 3V are applied across two individual resistor combination producing currents of 2A and 1A in the respective circuits. The power dissipated in the 1Ω resistor for Case A is then 4W and in Case B is 1W. Now we superpose the two excitations of Case A and Case B on the same circuit. The first goal is to demonstrate that superposition applies only to the voltages and the currents and not to power even for the case of resistive circuits. The second goal is to illustrate that when superposition is applied there are two possibilities and here the vector nature of the problem shows up. The two superposed voltages can be in phase

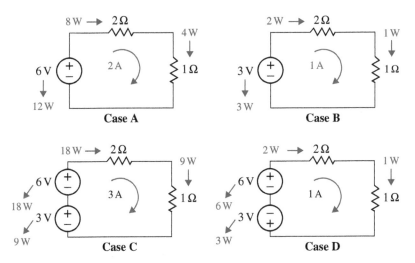

Figure 1.4 Illustration of the principle of superposition for a purely resistive circuit.

cohesion as illustrated in Case C or may be in phase opposition resulting in Case D. For case C we observe that the currents and the voltages superpose and the resultant power dissipated in the 1 ohm resistor is now 9W and not 4W+1W = 5W. Does that mean we have a power gain over the two cases A and B? No, definitely not as in electrical engineering the power defined through the principle of correlation states that individual powers cannot be superposed but the voltages and the currents can be combined. Now observe in Case D we have phase opposition between the two superposed voltages and here the 3V source is acting as a load on the 6V source, and instead of supplying useful power it is absorbing it. Hence it is extremely important how in practice generators are connected with the appropriate phase even in the DC situation. As is observed in many practical devices sources are generally connected in series and never in parallel. In a parallel connection of the sources, if there is a slight imbalance of the terminal voltages between the two sources then there will be a nonproductive circulating current. This leads to power loss and also not to minimize the situation that this circulating current may even cause one of the sources to heat up and explode!

Now we extend the concept of power flow in such a way that power is now thought to flow throughout space and is not associated only with a flow of current into and out of the terminals. Just as the circuit laws can be combined to describe the flow of power between the circuit elements, so Maxwell's equations are the basis for a field theoretical view of power flow expressed through the Poynting's theorem. The complex power density in space is given by the Poynting vector $\mathbf{S}$ as $\mathbf{E} \times \mathbf{H}^*$, where the superscript * represents the complex conjugate. When this power density is integrated over a surface with the

Poynting vector aligned with the outward normal to the surface we get the Power flow through that surface into space. Thus the Poynting's theorem states that the integral of the inward component of the Poynting vector over the surface of any volume V equals the sum of the power dissipated and the rate of energy storage increases inside that volume. We can now interpret the physical significance of the complex Poynting vector **S** by restating that the real part of it as the time-average quantity, with the time-average total power is radiated outward across the surface area A. The imaginary part of the Poynting vector is the reactive power.

In the near field of a radiator then in order to calculate the power flow one needs a measure of both the electric field intensity and the magnetic field intensity just like requiring both the voltages and the currents to calculate the power flow in an AC circuit. However in the far field of an antenna where the wave fronts become planar, then similar to a resistive circuit, the power flow can be obtained either from the Electric field or the Magnetic field alone as then they are related by the impedance of free space which is 377 ohms.

In summary, the principle of superposition does not apply to power. So in a power harvesting scenario when multiple antennas are used to generate useful power, it is the voltages induced in each of the receiving antennas that need to be added up with the proper orientation of the phase and not use the principle of superposition of power which does not exist in electrical engineering! Superposition holds only for linear systems, as mathematically defined. Power is not a linear system; hence, we cannot use superposition in power.

1.7   Radiation and Reception Properties of a Point Source Antenna in Frequency and in Time Domain

1.7.1   Radiation of Fields from Point Sources

In this section we first define what is meant by the term *radiation* and then observe the nature of the fields radiated by point sources and the temporal nature of the voltages induced when electromagnetic fields are incident on them. In contrast to the acoustic case (where an isotropic source exists), in the electromagnetic case there are no isotropic omnidirectional point sources. Even for a point source, which in the electromagnetic case is called a *Hertzian dipole*, the radiation pattern is not isotropic, but it can be omnidirectional in certain planes. We describe the solution in both the frequency and time domains for such classes of problems.

Any element of current or charge located in a medium will produce electric and magnetic fields. However, by the term *radiation* we imply the amount of finite energy transmitted to infinity from these currents. Hence, radiation is related to the far fields or the fields at infinity. This will be discussed in detail in Chapter 2. A static charge may generate near fields, but it does not produce radiation, as the field at infinity due to this charge is zero and moreover it does not produce any power flow to infinity. Therefore, radiated fields or far fields are synonymous. We will also explore the sources of a radiating field.

1.7.1.1 Far Field in Frequency Domain of a Point Radiator

If we consider a delta element of current or a Hertzian dipole located at the origin represented by a constant J_i times a delta function $\delta\,(0, 0, 0)$, the magnetic vector potential from that current element is given by

$$\mathbf{A}(x, y, z) = \frac{\mu}{4\pi} \frac{e^{-jkR}}{R} \mathbf{J}_i \tag{1.80}$$

where

$$R = \sqrt{x^2 + y^2 + z^2} \tag{1.81}$$

Here we limit our attention to the electric field. The electric field at any point in space is then given by

$$\mathbf{E}(x, y, z) = -j\omega\mathbf{A} - \nabla\psi = -j\omega\mathbf{A} + \frac{\nabla(\nabla\cdot\mathbf{A})}{j\omega\mu\varepsilon} = \frac{1}{j\omega\mu\varepsilon}\left[k^2\mathbf{A} + \nabla(\nabla\cdot\mathbf{A})\right] \tag{1.82}$$

In rectangular coordinates, the fields at any point located in space will be

$$\mathbf{E}(x, y, z) = \frac{1}{j\omega\mu\varepsilon}\left[\begin{array}{l} k^2\mathbf{A}(x, y, z) + \left\{\hat{x}\dfrac{\partial}{\partial x} + \hat{y}\dfrac{\partial}{\partial y} + \hat{z}\dfrac{\partial}{\partial z}\right\} \\ \times\left\{\dfrac{\partial A_x(x, y, z)}{\partial x} + \dfrac{\partial A_y(x, y, z)}{\partial y} + \dfrac{\partial A_z(x, y, z)}{\partial z}\right\} \end{array}\right] \tag{1.83}$$

However, some simplifications are possible for the far field (i.e., if we are observing the fields radiated by a source of finite size at a distance of $2D^2/\lambda$ from it, where D is the largest physical dimension of the source and λ is the wavelength – the physical significance of this will be addressed in Chapter 2). For a point source, everything is in the far field. Therefore, for all practical purposes, observing the fields at a distance $2D^2/\lambda$ from a source is equivalent to observing the fields from the same source at infinity. In that case, the far fields can be obtained from the first term only in (1.82) or (1.83). This first term due to the magnetic vector potential is responsible for the far field and there is no contribution from the scalar electric potential ψ. Hence,

$$\mathbf{E}_{\text{far}}(x, y, z) = -j\omega\mathbf{A} = -j\frac{\omega\mu}{4\pi}\mathbf{J}_i\frac{e^{-jkR}}{R} \tag{1.84}$$

and one obtains a radiated spherical wavefront and in the far field and it becomes planar (a sphere with an infinite radius becomes a plane) for a point source. However, the power density radiated is proportional to E_θ and that is clearly zero along $\theta = 0°$ and is maximum in the azimuth plane where $\theta = 90°$. The characteristic feature is that the far field is polarized and the orientation of the field is along the direction of the current element. It is also clear that one obtains a planar wavefront in the far field generated by spherical wavefronts radiated by a point source.

The situation is quite different in the time domain, as the presence of the term ω in the front of the expression of the magnetic vector potential in (1.84) represents a differentiation as we will illustrate later.

1.7.1.2 Far Field in Time Domain of a Point Radiator
We consider a delta current source at the origin of the form

$$\mathbf{J}_i(0, 0, 0, t) = \hat{z}\delta(0, 0, 0)f(t) \tag{1.85}$$

where $\hat{z}$ is the direction of the orientation of the elemental current element and $f(t)$ is the temporal variation for the current fed to the point source located at the origin. The magnetic vector potential in this case is given by

$$\mathbf{A}(\mathbf{r}, t) = \frac{\mu}{4\pi}\frac{\hat{z}f(t - R/c)}{R} \tag{1.86}$$

There will be a time retardation factor due to the space-time connection of the electromagnetic wave that is propagating, where R is given by (1.81).

Now the transient far field due to this impulsive current will be given by

$$\mathbf{E}(\mathbf{r}, t) = -\frac{\partial\mathbf{A}(\mathbf{r}, t)}{\partial t} = -\frac{\mu\hat{z}}{4\pi R}\frac{\partial f(t - R/c)}{\partial t} \tag{1.87}$$

Hence, the time domain field radiated by a point source is given by the time derivative of the transient variation of the elemental current. Therefore, a time-varying current element will always produce a radiated field and hence will cause radiation. However, if the current element is not changing with time, there will be no radiative fields emanating from it.

Equivalently, the current density $\mathbf{J}_i$ can be expressed in terms of the flow of charges; thus the current flow is equivalent to $\rho\mathbf{v}$, where ρ is the charge density and v is its velocity. Therefore, radiation from a time-varying current element in (1.87) can occur if any of the following three scenarios occur:

1) The charge density ρ is changing as a function of time.
2) The direction of the velocity vector v is changing as a function of time.
3) The magnitude of the velocity vector v is changing as a function of time, or equivalently, the charge is accelerated or decelerated.

Therefore, in theory any one of these three scenarios can cause radiation. For example, for a dipole element the current goes to zero at the ends of the wire structure and hence the charges decelerate when they come to the end of a wire. That is why radiation seems to emanate from the ends of the wire and also from the feed point of a dipole where a current is injected or a voltage is applied and where the charges are induced and hence accelerated. Current flowing in a loop of wire can also radiate as the direction of the velocity is changing as a function of time even though its magnitude remains constant. So a current flowing in a loop of wire may have a constant angular velocity, but the temporal change in the orientation of the velocity vector may cause radiation. To maintain the same current along a cross-section of the wire loop, the charges located along the inner circumference of the loop have to decelerate, whereas the charges on the outer boundary have to accelerate. This will cause radiation. In a klystron, by modulating the velocity of the electrons, one can have bunching of the electrons or a change of the electron density with time. This also causes radiation. In summary, if any one of the three conditions described above occurs, there will be radiation.

By observing (1.87), we see that a transmitting antenna acts as a differentiator of the transient waveform fed to its input. The important point to note is that an antenna impulse response on transmit is a differentiation of the excitation source. Therefore in all baseband broadband simulations the differential nature of the point source must be taken into account. This implies that if the input to a point radiator is a pulse, it will radiate two impulses of opposite polarities – a derivative of the pulse. This principle will be seen later in Figure 1.11. Therefore, when a baseband broadband signal is fed to an antenna, what comes out is the derivative of that pulse. It is rather unfortunate that very few simulations dealing with baseband broadband signals really take this property of an isotropic point source antenna into account in analyzing systems.

1.7.2 Reception Properties of a Point Receiver

On receive, an antenna behaves in a completely different way than on transmit. We observed that an isotropic point antenna acts as a differentiator on transmit. On receive, the voltage received at the terminals of the antenna is given by [14]

$$V = \int E \cdot d\ell \tag{1.88a}$$

where the path of the integral is along the length of the antenna. Equivalently, this voltage, which is called the *open-circuit voltage* V_{oc}, is equivalent to the dot product of the incident field vector and the effective height of the antenna and is given by [14, 15]

$$V_{oc} = \mathbf{E} \cdot \mathbf{H}_{eff} \tag{1.88b}$$

The effective height of an antenna is defined by [14]

$$\mathbf{H}_{eff} = \int_0^H \mathbf{I}(z)\, dz = H\mathbf{I}_{av} \tag{1.89}$$

where H is the length of the antenna and it is assumed that the maximum value of the current along the length of the antenna $I(z)$ is unity. I_{av} then is the average value of the current on the antenna. This equation is valid at a single frequency only. Therefore, when an electric field E^{inc} is incident on a small dipole of total length L from a broadside direction, it induces approximately a triangular shaped current on the structure [17]. Therefore, the effective height in this case is $L/2$ and the open-circuit voltage induced on the structure in the frequency domain is given by

$$V_{oc}(\omega) = -\frac{LE^{inc}(\omega)}{2} \tag{1.90}$$

and in the time domain as the effective height now becomes an impulse-like function, which is

$$V_{oc}(t) = -\frac{L\,E^{inc}(t)}{2} \tag{1.91}$$

Therefore, in an electrically small receiving antenna called a *voltage probe* the induced waveform will be a replica of the incident field provided that the frequency spectrum of the incident electric field lies mainly in the low-frequency region, so that the concept of an electrically small antenna is still applicable.

In summary, the impulse response of an antenna on transmit given by (1.84) is the time derivative of the impulse response of the same antenna when it is operating in the receive mode as expressed in (1.91). In the frequency domain, as we observe in (1.84), the term $j\omega$ is benign as it merely introduces a purely imaginary scale factor at a particular value of ω. However, the same term when transferred to the time domain represents a time derivative operation. Hence, in frequency domain the transmit radiation antenna pattern is identical to the antenna pattern when it is operating in the receive mode. In time domain, the transmit impulse response of the antenna is the time derivative of the impulse response in the receive mode for the same antenna. At this point, it may be too hasty to jump to the conclusion that something is really amiss as it does not relate quite the same way to the reciprocity theorem which in the frequency domain has shown that the two patterns in the transmit–receive modes are identical. This is because the mathematical form of the reciprocity theorem is quite different in the time and in the frequency domains. Since the reciprocity theorem manifests itself as a product of two quantities in the frequency domain, in the time domain it transforms to a convolution. It is this phenomenon that makes the impulse response of the transmit and the receive modes different. We use another example, namely of a dipole, to illustrate this point further.

1.8 Radiation and Reception Properties of Finite-Sized Dipole-Like Structures in Frequency and in Time

In this section we describe the impulse responses of transmitting and receiving dipole-like structures whose dimensions are comparable to a wavelength. Therefore, these structures are not electrically small. In this section, the main results are summarized. The reason for choosing finite-sized structures is that the impulse responses of these wire-like structures are quite different from the cases described in the preceding section. For a finite-sized antenna structure, which is comparable to the wavelength at the frequency of operation, the current distribution on the structure can no longer be taken to be independent of frequency. Hence the frequency term must explicitly be incorporated in the expression of the current.

1.8.1 Radiation Fields from Wire-Like Structures in the Frequency Domain

For a finite-sized dipole, the current distribution that is induced on it can be represented mathematically to be in the form [14, 15]

$$I(z) = \sin\left[k\left(L/2 - |z|\right)\right] \tag{1.92}$$

where L is the wire antenna length. We here assume that the current distribution is known. However, in a general situation we have to use a numerical technique to solve for the current distribution on the structure before we can solve for the far fields as illustrated in [18]. This is particularly important when mutual coupling effects are present or there are other near-field scatterers. For a current distribution given by (1.92), the far fields can be obtained [14, 15] as

$$E_\theta = \frac{I_0 \eta e^{j\omega\left[t - r/c\right]}}{2\pi r} \left[\frac{\cos\left(\dfrac{kL}{2}\cos\theta\right) - \cos\dfrac{kL}{2}}{\sin\theta}\right] \tag{1.93}$$

where η is the characteristic impendence of free space and I_0 represents the maximum value of the current. Here L is the total length of the antenna. k is the free-space wavenumber and is equal to $2\pi/\lambda = \omega/c$, where c is the velocity of light in that medium. It is important to note that only along the broadside direction and in the azimuth plane of $\theta = \pi/2 = 90°$ is the radiated electric field omni- directional in nature.

1.8.2 Radiation Fields from Wire-Like Structures in the Time Domain

When the current induced on the dipole is a function of frequency, the far-zone time-dependent electric field at a spatial location r is given approximately by [17] primarily along the broadside direction as

$$E_\theta(t) = \frac{\eta}{2\pi r \sin\theta} \left\{ \begin{array}{l} I\left(t - \dfrac{r}{c}\right) - I\left[t - \dfrac{r}{c} - \dfrac{L}{2c}(1 + \cos\theta)\right] \\[2mm] + I\left[t - \dfrac{r}{c} - \dfrac{L}{c}\right] - I\left[t - \dfrac{r}{c} - \dfrac{L}{2c}(1 - \cos\theta)\right] \end{array} \right\} \tag{1.94}$$

where $I(t)$ is the transient current distribution on the structure. It is interesting to note that for L/c small compared to the pulse duration of the transient current distribution on the structure, it has been shown in [17] the approximate far field can be written as

$$E_\theta \simeq \frac{\eta}{2\pi r}\left(\frac{L}{2c}\right)^2 \frac{\partial^2 I\left(t - \dfrac{r}{c}\right)}{\partial t^2} \sin\theta \tag{1.95}$$

that is, the far-field now is proportional to the second temporal derivative of the transient current on the structure.

1.8.3 Induced Voltage on a Finite-Sized Receive Wire-Like Structure Due to a Transient Incident Field

For a finite-sized antenna of total length L, the effective height will be a function of frequency and it is given by

$$H_{\text{eff}}(\omega) = \int_{-L/2}^{L/2} \sin\left[k\left(\frac{L}{2} - |z|\right)\right] dz = \frac{2c}{\omega}\left[1 - \cos\left(\frac{kL}{2}\right)\right] \tag{1.96}$$

Hence the induced voltage for a field incident along the broadside of the wire structure, it will be given approximately by

$$V_{oc}(\omega) = -H_{eff}(\omega)E^{inc}(\omega) \tag{1.97}$$

Hence the transient received voltage in the antenna due to an incident field will result in the following convolution (defined by the symbol ⊗) between the incident electric field and the effective height, resulting in

$$V_{oc}(t) = -E^{inc}(t) \otimes H_{eff}(t) \tag{1.98}$$

In the time domain, using (1.96) the effective height will be given by

$$H_{eff}(t) = jc \begin{cases} +1 & 0 < t < \dfrac{L}{2c} \\ -1 & \dfrac{-L}{2c} < t < 0 \end{cases} \tag{1.99}$$

This illustrates that when (1.99) is used in (1.98), the received open-circuit voltage will be approximately the derivative of the incident field when L/c is small compared to the initial duration of the transient incident field. In Chapter 2, we study the properties of arbitrary shaped antennas in the frequency domain using a general purpose computer code described in [18]. Furthermore, we focus on the implications of near and far fields. The near/far field concepts are really pertinent in the frequency domain as they characterize the radiation properties of antennas. However, in the time domain this distinction is really not applicable as everything is in the near field (as a finite time incident pulse represents a waveform in the frequency domain of infinite bandwidth) unless we have a strictly band limited signal!

1.8.4 Radiation Fields from Electrically Small Wire-Like Structures in the Time Domain

To calculate the transient radiated fields from a dipole structure with a pulsed excitation, we assume that the spatial and temporal variation of the current for the dipole is given by [18, 19]

$$J(z,t) = \hat{z} \cos\frac{\pi z}{L} P\{z\} f(t)P\{t\} \tag{1.100}$$

where the function P denotes a pulse function of the mathematical form

$$P\{z\} = 1 \quad \text{for} - L/2 \leq z \leq L/2$$
$$= 0 \quad \text{otherwise} \tag{1.101}$$

and

$$P\{t\} = 1 \quad \text{for} - T/2 \leq t \leq T/2$$
$$= 0 \quad \text{otherwise} \tag{1.102}$$

where $f(t)$ represents the temporal variation for the current. $L/2$ is the half-length of the dipole. Here it has been assumed that the length of the antenna is much smaller than the wavelength at the highest frequency of interest. This derivation is valid for an electrically small antenna. The duration of the excitation pulse is for a time T and it is centered at the origin. So the temporal function is not causal in a strict mathematical sense. However, we choose it in this way for computational convenience. The function $f(t)$ is next assumed to be a sinusoidal carrier, that is,

$$f(t) = \sin \omega_0 t \tag{1.103}$$

So for a digital signal, this carrier at frequency f_0 is on for a time duration T. Our goal is to find the radiated far-field wave shape from a dipole when such a pulsed current distribution is applied to the dipole.

We define

$$\zeta = \frac{L \cos \theta}{2c} \tag{1.104}$$

If θ is the angle measured from the broadside direction, then the transient far field at a distance R (distance to the center of the antenna) is given by

$$E_z(\theta,t) \approx -\frac{\partial A_z(\theta,t)}{\partial t} \approx -\frac{\mu}{4\pi R}\frac{\partial}{\partial t}\int_{-\infty}^{\infty} J\left(z',t-\frac{R-z'\cos\theta}{c}\right) dz'$$
$$= -\frac{\mu}{4\pi R}\frac{\partial}{\partial t}\int_{-\infty}^{\infty} J\left(z',\tau+\frac{2\zeta z'}{L}\right) dz' \tag{1.105}$$

where

$$\tau = t - \frac{R}{c} \tag{1.106}$$

Equivalently,

$$E_z(\theta,t) = -\frac{\mu}{4\pi R} \int_{-\infty}^{\infty} \cos\frac{\pi z'}{L} \; P\{z'\} \frac{\partial}{\partial t}\left\{\sin\left[\omega_0\left(\tau + \frac{2\zeta z'}{L}\right)\right]\right\} dz' \qquad (1.107)$$

Since

$$\frac{\partial}{\partial \tau}\left[f\left(\tau + \frac{2\zeta z'}{L}\right)\right] = \frac{L}{2\zeta}\frac{\partial}{\partial z'}\left[f\left(\tau + \frac{2\zeta z'}{L}\right)\right]$$

and after integrating by parts, one obtains

$$E_z(\theta,t) \approx -\frac{\mu}{8R\zeta} \int_{z_1}^{z_2} \sin\frac{\pi z'}{L} \; P(z') \sin\left[\omega_0\left(\tau + \frac{2\zeta z'}{L}\right)\right] dz' \qquad (1.108a)$$

where

$$z_1 = -\left(\tau + \frac{T}{2}\right)\frac{L}{2\zeta}; \qquad (1.108b)$$

and

$$z_2 = -\left(\tau - \frac{T}{2}\right)\frac{L}{2\zeta}. \qquad (1.108c)$$

The integral needs to be calculated with extreme caution.
    We consider two different distinct situations:

Case A: $\zeta > T/2$
Case B: $\zeta < T/2$

For each of these two cases there are five different situations, which need to be addressed separately [19].

Case A: $\zeta > 0; \zeta < T/2; z_2 - z_1 = \dfrac{TL}{2\zeta} \geq L$

i) $z_1 \geq \dfrac{L}{2}$. In this case we have $\tau \leq -\left[\dfrac{T}{2} + \zeta\right]$.

The transient fields in this region is going to be zero, so that $E_z(\theta, t) = 0$.

---

ii) $-\dfrac{L}{2} \le z_1 \le \dfrac{L}{2} \le z_2$. In this case we have $-\dfrac{T}{2} - \zeta \le \tau \le \left[-\dfrac{T}{2} + \zeta \right]$.

So the transient far field for this case is given by the expression

$$E_z(\theta,t) = -\frac{\mu}{8R\zeta} \int_{z_1}^{L/2} \sin\frac{\pi z'}{L} \sin\left[\omega_0\left(\tau + \frac{2\zeta z'}{L} \right) \right] dz'$$

$$= -\frac{\mu L \omega_0}{4R} \frac{\cos\left[\omega_0(\tau + \zeta) \right]}{\pi^2 - 4\zeta^2\omega_0^2}$$

$$+ \frac{\mu L}{16 R\zeta} \left[\frac{\sin\left\{ \dfrac{\omega_0 T}{2} - \dfrac{\pi}{2\zeta}\left(\tau + \dfrac{T}{2} \right) \right\}}{\pi - 2\zeta\omega_0} + \frac{\sin\left\{ \dfrac{\omega_0 T}{2} + \dfrac{\pi}{2\zeta}\left(\tau + \dfrac{T}{2} \right) \right\}}{\pi + 2\zeta\omega_0} \right]$$

(1.109)

iii) $z_1 \le -\dfrac{L}{2} \le \dfrac{L}{2} \le z_2$. In this case we have $-\dfrac{T}{2} + \zeta \le \tau \le \dfrac{T}{2} - \zeta$.

The transient far field is then given by

$$E_z(\theta,t) = -\frac{\mu}{8R\zeta} \int_{-L/2}^{L/2} \sin\frac{\pi z'}{L} \sin\left[\omega_0\left(\tau + \frac{2\zeta z'}{L} \right) \right] dz'$$

(1.110)

$$= \frac{\mu L \omega_0}{2R} \frac{\cos\left(\omega_0 \zeta \right)}{\pi^2 - 4\zeta^2\omega_0^2} \cos(\omega_0 \tau).$$

This is the steady-state region of the response.

iv) $z_1 \le -\dfrac{L}{2} \le z_2 \le \dfrac{L}{2}$. In this case $\dfrac{T}{2} - \zeta \le \tau \le \dfrac{T}{2} + \zeta$.

So the transient field can be expressed as

$$E_z(\theta,t) = -\frac{\mu}{8R\zeta} \int_{-L/2}^{z_2} \sin\frac{\pi z'}{L} \sin\left[\omega_0\left(\tau + \frac{2\zeta z'}{L} \right) \right] dz' = \frac{\mu L \omega_0}{4R} \frac{\cos\left[\omega_0(\tau - \zeta) \right]}{\pi^2 - 4\zeta^2\omega_0^2}$$

$$- \frac{\mu L}{16R\zeta} \left[\frac{\sin\left\{ \dfrac{\omega_0 T}{2} + \dfrac{\pi}{2\zeta}\left(\tau - \dfrac{T}{2} \right) \right\}}{\pi - 2\zeta\omega_0} + \frac{\sin\left\{ \dfrac{\omega_0 T}{2} - \dfrac{\pi}{2\zeta}\left(\tau - \dfrac{T}{2} \right) \right\}}{\pi + 2\zeta\omega_0} \right]$$

(1.111)

and finally,

v) $z_2 \le -\dfrac{L}{2}$ or $\tau \ge \zeta + \dfrac{T}{2}$.

In this case

$$E_z(\theta,t) = 0$$

Thus, the region defined by **case iii**, $-\zeta \le \tau \le +\zeta$, is the steady-state region and the other two regions where the fields are finite are the transient regions.

Next we consider **case B.**

Case B: $\zeta > T/2$

In this case, all the situations are exactly the same as in **case A**, except for the situation **iii a**. As this situation no longer exists for this case, the new conditions are

iii) **b.** $-\dfrac{L}{2} \le z_1 \le z_2 \le \dfrac{L}{2}$, or equivalently, $\dfrac{T}{2} - \zeta \le \tau \le \zeta - \dfrac{T}{2}$.

The transient field for this case is given by

$$E_z(\theta,t) = -\frac{\mu}{8R\zeta} \int_{z_1}^{z_2} \sin\frac{\pi z'}{L} \sin\left[\omega_0\left(\tau + \frac{2\zeta z'}{L}\right)\right] dz'$$

$$= -\frac{\mu L}{8R\zeta}\left[\frac{\sin\left\{\dfrac{T}{4\zeta}(\pi - 2\omega_0\zeta)\right\}}{\pi - 2\omega_0\zeta} - \frac{\sin\left\{\dfrac{T}{4\zeta}(\pi + 2\omega_0\zeta)\right\}}{\pi + 2\omega_0\zeta}\right]\cos\left(\frac{\pi\tau}{2\zeta}\right)$$

(1.112)

In contrast to **iii a** this does not represent a steady state situation as this is never reached for $\zeta \ge \dfrac{T}{2}$. For this case, the transient radiation is caused by an aperture of a reduced width of dimension $z_2 - z_1 = TL/(2\zeta)$, and the transient response lasts for a duration equal to $(2\zeta - T)$, during which time the effective part of the radiation shifts from one end of the physical aperture to the other.

The final regions of response are summarized in Figure 1.5.

As an example, consider several cycles of a sine wave of the form shown in Figure 1.6 We consider a dipole of length $L = 1$ m. We consider f_0 to be 2 GHz. If we consider the far field along the angle $\theta = 30°$, we see that $\zeta = 8.33\times 10^{-10}$. We now choose T to be 7.0 ns. The z-component of the electric field will then be given by case A as shown in Figure 1.7. It is seen that initially there is a transient region and after that the steady-state region of the fields is reached. There is also a derivative operation going on, as the wave shape is more of the form of a cosine than a sine. If we now shorten the pulse by making $T = 1$ ns so that the pulse is much narrower in time, as shown in Figure 1.8, we will not have any steady-state regions. In this case Figure 1.9 gives the far field.

As a second example we consider the response of the dipole to a pulse excitation. If the transient excitation is a simple rectangular pulse rather than part of a sine wave, then in that case, we have

$$E_z(\theta,t) = -\frac{\mu}{8R\zeta} \int_{z_1}^{z_2} \sin\frac{\pi z'}{L} dz'$$

(1.113)

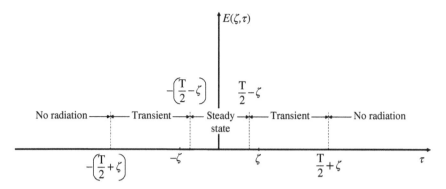

Figure 1.5 Regions of response for a current patch with a cosine distribution to rectangular pulses of sinusoidal carrier for $\dfrac{T}{2} \geq \zeta$.

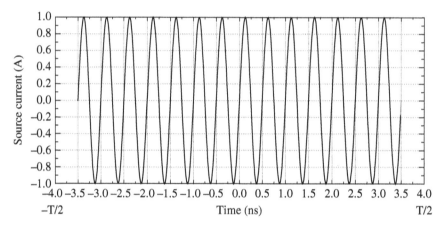

Figure 1.6 Sinusoidally modulated pulse.

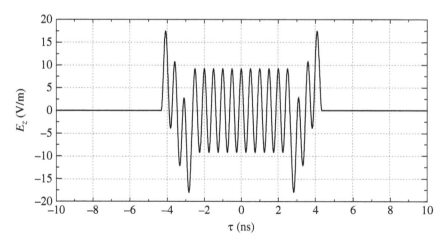

Figure 1.7 Far field along the elevation angle $\theta = 30°$ for case A.

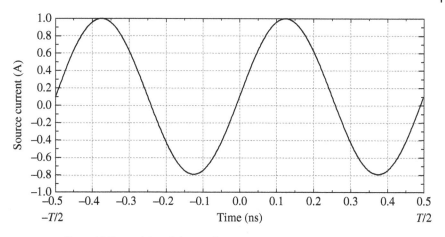

Figure 1.8 Sinusoidally modulated short pulse.

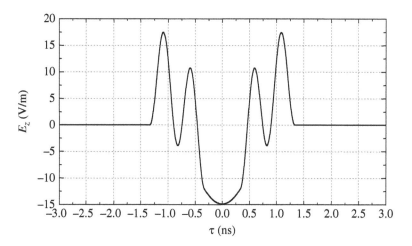

Figure 1.9 Far field along the elevation angle $\theta = 30°$ for case B.

where $z_1 = -\left[\tau + \dfrac{T}{2}\right]\dfrac{L}{2\zeta}$; and $z_2 = -\left[\tau - \dfrac{T}{2}\right]\dfrac{L}{2\zeta}$. In this case we have five different situations, as before.

For case B, we have the same expressions as in Case A, except for situation iii.

As an example we consider the pulse shown in Figure 1.10, having a width T of 0.67 ns. In this case, when $T \approx 4\zeta$, the radiated far field along the elevation angle $\theta = 30°$ for an input pulse is as shown in Figure 1.11. It is interesting to

Case A: $\zeta > 0; \zeta \leq T/2$

i) $z_1 \geq L/2, \tau \leq -\left[\dfrac{T}{2} + \zeta\right]$, and $E_z(\theta, t) = 0$.

ii) $-\dfrac{L}{2} \leq z_1 \leq \dfrac{L}{2} \leq z_2$.

In this case, $-T/2 - \zeta \leq \tau \leq -T/2 + \zeta$ and the transient fields are given by

$$E_z(\theta, t) = \frac{\mu}{8R\zeta} \int_{z_1}^{L/2} \sin\frac{\pi z'}{L} dz' = \frac{\mu L}{8\pi R\zeta} \cos\left[\frac{\pi}{2\zeta}\left(\tau + \frac{T}{2}\right)\right] \qquad (1.114)$$

iii) $z_1 \leq -\dfrac{L}{2} \leq \dfrac{L}{2} \leq z_2$ and $-\dfrac{T}{2} + \zeta \leq \tau \leq \dfrac{T}{2} - \zeta$.

The transient field in this case is $E_z(\theta, t) = 0$.

iv) $z_1 \leq -\dfrac{L}{2} \leq z_2 \leq \dfrac{L}{2}$ and $\dfrac{T}{2} - \zeta \leq \tau \leq \dfrac{T}{2} + \zeta$.

We have

$$E_z(\theta, t) = -\frac{\mu}{8R\zeta} \int_{-L/2}^{z_2} \sin\frac{\pi z'}{L} dz' = \frac{\mu L}{8\pi R\zeta} \cos\frac{\pi}{2\zeta}\left(\tau - \frac{T}{2}\right) \qquad (1.115)$$

v) $z_2 \leq -\dfrac{L}{2}$ or $\tau \geq \zeta + \dfrac{T}{2}$, this results in $E_z(\theta, t) = 0$.

So in this case, the steady-state response is zero. Therefore, when using digital modulation techniques for channel characterization, unless the transient response of the antennas is included in the model it is difficult to see how one can practically characterize the propagation channel without them.

Case B: $\zeta > T/2$

iii) $b - \dfrac{L}{2} \leq z_1 \leq z_2 \leq \dfrac{L}{2}$ or equivalently, $\dfrac{T}{2} - \zeta \leq \tau \leq \zeta - \dfrac{T}{2}$.

In this case

$$E_z(\theta, t) = -\frac{\mu L}{8\pi R\zeta}\left[\cos\left\{\frac{\pi}{2\zeta}\left(\tau + \frac{T}{2}\right)\right\} - \cos\left\{\frac{\pi}{2\zeta}\left(\tau - \frac{T}{2}\right)\right\}\right]$$

$$= \frac{\mu L}{4\pi R\zeta} \sin\frac{\pi\tau}{2\zeta} \sin\frac{\pi T}{4\zeta} \qquad (1.116)$$

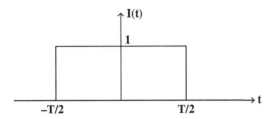

Figure 1.10 Rectangular pulse.

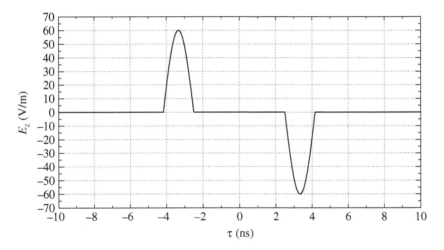

Figure 1.11 Far field along the elevation angle $\theta = 30°$ for case A.

observe that the steady-state response is zero, as the derivative of a constant is zero. However, there is a transient response. When $T = 1.6\zeta$, the radiated field wave shape is given by Figure 1.12, as there is no steady-state region. The important point to note is that the impulse response of the antennas needs to be taken into account in the design of systems, particularly when they are designed to transmit pulse-like waveforms.

In short, the transient responses from dipoles are quite complex, and except for a few simple cases, it is not possible in general to know *a priori* the wave shapes that come out of these structures. Hence, one needs to use numerical techniques for the solution of a general class of problems. The other important point is to note that the dipole on transmit actually differentiates the wave shape that is fed to it if the pulse width of the input waveform is larger than the transient time across the dipole.

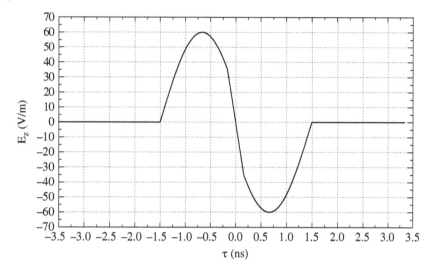

Figure 1.12 Far field along the elevation angle $\theta = 30°$ for case B.

1.9 An Expose on Channel Capacity

The evolution of the concept of Channel capacity has a long beginning. In 1861, James Clerk Maxwell [4, 7, 20] tried to determine the speed with which the smell fills the air when a perfume-bottle is opened while addressing the topic of Kinetic Theory of Gases. In order to address this problem he derived the Maxwell distribution for molecular velocities. The distribution turned out to be the well-known bell-shaped curve now known as the normal distribution. The objective of this section is to expose a different picture of electromagnetic wave propagation when the wave strength is much smaller than the ambient noise. The issue is how does one transmit and receive electromagnetic signals when its strength is comparable or lower than that of the ambient noise be it be thermal or that of an interfering signal. This section illustrates that through some modulations of the electromagnetic signals that carry the actual information such transmission and reception is possible. So ultimately we relate the electromagnetic wave transmission to the mechanism of a connecting link between the transmission and reception systems. A different dimension of physics related to statistics is introduced to characterize the deleterious effect of the environment when thermal noise is involved and not interference.

Maxwell's discovery opened up an entirely new approach to physics, which led to statistical mechanics, and to the use of probability distributions in quantum mechanics. No one before Maxwell ever applied a statistical law to a physical process. His work inspired the Austrian mathematician, physicist, and inventor of statistical mechanics Ludwig Boltzmann (1844–1906), and their works led to the *Maxwell-Boltzmann* distribution of molecular energies.

The logarithmic connection between entropy and probability was first stated by L. Boltzmann in his kinetic theory of gases. This famous formula for entropy $S = k_B\ Ln\ W$, where k_B is Boltzmann's constant, and *ln* is the natural logarithm and *W* is *Wahrscheinlichkeit*, a German word meaning the probability of occurrence of a macrostate or, more precisely, the number of possible microstates corresponding to the macroscopic state of a system — number of (unobservable) "ways" in the (observable) thermodynamic state of a system can be realized by assigning different positions and momenta to the various molecules. This formula was actually engraved on Boltzmann's tombstone [7].

Maxwell's paper in 1866 "On Dynamical Theory of Gases" produced the first statistical law of physics. In his subsequent work "On the Average Distribution of Energy" published in 1879 Maxwell introduced the concept of ensemble averaging [4, 7]. This enabled people to explain the properties of matter in terms of the behavior, en masse, of its molecules. One of the ideas in the paper was the method of ensemble averaging, where the whole system is much easier to analyze, rather than dealing with individual components. However, his result for the specific heat of air was off from the measurement. Instead of trying to explain this discrepancy in his theory by ingenious attempts, he said: *Something essential to the complete statement of the physical theory of molecular encounters must have hitherto escaped us, and that the only thing to do was to adopt the attitude of thoroughly conscious ignorance that is the prelude to every real advance in science.* He was right. The explanation came 50 years later from quantum theory.

Maxwell's subsequent research was on Theory of Heat conducted in 1871 where he claimed the second law of thermodynamics is a statistical law. So that if one throws a bucket of water into the sea, one cannot get the same bucket of water out again as the law applies to molecules in masses and not to individuals. The second law of thermodynamics states that the total entropy of an isolated system always increases over time, or remains constant in ideal cases where the system is in a steady state or undergoing a reversible process. The increase in entropy accounts for the irreversibility of natural processes, and the asymmetry between the future and the past. Historically, the second law of thermodynamics was an empirical finding that was accepted as an axiom of thermodynamic theory. Statistical thermodynamics, classical or quantum, explains the microscopic origin of the law. The second law has been expressed in many ways. Its first formulation is credited to the French scientist Sadi Carnot in 1824, who showed that there is an upper limit to the efficiency of conversion of heat to work in a heat engine [21].

Maxwell then generated a thought experiment in which he claimed the principles of the second law of thermodynamics might hypothetically be violated. In the thought experiment, a demon controls a small door between two chambers of gas, who has the capability to view each particle of the gas. As individual gas molecules approach the door, the demon quickly opens and shuts the door so that fast molecules pass into the other chamber, while slow molecules remain in the first chamber. Because faster molecules are hotter, the demon's behavior causes one chamber to warm up as the other cools, thus decreasing entropy and violating the Second Law of Thermodynamics.

Several physicists have presented calculations that show that the second law of thermodynamics will not actually be violated, if a more complete analysis is made of the whole system including the demon. The essence of the physical argument is to show, by calculation, that any demon must "generate" more entropy segregating the molecules than it could ever eliminate by the method described. That is, it would take more thermodynamic work to gauge the speed of the molecules and selectively allow them to pass through the opening than the amount of energy gained by the difference of temperature caused by the demon in this thought experiment of Maxwell. One of the most famous responses to this Maxwell's demon problem was addressed in 1929 by Leó Szilárd [22]. Szilard is perhaps best known for patenting particle accelerators and nuclear reactions in the 1930's and then preparing a letter to President Roosevelt delivered by Albert Einstein urging the United States to develop atomic weapons.

But Szilard's great contribution to information philosophy is his connecting an increase in thermodynamic (Boltzmann) entropy with any increase in information that results from a measurement. Szilárd pointed out that a real-life Maxwell's demon would need to have some means of measuring molecular speed, and that the act of acquiring information would require an expenditure of energy. Since the demon and the gas are interacting, we must consider the total entropy of the gas and the demon combined. The expenditure of energy by the demon will cause an increase in the entropy of the demon, which will be larger than the lowering of the entropy of the gas. Now Szilard argued that a gain in information in one place must also result in an increase in entropy in another place. Szilard argued that this demon made a "measurement" of the molecule's position, stored that measurement's information in memory, and then used the information to attach the moving partition to a weight in order to get useful work.

Because the measurement was a simple binary decision between the hotter and the cooler parts of the chamber, Szilard anticipated the idea in Claude Shannon's information entropy that a single "bit" (a binary digit) of information was involved. Szilard calculated the mean value of the quantity of entropy produced by a 1-bit measurement as the entropy $ε = k \log 2$, where k is Boltzmann's constant. The base-2 logarithm reflects the binary decision. The amount of entropy generated by the measurement may, of course, always be greater than this fundamental amount, but not smaller, or the second law of thermodynamics would be violated. Szilard connected the act of measurement to the acquisition of information, which was stored in the memory of the observing demon. This memory was then used to decide how to extract useful work. His discussion parallels in many respects the problem of measurement in quantum mechanics.

1.9.1 Shannon Channel Capacity

Claude Shannon developed an alternate methodology along similar lines 20 years later in 1949 through the introduction of the most popular form of the channel capacity theorem [23, 20]. As he pointed out: *"The ordinary capacity C of a noisy channel may be thought of as follows. There exists a sequence of codes for the channel of increasing block length such that the input rate of transmission approaches C and the probability of error in decoding at the receiving point approaches zero. Furthermore this is not true for any value higher than C. In some situations it may be of interest to consider, rather than codes with probability of error approaching zero, codes for which the probability is zero and to investigate the highest possible rate of transmission for these codes. This rate C_0 is the main object of investigation of the paper."* The point here is that if there is no coding involved at the final transmission stage then the word *channel capacity* carries little or no information.

This point was very well illustrated by Viterbi [24]. As he wrote: *"From the beginning, the additive white Gaussian noise (AWGN) channel was a favorite vehicle of information theorists. But the launch of the first artificial satellite generated a need for conserving transmitted power over a real channel that is closely approximated by this model, and thus served as a powerful stimulus to further theoretical work, particularly toward the goal of minimum-complexity decoding. In the sixties, the demands of satellite networks carrying large amount of data traffic increased the need for coding. In fact, by the end of sixties the state of the art of digital technology, together with the evolution of the theory, rendered feasible coding for the AWGN channel with an error probabilities of 10^{-5}, could conserve 5–6 dB of power relative to the uncoded operation, even at data rates in megabits per second. To gauge the economic advantage, consider that an alternative method of gaining 5 dB power would be to more than triple the area of the receiving antenna".*

Hence, the contribution of Shannon made satellite communication possible where the signal received can be well below the level of thermal noise and the information can still be extracted in such an environment if adequate redundancy is introduced in the form of a code at the radio frequency (RF) stage of transmission. So, in the operation of a real system besides the concept of electromagnetics we need to add another dimension of physics which will characterize the environment and hence it will be possible to extract the information of interest in the presence of other interfering signals. It is important to note that according to Shannon the transmitter is the encoder [25]. Shannon iterated: "*An encoding or Transmitting element. Mathematically this amounts to a transformation applied to a message to produce the signal, i.e., the encoded message*". Shannon also talks about Pulse Code Modulation and Pulse Position Modulation where the coding is done at the transmitter stage at the RF or the carrier frequency. Currently, there are two popular systems which does coding of the signal at the final radio frequency (RF) stage before transmission, namely satellite communication and also in the global positions system (GPS). In wireless communication there is coding, but that is done in the base band and not at the final RF stage, unless one uses totally digital transmission as advocated by Viterbi [26]. Some radars (pulse compression radars) nowadays use the Barker code (A Barker code is a finite sequence of N values of $+1$ and -1. The advantage is that it has a low autocorrelation property. The sidelobe level of a Barker code is typically $1/N$) during transmission and this can be factored into this group.

For those who ignored this subtle but important point made by Shannon – namely the coding at the transmitter or at the final transmission RF stage- Shannon had something to say as outlined in [20, 27]. Shannon wrote: *Information theory has, in the last few years, become something of a scientific bandwagon. Starting as a technical tool for the communication engineer, it has received an extraordinary amount of publicity in the popular as well as the scientific press. In part, this has been due to connections with such fashionable fields as computing machines, cybernetics, and automation; and in part, to the novelty of its subject matter. As a consequence, it has perhaps been ballooned to an importance beyond its actual accomplishments. Our fellow scientists in many different fields, attracted by the fanfare and by the new avenues opened to scientific analysis, are using these ideas in their own problems. Applications are being made to biology, psychology, linguistics, fundamental physics, economics, the theory of organization, and many others. In short, information theory is currently partaking of a somewhat heady draught of general popularity... ......While we feel that information theory is indeed a valuable tool in providing fundamental insights into the nature of communication problems and will continue to grow in importance, it is certainly no panacea for the communication engineer or, a fortiori, for anyone else. Seldom do more than a few of nature's secrets give way at one time. What can be done to inject a note of moderation in this situation? In the first place, workers in other fields should realize that the basic results of the subject are aimed in a very specific direction, a direction that is not necessarily relevant to such fields as psychology, economics, and other social sciences. Indeed, the hard core of information theory is, essentially, a branch of mathematics, a strictly deductive system. A thorough understanding of the mathematical foundation and its communication application is surely a prerequisite to other applications. I personally believe that many of the concepts of information theory will prove useful in these other fields-and, indeed, some results are already quite promising-but the establishing of such applications is not a trivial matter of translating words to a new domain, but rather the slow tedious process of hypothesis and experimental verification.*

If applied in the proper context, the Shannon channel capacity C is defined by

$$C = B\log_2\left(1 + \frac{S}{N}\right) \tag{1.117}$$

where B is the one-sided bandwidth of the channel, and S/N is the signal to noise ratio of the received signal. Hence Shannon considered the number of pulse levels $\left(1 + \frac{S}{N}\right)$ that can be literally sent without any confusion. It is important to emphasize here, that the parameter B in (1.117) of the Shannon Channel capacity formula is the bandwidth of the system and it is in no way related to the value of the carrier frequency. In other words, the carrier frequency can be 1 GHz or 20 GHz if the bandwidth of the system remains the same, the channel capacity remains the same irrespective of the value of the carrier frequency. Another point to be made is that the maximum bandwidth the current devices are capable of handling is at most 500 MHz and therefore going to a higher carrier frequency of transmission only makes sense if some form of multiplexing is used as it is impossible to fabricate devices with a bandwidth greater than 500 MHz. Also, Shannon considered this formula (1.117) for a point to point communication where there is a transmitter and a receiver. Here, S is the power of the received signal. For a wireless system where there is a transmitting antenna involved it is necessary to consider the radiation efficiency of this antenna, in addition to coding. Secondly, an antenna does not radiate power, it radiates an electric field and is related to the power density through the radiating fields. The antenna thus is a voltage sensing device as it integrates the incident electric/magnetic field impinging on it and produces a voltage or a current that needs to be amplified at the first RF stage. It is also important to stress that in nature there are no power amplifiers, it is either a voltage or a current amplifier as an amplifier uses DC voltage to produce an AC signal and therefore power amplification is just a misnomer. An antenna does not radiate power and nor does it receive power! Therefore, without specifying what the transmitter and the receiving antennas are for the system and their associated terminating loads it is difficult to characterize the received power! It must also be pointed out that the noise here is purely random and so multipaths or other interference sources are not involved in the formula.

So the *first question* that arises is how to apply this formula in a near field environment, which is pertinent for operations in a micro or a pico cell environment. In the near field of an antenna the wavefronts are not planar and there are both real and reactive forms of the power. From a Maxwellian point of view, we know that the Poynting vector in the near field is a complex quantity. So what value of the Poynting vector do we put in the capacity formula of (1.117) for the quantity S (the signal power)? Is it the real part or the magnitude of the Poynting vector? This poses an interesting point. Another point is that the capacity also should vary as a function of the distance (the nature of the functional variation is discussed in Chapter 3) between the transmitter and the receiver, particularly if we use an array of antenna elements. So, we need to relate the S/N in the above formula with that of the Poynting vector related to the transmitting antenna and in addition Friis's transmission formula (discussed in the later chapters) to get the total picture. This implies that the channel capacity will be a function of the separation distance between the transmitter and the receiver, and Shannon's formula (1.117) is applicable only at the receiver terminals after the receiving antenna has received the signal. It does not say anything as to the transmit power of the system, but only to the received power which is characterized with respect to the background thermal noise. It is important to state that when performing comparison of performance between various systems the input power must remain the same otherwise the comparison between two systems makes no sense!

In most communication system modeling, one typically deals with the voltages induced in the antennas which are treated as point sources and then takes the square of the absolute value to estimate the power, by relating the autocorrelation function of the voltage to the power spectral density through the Fourier transform. The squared absolute value of any quantity, either the voltage or the current, provides an estimate for power spectral density only for a strictly resistive circuit. For circuits containing complex impedances which can be either inductive or capacitive, as is always the case for any realistic antennas, use of only the voltage or the current is not sufficient to give the value for the power [20]. However, the real part of the power can be evaluated by knowing either the voltage or the current for only the resistors in the circuit. One needs both the current and the voltage and the phase angles between them to compute the total power. The point is that electromagnetics is the basis of Electrical Engineering and one is dealing with a vector problem and not a scalar problem, when dealing with wireless systems. Hence, for a general transmit and receive system in Electrical Engineering, the total complex power can never be computed exclusively from the power spectral density of either only the voltage or the current! One needs both the voltage and the current to compute the complex power for a general circuit. However, only for a purely resistive circuit, power can be obtained from either the voltage or the current. This fundamental principle that one needs, is to use both the voltage and the current to compute the power for real antennas which is often overlooked by communication theory practitioners leading to erroneous designs and conclusions as illustrated in Chapters 2 and 3. For example, it is widely believed in the communication theory and signal processing literature that deployment of multiple antennas will provide a better signal to noise ratio at the receiver. This cannot be far from the truth in a vector problem, as the total field induced in a receiving antenna is the vector sum of all the different field components responsible for inducing currents in the system. So, the problem is how one connects the Shannon channel capacity formula to the Poynting vector which actually provides the radiation efficiency and the gain of an antenna, with the constraint on the input power as the principle of conservation of power should not only be introduced in the presentations but also must be adhered to. Unfortunately the principle of conservation of power is completely missing from most of the related analysis and contemporary signal processing and information theory literature. They are not included in the system design thereby making the systems nonphysical in nature!

However, in many situations, thermal noise may not be the limiting factor in preventing a communication channel from performing its job. For this type of environment Gabor introduced his own version of the Channel capacity theorem. The environment may be interference limited. In this situation perhaps multipaths may limit the detection capability of the signal even though the signal may be several times stronger than the thermal noise. Gabor developed a channel capacity for this scenario where the channel is interference limited and not thermal noise limited.

1.9.2 Gabor Channel Capacity

Gabor investigated very thoroughly the energy levels which can be used, in each cell of his *logon* concept, to represent distinguishable signals [28, 29]. Since, the concept of the Gabor channel capacity is not well known, we include here for completeness. In his analysis, Gabor introduced the phenomenon of beats between the signal and noise (which in this case is deterministic unwanted signals). These beats resulted in increased energy fluctuations. If we superimpose a signal of amplitude A_1 and an interfering signal of amplitude A_2 with a variable phase ψ, then we obtain

$$I = A_1 \cos \omega t + A_2 \cos(\omega t + \psi) \tag{1.118}$$

Therefore,

$$I^2 = A_1^2 \cos^2 \omega t + A_2^2 \cos^2(\omega t + \psi) + 2A_1 A_2 \cos \omega t \cos(\omega t + \psi) \tag{1.119}$$

Now, if we perform an average over a few periods of the oscillation, represented by an over bar on the top and assuming a slow variation of A_2 and ψ, we get

$$\overline{I^2} = A_1^2/2 + A_2^2/2 + A_1 A_2 \cos \psi \tag{1.120}$$

The power over such a relatively short interval of time is

$$\overline{P} = P_1 + \overline{P_2} + 2\left(P_1 \overline{P_2}\right)^{1/2} \cos \psi \tag{1.121}$$

The average over a long interval of time will contain only the first two terms, since the average of cos ψ will be zero [28, 29] resulting in

$$\overline{\overline{P}} = P_1 + \overline{P_2} \tag{1.122}$$

The third term of (1.121) represents beats between the signal and the undesired one called noise. It plays an important role in the power fluctuations as

$$\overline{P} = \overline{\overline{P}} + \Delta p \tag{1.123}$$

And the average of the fluctuations will be given by

$$\overline{\Delta p^2} = \overline{P_2^2} + 4P_1 \overline{P_2} \, \overline{\cos^2 \psi} = \left(\overline{P_2}\right)^2 + 2P_1 \overline{P_2} \tag{1.124}$$

Since for thermal noise we have $(\overline{P_2})^2 = \overline{P_2^2}$ and $\overline{\cos^2 \psi} = 0.5$ we observe that the fluctuations increase with the signal power P_1.

These results can be used for the computation of the number m of distinguishable power levels to evaluate the channel capacity. Let us choose the step of size $\sqrt{2\left(\overline{\Delta p^2}\right)}$ to be consistent with Shannon's formula [28]. This is twice the step size originally proposed by Gabor. Let us say that P_M is the maximum power used in the signal transmission. The number of steps (including one at $P = 0$) available for signaling is

$$m_G = 1 + \int_0^{P_M} \frac{dP}{2\sqrt{\overline{\Delta p^2}}} = \frac{\sqrt{1 + 2\dfrac{P_M}{\overline{P_2}}} + 1}{2} = \frac{\sqrt{1 + 4\dfrac{P_{av}}{\overline{P_2}}} + 1}{2} \tag{1.125}$$

As illustrated in [28], for small values of the ratio $P_{av}/\overline{P_2}$, the number m_G reduces to $1 + (P_{av}/\overline{P_2})$, which is identical to that of Shannon's channel capacity. For large values of signal-to-noise ratio, m_G increases more slowly with P_{av}, the average value, than the value corresponding to Shannon's formula.

According to the actual conditions of observation, it will be necessary to investigate whether the additional fluctuations due to beats between the signal and noise may be of importance or not. That discussion will determine whether to use Shannon's or Gabor's channel capacity formula. Gabor's channel capacity formula then is given by

$$C_G = \Delta B \log_2 \left[\frac{1 + \sqrt{1 + \dfrac{2P_M}{\overline{P_2}}}}{2} \right] = \Delta B \log_2 \left[\frac{1 + \sqrt{1 + \dfrac{4P_{av}}{\overline{P_2}}}}{2} \right] \tag{1.126}$$

When a long time of transmission is selected, the cos ψ term averages to zero and the beats between the signal and the noise can be ignored, a situation which justifies the use of Shannon's formula in the limit.

Next, we deal with the Tuller channel capacity as it directly interfaces with the physics of electromagnetics. It is interesting to note that Tuller, a contemporary of Shannon, was an electrical engineer in contrast to Shannon who was a mathematician and Gabor who was a physicist. Tuller being an engineer actually designed a system using this type of theoretical analysis and it worked exactly as predicted [20].

1.9.3 Hartley-Nyquist-Tuller Channel Capacity

The third form of the channel capacity was developed by Tuller who actually built the first Private Line Air Traffic Control Communication System. We call it the Hartley-Nyquist-Tuller Capacity formula. As Tuller wrote [30–32] "*The general principles of the application of information theory to system design is discussed here and an example is given showing its use in studying the Private Line Air Traffic Control Communication System*". Tuller actually designed and built the first of the "private line" communication link between the aircraft traffic controller and the aircraft under their surveillance. In doing so it was clear to Tuller that power plays little role when dealing with a practical communication system but rather the sensitivity of the system depends on the intensity or equivalently the value of the fields. The sensitivity of a receiver is determined by the minimum electric field intensity (typically of the order of $1 \, \mu V/m$) it can receive and below that threshold, the receiver will not recognize any signal and therefore the system will not function at all. So Tuller used a formula for the channel capacity that dealt with the values of the electric fields and not in terms of power! He followed Nyquist [33] for the development of his channel capacity formula but unlike Nyquist, Tuller took the effect of noise into account. It was also important to note that Tuller was aware of Shannon's work and reference to it is made in his 1949 paper [30] and it is also interesting to observe that Shannon was also aware of Tuller's work [30] and also made references to other forms of channel capacity (developed by Norbert Weiner and H. Sullivan besides Tuller) theorems in his classical paper titled *A Mathematical Theory of Communication*.

A practical form of the channel capacity, which is more applicable to real systems, was first developed by Hartley [34–36] in 1928, where he generalized the earlier results. Among his conclusions, he stated, "*The total amount of information which may be transmitted over a system whose transmission is limited to frequencies lying in a restricted range is proportional to the product of the frequency range which it transmits by the time during which it is available for transmission*."

Hartley's treatment represented a first step in the direction of measuring a message and the message-transmitting capacity of a system [35]. Both Nyquist [33, 36] and Hartley [34, 35] showed, what must indeed have been common knowledge from the earliest days of telegraphy, namely, how the number of different signals available to a single receiver is limited by the relative magnitude of interference [35]. Also, as the signals traverse the transmission medium (or channel, as it is commonly called), they are distorted, noise and interfering signals are added, and it becomes a major task to interpret the signals correctly at the desired destination [37]. However, we deal with physical systems, and these systems do not allow us to increase the rate of signal change indefinitely or to distinguish indefinitely many voltage amplitudes or levels [37]:

1) *All of our systems have energy-storage devices present, such as inductances and capacitances, and changing the signal implies changing the energy content. Limits on making these changes are determined by the bandwidth of a particular system.*
2) *Every system provides inherent (even if small) variations or fluctuations in voltage, or whatever parameter is used to measure the signal amplitude. One cannot subdivide amplitudes indefinitely. These unwanted fluctuations of a parameter to be varied are called noise.*

The above two limitations are tied together by a simple expression developed for system capacity [37]: $C = 1/r \times \log_2 n$, where r is the minimum time required for the system to respond to signal changes, and n is the number of distinguishable signal levels. The minimum response time is proportional to the reciprocal of the system bandwidth [37]. Thus, system capacity can be written as $C = B \times \log_2 n$, where B is the system bandwidth in Hz. The information about the number of different amplitude levels can be sent by fewer numbers simply by binary encoding the different levels.

For example, if the signal amplitude can be separated into 16 different levels, then we require four bits to approximate it. If the signal amplitude has 32 different levels, we require five bits to characterize the signal. More generally, $\log_2 n$ bits are required for characterizing n levels of the signal. The Hartley capacity can be shown to be equivalent to the Shannon's definition for the capacity, when both the transmitting and the receiving antenna systems are conjugately matched. The basic form of Shannon's theorem states [23] that *to distinguish between M different signal functions of duration T on a channel, we can say that the channel can transmit* $\log_2 M$ *bits in time T. The rate of transmission is then* $\log_2 (M/T)$. *More precisely, the channel capacity, C, may be defined as*

$$C' = \lim_{T \to \infty} \frac{\log_2 M}{T} \tag{1.127}$$

By following Hartley, one can obtain a simplified expression for the capacity without delving into the concept of power, if one requires the received signal to be separated into 2Q distinct levels, so that the number of amplitudes that can be reasonably well distinguished is 2Q. Furthermore, as Shannon points out, *since in time T there are 2TB independent amplitudes, then the total number of distinct signals is*

$$M = \left(2^Q\right)^{2BT},$$

(1.128)

where B is the one-sided bandwidth of the signal. Hence, the number of bits that can be sent in this time T is

$$\frac{\log_2 M}{T} = 2BQ.$$

(1.129)

This expression is more pertinent in characterizing near-field environments, and is similar to the Hartley-Nyquist-Tuller form of the channel-capacity theorem, which has been elaborated in [30–32]

$$C_{HNT} = 2B\log_2\left(1 + \frac{A_S}{\Delta V}\right)$$

(1.130)

where A_s is the received root-mean-square (rms) amplitude voltage of the signal (and is related to the incident electric field of the signal), and ΔV is the rms level of the quantization noise voltage related to the discretization of the received signal. Using this formula for the channel capacity is simpler as it connects directly to the physical parameters of the system and may provide a different value for the channel capacity than that obtained when it is computed with the power relationship. This form dealing with the voltage and not power is ideally suited for wireless systems, since the capacity is directly related to the source, namely, the fields, and therefore is applicable both for near-field and far-field characterizations of the antennas. In this formula, the limiting factor is the lowest discretization level to which the voltages are quantized and can be discerned after propagation to the receiver.

It is important to note that these three formulas for channel capacity, namely Shannon, Gabor and Tuller do not represent the same transmission performance: as one is essentially error free; the others are error-free only in the statistical sense in that the average rate of errors can be made arbitrarily small. Therefore, if one assumes that the background noise characteristics are very similar in the two cases, then the two expressions might yield similar results for conjugately matched transmitting and receiving antennas as then they will be radiating the maximum power available from the source and the antennas will be resonant implying that the input power is real and directly related to the voltage.

In summary, in order to introduce the concept of channel capacity one needs to have coding at the RF stage as without the coding the discussion on use of the channel capacity quantifications is not very fruitful. This makes references

[26, 38, 39] quite relevant as it describes such scenarios. In other words the Maxwell-Poynting theory associated with the radiated fields from antennas need to be embedded in the channel capacity formulas to make them realistic and more amenable to system design.

1.10 Conclusion

The objective of this chapter has been to present the necessary mathematical formulations, popularly known as Maxwell's equations, which dictate the space-time behavior of antennas. Additionally, some examples are presented to note that the impulse response of antennas is quite complicated and the wave shapes depend on both the observation and the incident angles in azimuth and elevation of the electric fields. Specifically, the transmit impulse response of an antenna is the time derivative of the impulse response of the same antenna in the receive mode. This is in contrast to the properties in the frequency domain where the transmit antenna pattern is the same as the receive antenna pattern. Any broadband processing must deal with factoring out the impulse response of both the transmitting and receiving antennas. The examples presented in this chapter do reveal that the wave shape of the impulse response is indeed different for both transmit and receive modes, which are again dependent on both the azimuth and elevation angles. For an electrically small antenna, the radiated fields produced by it along the broadside direction are simply the differentiation of the time domain wave shape that is fed to it. While on receive it samples the field incident on it. However, for a finite-sized antenna, the radiated fields are proportional to the temporal double derivative of the current induced on it, and on receive, the same antenna differentiates the transient electric field that is incident on it. Hence all baseband broadband applications should deal with the complex problem of determining the impulse responses of the transmitting and receiving antennas. This is in contrast to spread spectrum methodologies where one deals with an instantaneous narrowband signals even during frequency hopping. For the narrowband case, determination of the impulse response is not necessary. The goal of this chapter is to outline the methodology that will be necessary to determine the impulse response of the transmit/receive antennas. This lays the foundation to comprehend what is actually fading and what is not. In addition the distinction between correlation and convolution has been delineated illustrating that superposition of power is not applicable in electrical engineering particularly when unmodulated and non-orthogonal carriers are considered. It is shown that superposition applies to convolution but not to correlation, which is power. Finally, it is illustrated that it is possible to combine the electromagnetic analysis with the signal-processing algorithms, particularly the concept of channel capacity. Such a connection is absolutely essential to be able not only to design better systems but also systems that work on the first design and no further tweaking is necessary. Also, the philosophical and the mathematical basis of various channel capacity concepts are described and they follow directly from the establishment of the Maxwellian concept of entropy.

References

1 J. C. Maxwell, "A Dynamical Theory of the Electromagnetic Field," *Philosophical Transactions*, Vol. 166, pp. 459–512, 1865 (reprinted in the *Scientific Papers of James Clerk Maxwell*, Vol. 1, pp. 528–597, Dover, New York, 1952).

2 H. Hertz, "On the Relations between Maxwell's Fundamental Equations of the Opposing Electromagnetics" (in German), *Wiedemann's Annalen*, Vol. 23, pp. 84–103, 1884. (English translation in [9, pp. 127–145].)

3 H. Hertz, "On the Fundamental Equations of Electromagnetics for Bodies at Rest," in H. Hertz (ed.), *Electric Waves*, Dover Publications Inc., New York, pp. 195–240, 1962.

4 T. K. Sarkar, M. Salazar-Palma, and D. L. Sengupta, "Who Was James Clerk Maxwell and What Was and Is His Electromagnetic Theory?," *IEEE Antennas and Propagation Magazine*, Vol. 51, No. 4, pp. 97–116, 2009.

5 C.-T. Tai and J. H. Bryant, "New Insights into Hertz's Theory of Electromagnetism,"*Radio Science*, Vol. 29, No. 4, pp. 685–690, July–Aug. 1994.

6 B. J. Hunt, *The Maxwellians*, Chap. 2, p. 24, Cornell University Press, Ithaca, NY, 1991.

7 T. K. Sarkar, R. J. Mailloux, A. A. Oliner, M. Salazar-Palma, and D. L. Sengupta, *History of Wireless*, Wiley/IEEE Press, Hoboken, NJ, 2006.

8 H. Hertz, *Electric Waves* (authorized English translation by D. E. Jones), Dover, New York, 1962.

9 J. H. Bryant, *Heinrich Hertz: The Beginning of Microwaves*, IEEE Service Center, Piscataway, NJ, 1988.

10 O. Heaviside, *Electromagnetic Theory*, Vols. I, II and III, The Electrician Series, Printing and Publishing Company Limited, London, 1893.

11 J. G. O'Hara and W. Pritcha, *Hertz and Maxwellians*, Peter Peregrinus, London, UK, 1987.

12 J. D. Kraus, *Electromagnetics*, McGraw-Hill, New York, 1980.

13 R. Nevels and C. Shin, "Lorenz, Lorentz, and the Gauge," *IEEE Antennas and Propagation Magazine*, Vol. 43, No. 3, pp. 70–72, June 2001.

14 J. D. Kraus, *Antennas*, McGraw-Hill, New York, 1988.

15 R. F. Harrington, *Time Harmonic Electromagnetic Fields*, Wiley-IEEE Press, New York, 2001.

16 R. E. Collin, *Foundations for Microwave Engineering*, 2nd edition, Wiley-IEEE Press, 2000.

17 D. L. Sengupta and C. T. Tai, "Radiation and Reception of Transients by Linear Antennas," in L. B. Felsen (ed.), *Transient Electromagnetic Fields*, Springer-Verlag, New York, Chap. 4, pp. 182–234, 1976.

18 T. K. Sarkar, W. Lee, and S. M. Rao, "Analysis of Transient Scattering from Composite Arbitrarily Shaped Complex Structures," *IEEE Transactions on Antennas and Propagation*, Vol. 48, No. 10, pp. 1625–1634, Oct. 2000 and also in B. H. Jung, T. K. Sarkar, S. Ting, Y. Zhang, M. Mei, Z. Ji, M. Yuan, A. De, M. Salazar-Palma, and S. M. Rao, *Time and Frequency Domain Solutions of EM Problems Using Integral Equations and a Hybrid Methodology*, IEEE Press/ John Wiley & Sons, 2010.

19 F. I. Tseng and D. K. Cheng, "Antenna Pattern Response to Arbitrary Time Signals," *Canadian Journal of Physics*, Vol. 42, No. 7, pp. 1358–1368, July 1964.

20 T. K. Sarkar, E. L. Mokole, and M. Salazar-Palma, "Relevance of Electromagnetics in Wireless System Design," *IEEE Aerospace and Electronics Systems Magazine*, Vol. 31, No. 10, pp. 8–19, 2016.

21 T. Sarkar, S. Burintramart, N. Yilmazer, Y. Zhang, A. De, M. Salazar-Palma, M. Lagunas, E. Mokole, and M. Wicks, "A Look at the Concept of Channel Capacity from a Maxwellian Viewpoint," *IEEE Antennas and Propagation Magazine*, Vol. 50, No. 3, pp. 21–50, 2008.

22 L. Szilard, "Über die Entropieverminderung in einem thermodynamischen System bei Eingriffen intelligenter Wesen (On the Reduction of Entropy in a Thermodynamic System by the Intervention of Intelligent Beings)," *Zeitschrift für Physik*, Vol. 53,pp. 840–856, 1929. English translation available as NASA document TT F-16723 published 1976 and English translation by A. Rapport and M. Knoller, *Behavioral Science*, Vol. 9, p. 301, 1964.

23 C. E. Shannon, "Communication in the Presence of Noise," *Proceedings of the IEEE*, Vol. 86, No. 2, pp. 447–457, Feb. 1998. (Reprinted from *Proceedings of IRE*, Vol. 37, No. 1, pp. 10–21, 1949.)

24 A. J. Viterbi, "Information Theory in the Sixties," *IEEE Transactions on Information Theory*, Vol. 19, No. 3, pp. 257–262, 1973.

25 C. E. Shannon, "General Treatment of the Problem of Coding," *Transactions of IRE on Information Theory*, Vol. 1, No. 1, pp. 102–104, 1953.

26 A. J. Viterbi, "The Evolution of Digital Wireless Technology from Space Exploration to Personal Communication Services," *IEEE Transactions on Vehicular Technology*, Vol. 43, No. 3, pp. 638–644, 1994.

27 C. E. Shannon, "The Bandwagon (Edtl)," *Transactions of IRE on Information Theory*, Vol. 2, No. 1, p. 3, 1956.

28 L. Brillouin, *Science and Information Theory*, 2nd edition, Academic Press, New York, 1962.

29 D. Gabor, Lectures in communication theory, Tech. Rep. 238. Massachusetts Institute of Technology, Research Laboratory of Electronics, Cambridge, MA, Apr. 1992.

30 W. G. Tuller, "Theoretical Limitations on the Rate of Transmission of Information," *Proceedings of the IRE*, Vol. 37, No. 5, pp. 468–478, 1949.

31 W. G. Tuller, "Application of Information Theory to System Design," *Electrical Engineering*, Vol. 70, No. 2, pp. 124–126, 1951.

32 W. G. Tuller, "Information Theory Applied to System Design," *Transactions of American Institute of Electrical Engineers*, Vol. 69, No. 2, pp. 1612–1614, 1950.

33 H. Nyquist, "Certain Factors Affecting Telegraph Speed," *Bell System Technical Journal*, Vol. 3, p. 324, 1924.

34 R. V. L. Hartley, "Transmission of Information," *Bell System Technical Journal*, Vol. 7, pp. 535–563, 1928.

35 H. S. Black, *Modulation Theory*, D. Van Nostrand Company Inc., New York, 1953.

36 H. Nyquist, "Certain Topics in Telegraph Transmission Theory," *AIEE Transactions*, 47, pp. 617–644, Apr. 1928.

37 M. Schwartz, *Information, Transmission, Modulation, and Noise*, McGraw Hill Book Company, New York, 1959.

38 A. J. Viterbi, "Wireless Digital Communication: A View Based on Three Lessons Learned," *IEEE Communications Magazine*, Vol. 29, No. 9, pp. 33–36, 1991.

39 A. Viterbi, "A Fresh Look at the Physical Layer of Terrestrial Mobile Multiple Access Networks," *IEEE International Conference on Personal Wireless Communications*, New Delhi, India, 1996, pp. 1–7.

2

Characterization of Radiating Elements Using Electromagnetic Principles in the Frequency Domain

Summary

In this chapter we are going to look at the fields produced by some simple antennas and their current distributions in the frequency domain. Specifically, we will observe the fields produced by a Hertzian dipole, a finite sized dipole antenna, and a small loop antenna. We will define the radiating regions of an antenna in terms of the near fields from the far fields and how that is tied to radiation. We will define what is meant by the term radiation and what does it signify and how does it relate to directivity and gain. Examples will be presented to illustrate what are valid near field/far field modeling including analysis of antennas over an imperfectly conducting earth as in a wireless communication environment.

In electrical engineering, we obtain maximum average power from a source with some internal impedance when the connected load impedance equals the complex conjugate of the internal source impedance. This is known as the maximum power transfer theorem. Applying this theorem means that the best we can do is to distribute the source power equally between the source internal impedance and the load impedance; i.e., the efficiency of such a system is at most 50%. Efficiency takes into account the ratio of the dissipated power in the load divided by the source power, while on the other hand the maximum power transfer considers only the magnitude of the dissipated power. If we increase the resistance of the load more than the internal resistance of the source then we will achieve better efficiency, however the magnitude of the dissipated power will be less since the total resistance in the circuit is increased. We will try to emphasize this fact in antenna problems and show that considering radiation efficiency is more appropriate than considering maximum power transfer. Analysis is performed on dipole antennas of various lengths to show

The Physics and Mathematics of Electromagnetic Wave Propagation in Cellular Wireless Communication, First Edition. Tapan K. Sarkar, Magdalena Salazar Palma, and Mohammad Najib Abdallah.
© 2018 John Wiley & Sons, Inc. Published 2018 by John Wiley & Sons, Inc.

that the maximum power transfer principle used in impedance matching is not the optimum solution in terms of efficiency of a radiating/receiving antenna.

The concept of an Electrically Small Antenna (ESA) is discussed next and it will be shown that contrary to popular belief, an ESA can perform well compared to larger antennas under matched conditions. In this chapter, we study the difficulties in achieving a good design of an ESA matched to a system. Methodologies are then explored to match an ESA.

The nature of the near and far field properties of an antenna are considered when deploying them in free space and over Earth. Next the concept of diversity is introduced to illustrate that it has little effect on the pattern of an antenna array. Performance of antennas over different type of grounds and in the presence of obstacles in the LOS (line-of-sight) propagation is also considered. The fields from an antenna operating inside a room is considered. It is shown that the antenna is operating in a near field environment irrespective of the size of the room due to the generation of multiple images created by the various walls, floor and the ceiling. It is shown that the antenna does not radiate when placed inside a conducting room. The important part in this case is that there may be some fields coming out of the antenna but they are not radiating fields. Hence, it is difficult to make sense out of the myriads of experiments carried out to characterize a rich multipath environment by placing antennas in a chamber made of conductors. The mathematics and physics of an antenna array is also discussed.

The concepts and the philosophy of the use of Multiple-Input Multiple-Output (MIMO) antenna principles are presented. Even though from a theoretical point of view the concept of using the same spatial channel for simultaneous transmission of multiple signals is quite interesting, there are some major problems when it comes to the physical realization! It is also illustrated how the principle of reciprocity can be used to direct a signal to the receiving antenna of interest and simultaneously placing nulls along the undesired receivers without even characterizing the channel. This is equivalent to adaptivity on transmit. In the signal processing literature this is also known as the zero-forcing algorithm.

Finally, the Appendix describes where the far field of an antenna starts when it is operating in free space and/or over an Earth.

2.1 Field Produced by a Hertzian Dipole

Consider an element of current oriented along the z-direction as shown in Figure 2.1. The current element, a Hertzian dipole, is an infinitesimally small current element so that [1]

$$\iiint_{\substack{\text{volume encompassing} \\ \text{the source}}} J_z^i \, dv = I_d l \tag{2.1}$$

where J_z^i is the current density (amp/m³). This particular current distribution belongs to an antenna called an electric dipole. As shown in Figure 2.1, l is the length of the wire along which the current I_d flows. This element of current terminates and originates from two charges. The uniform current element I_d distributed along the entire length l gives rise to the charges $+Q$, and $-Q$, which in turn gives rise to a displacement current that flows through space. The current is related to the charge by $I_d = dQ/dt$ [1–4].

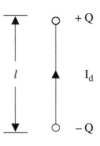

Figure 2.1 An equivalent circuit of a Hertzian dipole.

The magnetic vector potential produced by this elementary dipole located at (x',y',z') is given by

$$\mathbf{A}(x,y,z) = \hat{z}\frac{\mu_0 I_d l}{4\pi}\frac{\exp(-jkR)}{R} \tag{2.2}$$

where x, y, z are the coordinate points at which the potential is evaluated. In addition $k = 2\pi/\lambda = 2\pi f/c = 2\pi f\sqrt{\mu_0\varepsilon_0}$, where λ is the wavelength of the wave corresponding to a frequency f and c is the velocity of light in free space, and $\delta(\cdot)$ is the classical delta function. If the dipole is located at the origin, we have

$$R = \sqrt{(x-x')^2 + (y-y')^2 + (z-z')^2} = \sqrt{x^2 + y^2 + z^2} = r \tag{2.3}$$

It is now convenient to obtain the magnetic vector potential in spherical coordinates (r, θ, ϕ). The components of the magnetic vector potential are:

$$A_r = A_z\cos\theta = \frac{\mu_0 I_d l e^{-jkr}\cos\theta}{4\pi r} \tag{2.4}$$

$$A_\theta = -A_z\sin\theta = \frac{-\mu_0 I_d l e^{-jkr}\sin\theta}{4\pi r} \tag{2.5}$$

$$A_\varphi = 0 \tag{2.6}$$

Since, the magnetic field intensity is given by, $\mathbf{H} = (\nabla \times \mathbf{A})/\mu_0$, we obtain

$$H_\varphi = \frac{1}{\mu_0 r}\left[\frac{\partial}{\partial r}(rA_\theta) - \frac{\partial A_r}{\partial\theta}\right] = \frac{I_d l e^{-jkr}\sin\theta}{4\pi r^2} + \frac{jkI_d l e^{-jkr}\sin\theta}{4\pi r} \tag{2.7}$$

with $H_r = H_\theta = 0$. The first term in (2.7) is due to the induction field of a current element and can be obtained from Ampere's law of (1.8). This is also the induction field with a time retardation or phase delay. It is in phase with the exciting current I_d and decreases as the inverse square law of the distance. The second term in (2.7) is in phase quadrature with the excitation current and remains so, even if we come in close proximity to the current element. This second component does not occur in the application of Ampere's law. We shall see later that this term is related to the radiation of the fields generated by the dipole.

Next, we find the electric field intensity associated with this magnetic field. Since $E = (\nabla \times H)/(j\omega\varepsilon_0)$, therefore

$$E_r = \frac{-jI_d le^{-jkr}\cos\theta}{2\pi\omega\varepsilon_0\, r^3} + \frac{kI_d le^{-jkr}\cos\theta}{2\pi\omega\varepsilon_0\, r^2} \tag{2.8}$$

$$E_\theta = \frac{kI_d le^{-jkr}\sin\theta}{4\pi\omega\varepsilon_0\, r^2} - \frac{jI_d le^{-jkr}\sin\theta}{4\pi\omega\varepsilon_0\, r^3} + \frac{jk^2 I_d le^{-jkr}\sin\theta}{4\pi\omega\varepsilon_0\, r} \tag{2.9}$$

with $E_\varphi = 0$. There are two different variations of the field components with respect to the distance for E_r. The $1/r^3$ variation of the field intensity is due to the fields produced by a charge and it represents the static fields from a dipole and the $1/r^2$ variation of the fields is due to electromagnetic induction or is often referred to as transformer action. Therefore, very close to the current element, the **E** field reduces to that of a static charge dipole and the **H** field reduces to that produced by a constant current element. So, the fields are said to be quasi-static. At intermediate values of r the field is said to be the induction field.

Next, we look at the direction and the amplitude of the flow of the power density (power per unit area) from the dipole. That is given by the Poynting vector S_{dipole} characterizing the complex power density and can be mathematically expressed as [1–4]

$$S_{dipole} = \frac{1}{2}\left(E \times H^*\right) = \frac{1}{2}\left(\hat{r}E_\theta H_\varphi^* - \hat{\theta}E_r H_\varphi^*\right)$$

$$= \hat{r}\frac{\eta k^2 |I_d l|^2 \sin^2\theta}{32\pi^2 r^2}\left[1 - \frac{j}{(kr)^3}\right] + \hat{\theta}\frac{j\eta k |I_d l|^2 \cos\theta\sin\theta}{16\pi^2 r^3}\left[1 + \frac{1}{(kr)^2}\right] \tag{2.10}$$

Here the superscript * denotes the complex conjugate. The factor ½ will not be there if we use the root mean square (rms) values rather than the peak values for the fields. In the later chapters we shall be using the rms values and therefore the factor of ½ will not be there. And $\eta = k/(\omega\varepsilon_0) = \sqrt{\mu_0/\varepsilon_0} = 120\pi$

defines the characteristic impedance of free space and is approximately equal to 377 ohms. $\hat{r}$ is the unit vector in the radial direction and $\hat{\theta}$ is the unit vector along the elevation angle, or formally called the polar angle. The outward directed power along the radial direction is obtained by integrating the Poynting vector over a sphere of radius r

$$P_{dipole} = \oiint S_{dipole} \cdot dS = \oiint S_{dipole} \cdot \left(\hat{r} \, r^2 \sin \theta \, d\theta \, d\phi \right)$$

$$= \int_0^{2\pi} d\phi \int_0^{\pi} d\theta \left(r^2 \sin \theta \, E_\theta H_\phi^* \right) = \frac{\pi \eta}{3} \left| \frac{I_d l}{\lambda} \right|^2 \left[1 - \frac{j}{(kr)^3} \right]$$

$$(2.11)$$

The second term, or the θ-component of the Poynting vector of (2.10) does not contribute to the radiated power. Therefore the entire outward power flow comes from the first term. This real part of the power, when the Poynting vector, related to the power density, is integrated over a sphere and is independent of the distance r. It represents the power crossing the surface of a sphere, even if the radius of the sphere is infinite. The reactive part of the power diminishes with the distance and vanishes when $r = \infty$. Since the reactive power is negative, it indicates that there is an excess of electric energy over the magnetic energy in the near field, at a finite distance r from the dipole.

2.2 Concept of Near and Far Fields

An alternating current in a circuit has a near field and a far field. In the near field, it is assumed that the fields are concentrated near the source (http://www.majr.com/docs/Understanding_Electromagnetic_Fields_And_Antenna.pdf). The radiating field is referred to as the far field as its effect extends beyond the source, particularly, it is the received power at infinity. To illustrate this point, we observe that the scalar power density S_{dipole} characterized by (2.10) from the dipole at a finite distance from a source can be represented by

$$S_{dipole} = \frac{C_1}{r^2} + \frac{C_2}{r^3} + \frac{C_3}{r^4} + \cdots$$

$$(2.12)$$

Now consider a sphere with radius r, centered at the source. Then the total power passing the surface of the sphere will be given by

$$P = \left(4\pi r^2 \right) S_{dipole} = 4\pi \left(C_1 + \frac{C_2}{r} + \frac{C_3}{r^2} + \cdots \right)$$

$$(2.13)$$

It is seen that the first term is a constant. So for this term no matter what size of the sphere is chosen in space, the same amount of power flows through it and it demonstrates that power is flowing away from the source and is called the radiation field. The other terms become negligible as r gets large. Consequently at a small distance r, the other terms become much larger compared to the radiative field. These other terms taken together represent the power in the near field or termed as the reactive field.

Therefore in the far field there are only two components of the field for the dipole. They are

$$H_\varphi \approx \frac{jk I_d l e^{-jkr} \sin\theta}{4\pi r} = \frac{je^{-jkr} \sin\theta I_d l}{2r}\frac{}{\lambda} \tag{2.14}$$

$$E_\theta \approx \frac{jk^2 I_d l e^{-jkr} \sin\theta}{4\pi\omega\varepsilon_0 r} = \frac{j\eta e^{-jkr} \sin\theta I_d l}{2r}\frac{}{\lambda} \tag{2.15}$$

It is now seen that their ratio is given by $E_\theta / H_\varphi = k/(\omega\varepsilon_0) = \sqrt{\mu_0/\varepsilon_0} = \eta = 120\pi \approx 377\Omega$ where η is the characteristic impedance of free space. Therefore, in the far field the electric and the magnetic fields are orthogonal to each other in space, but coherent in time. In addition, they are related by the characteristic impedance of free space. The emanating wave generated from the source in the far field is a plane wave. This implies that there is also no radial component in the far fields. The factor $\sin\theta$ in the two expressions represents the radiation pattern. It is the field pattern at $r \to \infty$. This field pattern is independent of the distance r and is associated with the far field component. The Poynting vector and the power flow can then be approximated in the far field as

$$S_{dipole} = \hat{r}\frac{\eta k^2 |I_d l|^2 \sin^2\theta}{32\pi^2 r^2} = \hat{r}\frac{15\pi \sin^2\theta |I_d l|^2}{r^2}\left|\frac{I_d l}{\lambda}\right|^2 \tag{2.16}$$

$$P_{dipole} = \frac{2\pi\eta}{3}\left|\frac{I_d l}{\lambda}\right|^2 = 80\pi^2\left|\frac{I_d l}{\lambda}\right|^2 \tag{2.17}$$

In the far field the power flow is finite and is independent of the distance r.

So the differences between the properties of the reactive near field and the radiated far field can be summarized as follows:

1) In the far field there is always a transmission of energy through space whereas in the reactive near field the energy is stored in a fixed location.
2) In the far field, the energy is radiated outward from the source and is always a real quantity as the waves of **E** & **H** fields move through space, whereas, in the reactive near field, energy is recoverable as they oscillate in space and time.

3) In the far field, the **E** & **H** fields decrease as the inverse of the distance, whereas in the near fields they follow an inverse squared law or higher.

4) The electric and the magnetic far fields are in space quadrature as seen from (2.14) and (2.15) but they are coherent in time. The fields form a plane wave. Whereas in the reactive near field, the electric and the magnetic fields can have any spatial or temporal orientation with respect to each other.

5) The radiated far field is related to the real part of the antenna impedance, whereas the reactive part of the impedance is related to the reactive near field.

6) In the far field, the antenna field pattern is defined independent of the spatial distance from the antenna but is related only to the spatial angles as the r terms in the field expressions are generally not deemed relevant.

7) In the far field, the power radiated from the antenna is always real and is considered to exist even at infinity whereas the reactive power component becomes zero.

We will define a rule of thumb by placing some mathematical constraints on the distance r later on to demarcate the near and the far field regions. A detailed discussion of this topic is carried out in the Appendix 2A.

At this point it might be useful to introduce the concept of a point source [2] instead of an infinitesimal current element of a finite length l. If we focus our attention only to the far field region of the dipole, we observe that the electric and the magnetic fields are transverse to each other and the power flow or the Poynting vector is oriented along the radial direction. It is convenient in many analyses to assume that the fields from the antenna are everywhere of this type. In fact, we may assume, by extrapolating inward along the radii of the circle along the direction r that the waves originate at a fictitious volumeless emitter, or a point source at the center of the observation circle of radius r. The actual field variation near the antenna, or the near fields is ignored and we describe the source of the waves only in terms of the far field, it produces. Provided that the observations are made at a sufficient distance, any antenna, regardless of its size or complexity, can be represented in this way by a point source [2]. A complete description of the far field of a source requires knowledge of the electric field as a function of both space and time. For many purposes, however, such a complete knowledge is not necessary. It may be sufficient to specify merely the variation with angle of the power density from the antenna. In this case the vector nature of the field is disregarded, and the radiation is treated as a scalar quantity. When polarization of the fields is of interest then the variation of the nature of the fields must be specified as a function of time. ***Hence, we lose the vector nature of the problem when we approximate an antenna by a point source*** [2]. An accurate characterization of the near and far fields is given in Appendix 2A to illustrate the limitations of the currently used rule of thumb in the characterization of wireless systems.

Next, we consider the electric and magnetic fields produced by a small circular loop of current.

2.3   Field Radiated by a Small Circular Loop

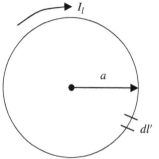

Consider a small loop antenna of radius a placed symmetrically on the x-y plane at $z = 0$ as shown in Figure 2.2. The wire loop is considered to be very thin and is assumed to have a constant current I_ℓ distribution along the circumference.

To calculate the field radiated by the loop, we first need to find the magnetic vector potential **A** as [2–4]

Figure 2.2 A loop made of very thin wire.

$$\mathbf{A}(x, y, z) = \frac{\mu}{4\pi} \int_c \mathbf{J}(x', y', z') \frac{e^{-jkR}}{R} dl' \qquad (2.18)$$

$$R = \sqrt{r^2 + a^2 - 2ar\sin\theta\cos(\phi - \phi')}, \qquad (2.19)$$

and

$$dl' = ad\phi \qquad (2.20)$$

We have,

$$\mathbf{J} = \hat{r} I_\ell \sin\theta \sin(\phi - \phi') + \hat{\phi} I_\ell \cos\theta \sin(\phi - \phi') + \hat{\theta} I_\ell \cos(\phi - \phi') \qquad (2.21)$$

$$
\begin{aligned}
x &= r\sin\theta\cos\theta & \qquad x' &= a\cos\phi' \\
y &= r\sin\theta\sin\phi & \qquad y' &= a\sin\phi' \\
z &= r\cos\theta & \qquad z' &= 0 \\
x^2 + y^2 + z^2 &= r^2 & \qquad x'^2 + y'^2 + z'^2 &= a^2
\end{aligned}
\qquad (2.22)
$$

Therefore,

$$A_\varphi \cong \frac{a\mu}{4\pi} \int_0^{2\pi} I_\ell \cos(\phi - \phi') \frac{e^{-jkR}}{R} d\phi' \qquad (2.23)$$

If we set $\phi = 0$ for simplicity, then the radiated field can be obtained as

$$A_\varphi \cong \frac{a\mu I_\ell}{4\pi} \int_0^{2\pi} \cos\phi' \left[\frac{1}{r} + a\left(\frac{jk}{r} + \frac{1}{r^2} \right) \sin\theta \cos\phi' \right] e^{-jkr} d\phi$$

$$\cong \frac{a\mu I_\ell}{4} e^{-jkr} \left(\frac{jk}{r} + \frac{1}{r^2} \right) \sin\theta$$

$$(2.24)$$

The integration for A_r and A_θ becomes zero. Therefore

$$H_r = j\frac{ka^2 e^{-jkr} I_\ell \cos\theta}{2r^2}\left[1+\frac{1}{jkr}\right] \tag{2.25}$$

$$H_\theta = -\frac{(ka)^2 I_\ell \sin\theta}{4r}\left[1+\frac{1}{jkr}\right]e^{-jkr} \tag{2.26}$$

$$E_\varphi = \frac{\eta(ka)^2 e^{-jkr} I_\ell \sin\theta}{4r}\left(1+\frac{1}{jkr}\right) \tag{2.27}$$

with $H_\varphi = E_r = E_\theta = 0$ \hfill (2.28)

It is interesting to note that the complex power density **S** radiated by the loop is given by

$$\mathbf{S}_{loop} = \frac{1}{2}\left(\mathbf{E}\times\mathbf{H}^*\right) \tag{2.29}$$

(Here we are using the peak values. For root mean square (rms) values the factor ½ will not be there) which yields

$$\mathbf{S}_{loop} = \hat{r}\frac{(ka)^4}{32}|I_\ell|^2\frac{\sin^2\theta}{r^2}\left[1+j\frac{1}{(kr)^3}\right] \tag{2.30}$$

and the complex power P_r is given by

$$P_r = \oiint_S \mathbf{S}_{loop}\cdot d\mathbf{S} = \eta\left(\frac{\pi}{12}\right)(ka)^4|I_\ell|^2\left[1+\frac{j}{(kr)^3}\right] \tag{2.31}$$

Therefore, for a small loop of area $A_{loop} = \pi a^2$ carrying a constant current I_l, the far or the radiation fields are given by

$$H_\theta = -\frac{(ka)^2 e^{-jkr} I_\ell \sin\theta}{4r} = -\frac{\pi e^{-jkr} I_\ell \sin\theta}{r}\frac{A_{loop}}{\lambda^2} \tag{2.32}$$

$$E_\varphi = \frac{\eta(ka)^2 e^{-jkr} I_\ell \sin\theta}{4r} = \frac{120\pi^2 e^{-jkr} I_\ell \sin\theta}{r}\frac{A_{loop}}{\lambda^2} \tag{2.33}$$

It is now of interest to compare the far field expressions for a small loop with that of a short dipole. The presence of the factor j in the dipole expressions (2.14) and (2.15) and its absence in the field expressions for the loop indicate that the fields of the electric dipole and of the loop are in time-phase quadrature, the current I being assumed to be the same phase in both cases. The dipole is considered to be oriented parallel to the z-axis and the loop is located in the x-y plane. Therefore if a short electric dipole is mounted inside a small loop antenna and both the dipole and the loop are fed in phase with equal power, then the radiated fields are circularly polarized with the doughnut-shaped field pattern of the dipole.

2.4 Field Produced by a Finite-Sized Dipole

In order to calculate the fields from antennas, it is necessary to know the current distribution along the length of the antenna. For that we need to solve Maxwell's equations, subject to the appropriate boundary conditions along the antenna. In the absence of a known antenna current, it is possible to assume certain current distribution and from that calculate the approximate field distribution. The accuracy of the calculated fields will depend on how good an assumption was made for the current distribution. It turns out that for a thin linear wire antenna; the sinusoidal current distribution is a very good approximation (the thinner the antenna, i.e., radius $\rightarrow$ 0). When greater accuracy is desired, and for the cases where the sinusoidal approximation breaks down (for thick or short dipole antennas, where the diameter of the wire is greater than one-tenth of its length) it is necessary to use a distribution closer to the true one. Our objective in this section is to bring out certain characteristic properties of antennas as they relate to the field distribution and hence applicable to beamforming.

Consider a center-fed dipole antenna of length $L = 2H$. It will be assumed for convenience that the current distribution on the structure [2–4] is sinusoidal and symmetrical. Therefore it can be written as

$$\mathbf{J}(z') = \hat{\mathbf{z}} I_m \sin k(H - z') \quad \text{for } z' > 0 \tag{2.34}$$
$$= \hat{\mathbf{z}} I_m \sin k(H + z') \quad \text{for } z' < 0$$

where I_m is the value of the maximum value of the current. The vector potential at a point $P(x, y, z)$ due to the current element will be

$$A_z = \frac{\mu_0}{4\pi} \left[\int_{-H}^{0} \frac{I_m \sin k(H + z') e^{-jkR}}{R} \, dz' + \int_{0}^{H} \frac{I_m \sin k(H - z') e^{-jkR}}{R} \, dz' \right]$$

$$\tag{2.35}$$

where $R = \sqrt{(z-z')^2 + y^2 + x^2}$, with $x' \equiv y' \equiv 0$. In cylindrical coordinates, the magnetic field in the Y-Z plane, H_ϕ at the point P can be obtained as [2–4]

$$H_\phi = -\frac{I_m}{4\pi j}\left(\frac{e^{-jkR_1}}{y} + \frac{e^{-jkR_2}}{y} - \frac{2\cos kH e^{-jkr}}{y}\right) \tag{2.36}$$

The electric field in the $x = 0$ plane is given by

$$E_z = -j30I_m\left(\frac{e^{-jkR_1}}{R_1} + \frac{e^{-jkR_2}}{R_2} - \frac{2\cos kH e^{-jkr}}{r}\right) \tag{2.37}$$

and

$$E_y = j30I_m\left(\frac{z-H}{y}\frac{e^{-jkR_1}}{R_1} + \frac{z+H}{y}\frac{e^{-jkR_2}}{R_2} - \frac{2\cos kH}{y}\frac{e^{-jkr}}{r}\right) \tag{2.38}$$

The parameters R_1, R_2, and r are shown in Figure 2.3. These equations provide the electric and the magnetic fields both near and far from an antenna carrying a sinusoidal current distribution. This is an exact analytic expression for the assumed sinusoidal current distribution on the antenna.

It is interesting to observe that E_z represents a juxtaposition of three spherical waves. One each from the ends of the antenna and the last term

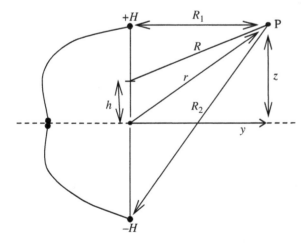

Figure 2.3 The geometry of a dipole antenna and the location of the field point. The curve on the left represents the current distribution on the wire antenna.

is originating from the feed point. However, for a half wavelength long dipole $H = \lambda/4$, the third term in (2.38) disappears and the total field is a combination of two spherical waves originating from two isotropic sources located at the ends of one dipole. **The other important point to note is that the total field can never be zero at any finite distance from the dipole** (This point has been illustrated in the Appendix 2A). **The only place where the field can be zero is when the distance** $y \to \infty$ **and we talk about the radiation field.** The electric field is a complex quantity and the real part of E_z is zero when the imaginary part is a maximum and vice-versa. The absolute value of the fields decays monotonically as we recede from the antenna.

Therefore, **the fields from a finite-sized dipole actually never become zero as a function of the angle and also independent of the distance from the antenna. In addition, in the near field, an antenna field distribution cannot be defined as independent of distance such as an antenna pattern and will always be range dependent. It will have no nulls except at a few isolated points. In the near field, even the fields radiated by a highly directive antenna is omnidirectional.** The Figure 2A.1 in the Appendix further illustrates this point. We now look at the radiation fields from any finite arbitrary shaped antenna radiating in free space as in real life there are no elementary Hertzian dipoles.

2.5 Radiation Field from a Finite-Sized Dipole Antenna

To calculate the radiated far field from a linear dipole antenna certain approximations can be made. Even though these mathematical approximations are independent of the nature of the antenna and are universal in nature, here we apply them to a simple dipole antenna. The magnetic vector potential for a z-directed straight wire in free space of length L and carrying a current $I(z)$ can be written as [2–4]

$$A_z = \frac{1}{4\pi} \int_{-L/2}^{L/2} I(z') \frac{e^{-jkR}}{R} dz' \tag{2.39}$$

$$R = |r - r'| = \sqrt{r^2 + z'^2 - 2rz'\cos\theta} \tag{2.40}$$

where $r^2 = x^2 + y^2 + z^2$ and $z = r\cos\theta$. If we consider that $r \gg z$, i.e., we are quite far away from the antenna, then one can simplify the expression for R using a binomial expansion as shown in the Appendix 2A

$$R = r\left[1 + \frac{z'^2 - 2rz'\cos\theta}{r^2}\right]^{\frac{1}{2}}$$

$$\approx r - z'\cos\theta + \frac{1}{2}\frac{z'^2}{r}\sin^2\theta + \frac{1}{2}\frac{z'^3}{r^2}\cos\theta\sin^2\theta + \text{other terms}$$
(2.41)

while neglecting the remaining of the higher order terms. It is important to note that when the third term becomes maximum at $\theta = \pi/2$, the fourth term becomes zero.

Now "R" appears in both the denominator of (2.39) as an amplitude term and in the numerator in the form of a phase term. It is sufficient to approximate $R \approx r$ for all practical purposes for the denominator. For the phase term we need to consider the function kR, by neglecting the last term of (2.41)

$$kR = k\left[r - z'\cos\theta\right] + \frac{1}{2}\frac{kz'^2}{r}$$
(2.42)

If we consider a three bit phase shifter, then it has eight phase states between $0°$ and $360°$ in steps of $45°$. Therefore, the maximum phase error that can be tolerated across the aperture in this case is a maximum of $45°$. Over half of the aperture the acceptable phase error is $22.5°$. [It is not clear what the origin of this choice was. Here, we explained the choice by an example. A more accurate and complete analysis is given in the Appendix 2A]. If we bound the error by this magical number, and find the value of r (it is a rule of thumb for far field approximation) at which the following equality is satisfied, i.e.,

$$\frac{1}{2}\frac{k(L^2/2)}{r} \leq \frac{\pi}{8} \quad \text{or} \quad \frac{1}{2}\frac{kL^2}{4r} \leq \frac{\pi}{8}; \quad \text{therefore,} \quad r \geq \frac{2L^2}{\lambda}$$
(2.43)

Here L is the entire length of the antenna. The rules of thumb for small ($L < 0.33\lambda$) and intermediate ($0.33\lambda < L < 2.5\lambda$) values of L, dictate that the far field commence at $r > 5D$ and $r > 1.6\lambda$, respectively, instead of (2.43) and as described in Appendix 2A.

The question that is now addressed is if this dipole is placed over a perfect ground plane, what will be the size of the effective aperture. In that case, we assert that this length L must be an equivalent length as it should include the image of the antenna below the ground plane. Equivalently, L is the diameter of the circle that encompasses the entire source antenna along with all the images produced by its operational environment, i.e., this circle must encompass both the original source and its image. For illustration purposes, consider a center-fed half wave dipole antenna operating in free space.

From (2.36) to (2.38), it is observed that only two spherical waves emanate from the two ends of the dipole and nothing contributes from the middle section of the dipole. The field from a center-fed dipole is then equivalent to two point sources separated by $\lambda/2$. The far field from such a configuration starts at $2L^2/\lambda$. This is also equivalent to a $\lambda/4$ monopole radiating over a ground plane where the field is produced by a single spherical wave source over a perfect ground plane. One can then use this observation to predict the far field from center fed dipoles that are located on top of towers. For antennas located on top of high towers operating over earth, the value for L in (2.43) should encompass all the antenna sources along with their images. This point is also illustrated in Appendix 2A.

Alternately, the Fresnel number, N_B has been introduced in the context of diffraction theory to arbitrarily demarcate the separation between the regions of the near and the far fields. It is a dimensionless number and it is given in optics in the context of an electromagnetic wave propagation through an aperture of size ξ (e.g. radius) and then the wave propagates over a distance W to a screen. The Fresnel number is given by

$$N_F = \frac{\xi^2}{W\lambda} \tag{2.44}$$

where λ is the wavelength. For values of the Fresnel number well below 1, one has the case of Fraunhoffer diffraction where the screen essentially displays the spatial Fourier transform of the complex amplitude distribution of the fields after the aperture. In Fraunhoffer diffraction, $N_F << 1$; and the diffraction pattern is independent of the distance between the aperture and the screen, depending only on the angles to the screen from the aperture. For Fresnel number around 1 or larger, i.e., $N_F \geq 1$, one has the case of Fresnel diffraction (near-field diffraction) where mathematical description of the fields is somewhat complicated. If we assume the Fresnel number to be $N_F = 0.125$ which is $<< 1$, i.e., we are in the far field Fraunhoffer region, then (2.44) translates to (2.43) as $W > 8\,\xi^2/\lambda$ and ξ is the radius and so we need to consider $L = 2\xi$ providing essentially the same approximation for the far field condition.

2.6 Maximum Power Transfer and Efficiency

Next we look at the concept of maximum power transfer and efficiency of power conversion from the source to the load. The question now is which of the two principles to use for an antenna problem so that one can radiate/receive the maximum electrical field from an antenna for a fixed input power.

2.6.1 Maximum Power Transfer

The maximum power transfer theorem states that for a source network, the maximum amount of power will be dissipated by a load resistance when that load resistance is equal to the Thevenin or Norton resistance of the network supplying the power. If the load resistance is not equal to the Thevenin/Norton resistance of the source network, the network's dissipated power will be less than maximal. This theorem was perhaps generated in 1839 by Moritz Herman von Jacobi, brother of mathematician Carl Gustav Jacob Jacobi, when he constructed an electric motor boat powered by battery cells which carried 14 people and travelled in the Neva River against the current at 3 miles per hour. In the course of these experiments, he considered how much power he could get out of a battery. Jacobi observed that the transfer of maximum power from a source with a fixed internal resistance to a load, will occur when the resistance of the load is same as that of the source. This law is of use when driving a load such as an electric motor from a battery. Maximum power transfer theorem states that the maximum load power is achieved when the load impedance equals the complex conjugate of the internal impedance of the source [5]. So when the principle of maximum power transfer is evoked, one can deliver at the most 50% of the useful power to the load. However, this mythology continued and researchers took it as the best they can do till Edison destroyed that mythical concept.

Maximum power transfer yields 50% of energy transfer and thus is achieved at the expense of reduced efficiency. In 1880 the assumption that the best we can do is 50% efficiency was shown to be false by Thomas Edison who realized that the maximum efficiency was not the same as maximum power transfer [6]. It is important to note that the maximum power transfer theorem finds the load which will receive the maximum power from the source without considering the total input power of the source. The total input power delivered by the source varies as we change the load. The maximum power that we obtain at the load in the case of the maximum power transfer is associated with a high value for the input power compared to the other situations! Efficiency on the other hand considers both the magnitude of the load power and the source input power.

The application of maximum power theorem is done only under the conditions when the maximum performance is desired over the overall efficiency of the circuit because the efficiency of a circuit under maximum power transfer condition is only 50%. This means that the maximum power transfer concept is most appropriate in cases where the voltage and internal resistance properties of the source circuit are fixed, and the aim is to maximize the amount of power transferred, even at the expense of reduced efficiency. The maximum power transfer theorem has a wide range of usage in real life situation, for example to match an amplifier with a loudspeaker to yield maximum power to the speaker and thus produce maximum sound.

However, if the source circuit is also under the designer's control, then efficiency may be improved by making the load resistance higher than the source resistance, while power may be boosted by increasing the source voltage. This could be applicable for the transmitting antenna on top of the tower of any cellular network. It is obvious that efficiency in such an application is more desired than the output power since we can basically increase the output power by increasing the input voltage (which we have access to) and this will increase the output power and maintain reasonable efficiency rate. At the end of the day, the expenses of the power usage that feeds the tower will decrease for a given radiated field intensity and that is what matters in this commercial application.

For antenna systems we need to consider radiation efficiency, or maximum conversion of source power to the load. Also, note that an antenna is a radiator of fields that are excited by a voltage source. In an antenna system, the goal should be to improve the radiation efficiency of an antenna and NOT to focus on the maximum power transfer, just like what is done in simple household appliances where increased efficiency is the objective and not maximum power transfer. If the appliance industry had sought to maximize power transfer, it would have achieved a mere 50% efficiency, with half of the power lost (wasted) because it is not absorbed by the desired load and is equivalent to the amount radiated by the antenna. Therefore, it is necessary to pay attention to the efficiency of power or energy transfer and not to maximum power transfer. The subtlety between these two concepts is illustrated next.

The best load in terms of efficiency is not the same as for the case of the maximum power transfer. This principle of achieving maximum efficiency may then lead to change our way of thinking that maximum power transfer matching is the optimum solution in all cases. It seems that we can think of easier matching procedures that will give better efficiency rates.

Next, we formulate the output power at the load side and also evaluate the efficiency of the system. In addition, a brief comparison is provided resulting in some concrete conclusions. Also we give an example of applying the efficiency concept to antenna problems, where we illustrate that the usual way of complex conjugate matching is not the best procedure to follow, instead achieving a high efficiency is the goal.

In electrical engineering, we obtain maximum average power from a source with some internal impedance when the connected load impedance equals the complex conjugate of the internal source impedance. This is known by the maximum power transfer theorem. Applying this theorem means that the best we can do is to distribute the source power equally between the source internal impedance and the load impedance; i.e., the maximum power efficiency that can be achieved is ONLY 50%. Efficiency takes into account the ratio of the dissipated power in the load divided by the source power.

It is important to recognize that the maximum power transfer considers only the magnitude of the dissipated power. If we increase the resistance of the load more than the internal resistance of the source then we will achieve better efficiency as more of the source power will be delivered to the load than to the power dissipated in the source impedance. The magnitude of the dissipated power will be less since the total resistance in the circuit will be increased. We will try to emphasize this fact in antenna problems and show that considering efficiency is more appropriate than considering maximum power transfer. Analysis is performed on half wave dipole to show that maximum power transfer impedance matching is not the optimum solution in terms of efficiency.

2.6.2 Analysis Using Simple Circuits

Consider the electrical circuit of a voltage divider in Figure 2.4, where the antenna is characterized by the load R_L. Recall that an antenna is excited by a voltage source and not by a power source. The total input power to the antenna is calculated by multiplying the excitation voltage V_S and the input current I. So what is required here is the efficiency in transferring most of the power from the source to the load. The efficiency η of the circuit is the proportion of the power dissipated in the load to all the power dissipated in the circuit (i.e., supplied by the source).

Therefore, for the circuit in Figure 2.4b, let us calculate the output power P_{out} and the efficiency η. The output power is the dissipated power in the load Z_L and the efficiency of power transfer from the source to the load are

$$P_{out} = \mathrm{Re}\left(V_L I_L^*\right) = \frac{|V_S|^2 R_L}{\left(R_S + R_L\right)^2 + X_L^2}, \quad P_{out-max} = \frac{|V_S|^2}{4R_L} \qquad (2.45)$$

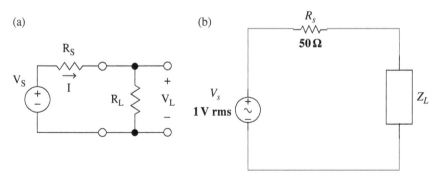

Figure 2.4 A simple circuit for a voltage divider. (a) Real value for the load and (b) Complex value for the load.

This principle becomes apparent when a comparison of two systems is made by keeping the input power to each system exactly the same. To calculate the input power for a general AC circuit, it is also apparent that one needs to have complex values of both the voltage and the current. Just considering the voltage and relating its correlation to power spectral density only applies to stationary systems where there are no inductors and capacitors in the system because they destroy the stationarity assumption – thus the assumption of stationarity is only valid for resistive circuits. Therefore, to compute the radiated power density in antenna theory, one needs to have the quantifications of both the electric and the magnetic fields. Unless one is operating in the far field of an antenna, where only one of the field components provides the power density, as the electric and the magnetic fields are related by the impedance of free space (377 Ω), both the electric and the magnetic field components are required to evaluate the radiated power density. This issue will be addressed later.

In radio circuits, the internal resistance of a source of power cannot usually be reduced to a small value, so the load resistance is made equal to it for maximum power transfer. The powers involved are very small, and losses are unimportant – only the amount of power that gets through is important. The situation is quite different for AC (alternating current) power transmission, where losses are important – or equivalently for the antenna radiation problem where efficiency is paramount! Clearly, maximum power transfer does not coincide with maximum efficiency. Application of the Maximum Power Transfer Theorem to AC power distribution will result in neither a maximum nor high efficiency. The goal of high efficiency is more important for AC power distribution, which dictates a relatively low generator impedance compared to load impedance and which consequently should be the approach when deploying antennas. For example, similar to AC power distribution, high-fidelity audio amplifiers are designed for relatively low output impedances and relatively high load impedances of the speakers. The ratio of "output impedance" to "load impedance" is known as the *damping factor*, and is typically in the range of 100 to 1000. Also, maximum power transfer does not coincide with the goal of lowest noise. For example, the low-level RF amplifier between an antenna and a radio receiver is often designed to have the lowest possible noise. This design often requires a mismatch of the amplifier's input impedance to the antenna instead of that dictated by the Maximum Power Transfer Theorem.

In summary, the *Maximum Power Transfer Theorem* states that the maximum amount of power will be dissipated by a load resistance if it is equal to the Thevenin or Norton resistance of the network supplying the power resulting in a 50% efficiency as ONLY half of the input power is dissipated in the load. This theorem does NOT satisfy the goal of maximum efficiency (a transformer or a heater in a power system needs to be >90% efficient), where the supplied power remains constant. It is important to note that under conditions of maximum power transfer, as much power is dissipated in the source as in the load.

The maximum power transfer condition is used in communication systems (usually at RF) where the source resistance cannot be made low, the power levels are relatively low, and it is of paramount importance to supply as much signal power as possible to the receiving end of the system (the load).

2.6.3 Computed Results Using Realistic Antennas

We used a commercially available antenna analysis software called AWAS [7] to simulate the fields radiated by a half wave dipole at 1 GHz under various terminating conditions. We considered three cases, namely the unmatched, conjugately matched and the reactance only matched antenna through the following:

A) **Unmatched:** We simulate a z-oriented half-wave dipole with a 1 V excitation at the center point, i.e., center fed. The internal resistance of the excitation is 50Ω. The operating frequency is 1 GHz. The radius of the wire antenna is 1 mm. No matching network exists between the source and the antenna. This is illustrated in Figure 2.8, Case A

B) **Matching network** to match the half-wave dipole antenna to a 50Ω labelled as (**Matched**):
All the simulation parameters are the same as in case (A) but we included an L-section matching network to match the antenna to a 50Ωsource. This is shown in Figure 2.8, Case B

C) **Matching network** to cancel the effect of the imaginary part of the input impedance (i.e., the reactance) *only* of the half-wave antenna labelled as (**Xmatched**):

All the simulation parameters are the same as in case (A) but we include a capacitor to cancel the effect of the imaginary part of the input impedance of the antenna. This is shown in Figure 2.8, Case C.

All the simulation results are summarized in Table 2.1 for the three cases. Please note that the input power for all the three cases is not the same. So they

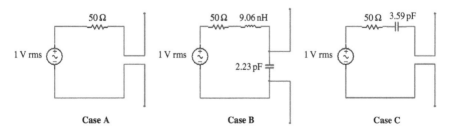

Figure 2.8 Performance of Antennas under different matching conditions.

Table 2.1 The simulation results for all the previous cases A, B, and C.

Case	$|P_{in}|$ in mVA	P_{out} in mW	η
Matched	10	5	0.5
Unmatched	6.64	4.14	0.62
Xmatched	6.95	4.54	0.65

are now all normalized so that the input power is restricted to be the same for all the three cases and fixed at 10 mVA. Figure 2.9(a) plots the magnitude of the electric field in dB as a function of the horizontal distance from the antenna for all the three cases, Figure 2.9(b) is just a zoom in of the plot of Figure 2.9(a). The fields are scaled so as to have the same source input power in all the three cases, in other words we are looking at the radiation efficiency rather than the output power of the system. Looking at Figures 2.9(a) and 2.9(b) we conclude that if we consider efficiency for the half wave dipole then a better way of thinking is to *xmatch* the half wave dipole antenna (cancel the effect of the imaginary part of the input impedance of the antenna *only*). Moreover, the unmatched half wave dipole has a better radiation efficiency than the matched dipole antenna as seen in Table 2.1. Also if we compare the performance of an unmatched $3\lambda/4$ dipole, where λ is the wavelength, with a matched half wave dipole then the radiation efficiency of the unmatched $3\lambda/4$ dipole will be much higher than that of the matched half wave dipole. The radiation efficiency of the unmatched $3\lambda/4$ dipole is in the range of about 88%. Furthermore, if we compare the performance of an unmatched 0.65λ dipole with a matched half wave dipole then the radiation efficiency of the unmatched 0.65λ dipole will be much higher than the matched half wave dipole as the efficiency of the unmatched 0.65λ dipole is in the range of about 69%. The simulation results are summarized in Table 2.2. Figure 2.10 shows a comparison of the performance of an unmatched $3\lambda/4$ dipole, unmatched 0.65λ dipole, unmatched half wave dipole, and matched half wave dipole. The fields are scaled such to have the same source input power in all the cases.

In summary, the maximum power transfer concept is most appropriate in cases where the total input power and the internal resistance of the source circuit are fixed, and the aim is to maximize the amount of power transferred from the source to the load, even at the expense of reduced efficiency. However, if the source circuit is also under the designer's control, then efficiency may be improved by making the load resistance much higher than the source resistance, while power may be boosted by increasing the source voltage. Efficiency could be applicable for the transmitting antenna on top of the tower of any cellular network.

(a)

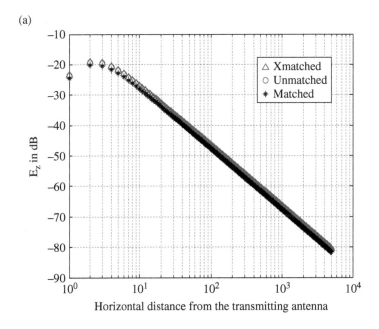

(b)

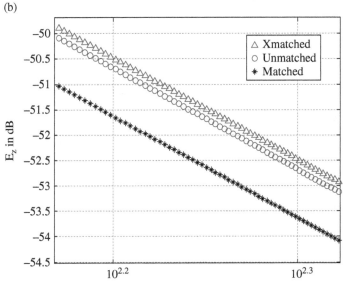

Figure 2.9 (a) The magnitude of the electric field in dB as a function of the horizontal distance from the antenna for all the three cases A, B, and C. (b) Zoom in of Figure 2.9(a).

Table 2.2 The simulation results for the unmatched $3\lambda/4$ dipole and the unmatched 0.65λ dipole cases.

Case	$|S_{in}|$ in mVA	P_{out} in mW	η
Unmatched $3\lambda/4$	1.16	1.02	0.88
Unmatched 0.65λ	2.16	1.48	0.69

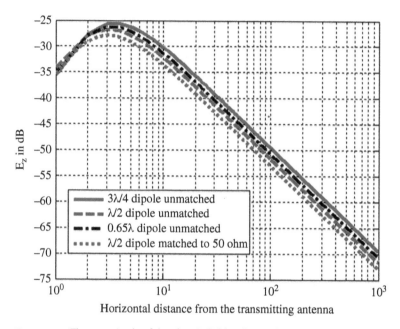

Figure 2.10 The magnitude of the electric field in dB as a function of the horizontal distance from an unmatched $3\lambda/4$ dipole, unmatched 0.65λ dipole, unmatched half wave dipole, and matched half wave dipole.

All the figures indicate that a conjugately matched dipole radiate the least electric field strength when the same amount of power is used to feed all the different antennas. Therefore, in contrast to popular beliefs of a conjugately matched antenna systems we should use the concept of radiation efficiency if we desire to radiate the maximum field strength for the same value of the input power.

2.6.4 Use/Misuse of the S-Parameters

The previous discussion leads to a related topic, the use of the S-parameters in analyzing antenna problems. Unfortunately, the S-parameters are not suitable for characterizing antenna problems where the impedance changes as a function of frequency. The point here is that treating the forward and the backward waves in the formulation of the S-parameters as a sum of weighted voltages and currents make it possible to apply the principle of superposition of power in the S-parameter formulation, because the waves representing power are related to the voltages and currents to which superposition applies! In antenna problems, one wants to increase the radiation efficiency just like in household electrical appliances and to avoid the concept of power transfer, as maximum power transfer implies an efficiency of only 50% with half of the generated power wasted! The crucial takeaway is that an antenna does NOT radiate power; rather it radiates power density, which is related to the radiated electric field. Therefore in antenna design, the goal should be to radiate the maximum electric field intensity as the input power to the system is fixed and not to dwell on power to which the principle of superposition does not apply as has been illustrated in Chapter 1 [8]!

As an example, consider a center fed dipole antenna of length 0.475 λ with a half-length of 0.2375 λ. This structure is resonant as for an input exciting voltage at the center fed delta gap of 1.0 V, the current is approximately real at the feed point and is 13.84 − *j* 0.0227 mA. The analysis is carried out using the commercially available general electromagnetic simulator called HOBBIES [9]. The broadside radiated field is 0.063 − *j* 0.82 V/m for this excitation. So for an input V-mA of 1.0 the maximum absolute value for the radiated field is 0.0594 mVA. Next consider a dipole of length 0.52 λ which is non resonant as its half-length is 0.26 λ. For a center fed applied voltage of 1.0 V the complex value of the input current in mA is 5.72 − *j*4.9. The value of the radiated field along the broadside direction is 0.39 − *j* 0.37 V/m. So for an input V-mA of 1.0, which is the same as that of the previous case, the maximum absolute value for the radiated field is 0.0714. Hence this non resonant antenna radiates a larger value for the electric far field along the broadside direction for the same input power in V-A. **Hence a resonant structure does not imply that it is going to radiate more signal – a fallacious assumption often used in the matching of antennas where it is assumed that if the current is resonant, the structure will radiate more power!** Finally, it is important to recollect that it is **the external resonance [10] that provides the maximum radiating fields** and the current at an internal resonance does not radiate. So, there is little connection between the current distribution on an antenna structure and the magnitude of the fields that it radiates [10]. Hence in antenna design, the emphasis should be on radiation efficiency and not on the maximum power transfer principles.

2.7 Radiation Efficiency of Electrically Small Versus Electrically Large Antenna

In this section, we describe the performance of Electrically Small Antennas (ESAs) – small in size compared to the wavelength – versus larger sized antennas in both transmitting and receiving modes. We conclude that the ESAs perform well compared to larger antennas under matched conditions where the input power to the antenna is real (input power is $P_{in} = V_{in}I_{in}$, V_{in} and I_{in} are the phasor domain rms input voltage and input current, respectively). The difficulties in achieving a good design of an ESA matched to a 50 Ω system will be discussed in the next section.

2.7.1 What is an Electrically Small Antenna (ESA)?

An Electrically Small Antenna (ESA) is an antenna whose maximum dimension can be enclosed in an imaginary sphere with radius a such that $ka < 0.5$, k is the wave number which equals $2\pi/\lambda$ [11], λ is the free space wavelength or we could say that an ESA is the antenna whose maximum dimension is less than the radian length which is given by $\lambda/(2\pi)$ [12]. In this section, we study the performance of the ESAs (satisfying the criterion of $ka < 0.5$) versus large antennas in both transmitting and receiving modes operating under different scenarios.

2.7.2 Performance of Electrically Small Antenna Versus Large Resonant Antennas

First, we consider the antenna to be in free space and not matched. We simulated a z-directed dipole antenna with four different lengths: 0.01λ, 0.1λ, 0.25λ, and 0.5λ. The frequency in all the simulations performed in this chapter is 1 GHz. We used the antennas in the transmitting mode where we center excited the antennas using a 1 V voltage source with a 50 Ω source resistance. The results are shown in Figure 2.11 which plots the magnitude of the z-component of the electric field in dB.

We use a commercially available simulation tool called Analysis of Wire Antennas and Scatterers (AWAS) [7]. This simulation tool uses the Method of Moments to calculate the current distribution on the antenna and then calculates the radiated fields. AWAS can be used to simulate wire antennas in transmitting or receiving mode where only the excitation differs but not the Z matrix computed in the Method of Moments.

One must be careful when generating such plots for the sake of comparison. When the antennas are not matched at the input port then the input power given by $S_{in} = V_{in}I_{in}$ (V_{in} and I_{in} are the phasor domain Root Mean Square (RMS) input voltage and input current, respectively), consists of a real part which is the average power and an imaginary part which is due to capacitive/inductive loads. In general, the input impedance of the antenna is complex so we have to make the complex input power magnitude, $|S_{in}|$, the same for all the cases in order to make a fair comparison between the fields radiated by antennas of different lengths. Instead, we calculated the ratio between the magnitude of the complex input powers for the different cases and then scale the fields by the

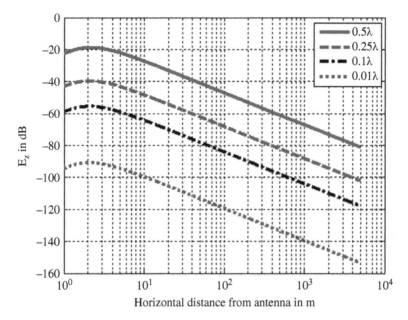

Figure 2.11 The magnitude of the z-component of the electric field in dB for the various dipole lengths (unmatched scenario).

square root of the ratio accordingly. In our simulations, the magnitude of the input complex power for the 0.01λ antenna was 35.083 µVA, the magnitude of the input complex power for the 0.1λ antenna was 363.3 µVA, the magnitude of the input complex power for the 0.25λ antenna was 1.131 mVA, and the magnitude of the input complex power for the 0.5λ antenna was 4.87 mVA. It makes sense that the input power for the half wave dipole is the largest as the antenna is almost self-resonant compared to the other lengths and so the real part of the input impedance is larger than the others but the input reactance is significantly smaller. As previously mentioned, we need to scale the fields' strengths according to the ratios of the square root of the magnitude of the input complex power. Basically we multiplied the magnitude of the z-component of the electric field for the 0.25λ by 2.075, for the 0.1λ by 3.66, for the 0.01λ by 11.782, and for the 0.5λ by 1. Applying the scaling so that the input power is the same for all cases, we can consider Figure 2.11 a fair comparison between the fields radiated by antennas of different lengths.

The second scenario we study is when all the antennas are matched to the real part of the input impedance using a lossless matching network (inductor or capacitor with zero resistance). We again used AWAS to perform the simulations. We considered the antenna to be radiating in free space. We took the same four z-directed dipole antenna with four different lengths: 0.01λ, 0.1λ,

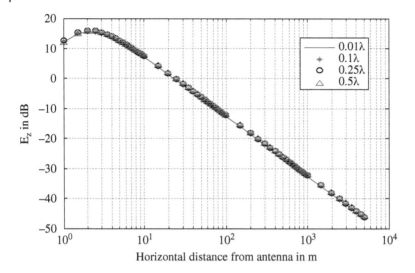

Figure 2.12 The magnitude of the z-component of the electric field in dB for the various dipole lengths (matched scenario).

0.25λ, and 0.5λ. The input power, in this case, is real so we made sure that the total input power is the same for the excitations of the dipoles of different lengths so as to be able to make a fair comparison. To achieve the goal that all the antennas are excited by the same power levels, we scale the electric field strengths by the square root of the ratio between the input powers as in the previous unmatched scenario. Figure 2.12 shows the results for the antennas when they are conjugately matched to the real part of the input impedance. We can clearly see that the field strength is almost the same for the antennas of various lengths when they are matched (the input power consists of real part only) and the difference between them does not exceed 0.3 dB. This means that the degradation in the antenna performance for an ESA is negligible, when operating in a matched condition. For these types of problems, the application of the maximum power transfer theorem provides a meaningful solution as in this case they all radiate similar electric field intensities.

We also study the performance of the antennas in the receiving mode operating in free space. We used AWAS to perform the simulations. We took four z-directed dipole antenna with four different lengths: 0.01λ, 0.1λ, 0.25λ, and 0.5λ. We excited the antenna with a plane wave where we set the theta component of the electric field to be 1 V/m incident at $\theta = 90°$ and we set the phi component of the electric field to be 0 V/m. the angles theta and phi are measured from the z-axis and x-axis, respectively. First, we study the scenario of the receiving mode of dipoles with various lengths while the antenna is connected

Table 2.3 Received power by a 50 Ω load connected to dipole antenna of various lengths.

Length	0.01λ	0.1λ	0.25λ	0.5λ
P_r (mW)	1.3×10^{-10}	1.43×10^{-6}	9.77×10^{-5}	0.026
P_r(dBm)	−98.86	−58.45	−40.1	−15.85

Table 2.4 Received power by a load, conjugately matched to the input impedance of the antenna, connected to dipole antenna with various lengths.

Length	0.01λ	0.1λ	0.25λ	0.5λ
P_r (mW)	0.0279	0.0286	0.0291	0.0312
P_r(dBm)	−15.54	−15.43	−15.36	−15.05

to a 50Ω load. Table 2.3 summarizes the received power delivered into a 50 Ω load, when all the antennas are unmatched.

From the results in Table 2.3, we conclude that the received power in a 50 Ω load is the largest in the case of the half wave dipole. This result is expected because the input impedance of the half wave dipole is best matched to 50 Ω compared to all other dipole lengths.

For the same previous setup, we study the scenario of the receiving mode for dipoles of different lengths while the connected load at the antenna port is conjugately matched to the antenna input impedance. Table 2.4 summarizes the received power by a conjugately matched load. From the results in Table 2.4, we conclude that received power by a conjugately matched load is almost the same for all different dipole lengths. The difference does not exceed 0.5 dB in the worst case. This means that the degradation in the antenna performance for an ESA is negligible, when operating in a matched condition.

As an example consider the transistor radios that were popular several decades back. That little device was able to receive transmitted signals from the other part of the world with ease operating at 1 MHz (λ = 300 m and so λ/4 = 75 m). The antenna used in those devices were very small in size and not of resonant lengths as the wavelength is extremely large. But since at the first stage of that radio there was a tuning capacitor to resonate the electrically small antenna, the antenna size did not matter. Unfortunately, in modern times this critical matching component is missing from the wireless phone systems resulting in poor signal reception! However, recent publications indicate that this well-known design principle known to the antenna engineers from time immemorial is finding its way again as the wireless system design engineers are now learning something about a subject called *electromagnetics* and reading topics on *antennas*.

In short, the radiation efficiency of an ESA is excellent compared to larger antenna as long as it is matched. Our simulation shows a maximum of 0.3 dB field strength degradation for the ESA in the transmitting mode as compared to the larger antennas. Also, our simulation shows a maximum of 0.5 dB received power degradation for the ESA in the receiving mode as compared to the larger antennas. Designing an ESA is not an easy task because the quality factor of the ESA is high in general and trying to obtain a design for an ESA with low enough quality factor is a tough task. That is the content of the next section as how to match an ESA.

2.8 Challenges in Designing a Matched ESA

In this section, we will try to highlight the challenges in designing an ESA. Let us study the following question: how can we match an ESA? One could do that externally using the traditional ways of impedance matching. We know that there exists a limit on the bandwidth of the impedance matching network. In general, there are two limits for the maximum bandwidth that one could achieve through the use of passive matching network in the case of ESA. One is the minimum quality factor limit which is proportional to the maximum bandwidth of the ESA [12, 13] and the other one is the Fano-Bode limit for the passive matching network [14–16].

If we want to match a load consisting of a resistor R and capacitor C by a lossless matching network then the question arises how well can this be achieved over a band of frequencies. Fano-Bode provided a theoretical limit on the value of the reflection coefficient Γ that can be achieved at the input of the matching network which is given by the following integral equation

$$\int_0^\infty \ln\left(\frac{1}{\Gamma(\omega)}\right) d\omega \le \frac{\pi}{RC} \tag{2.47}$$

If the objective is to attain a minimum reflection coefficient $\Gamma_{\min}$ at the input of the lossless matching networks trying to match the load consisting of R and C over a bandwidth of $\Delta\omega$ in the angular frequency domain, and the reflection coefficient is unity at the other frequencies outside the band, then one can approximate the integral of (2.47) by

$$\Gamma_{\min} \ge \exp\left(-\frac{\pi}{\Delta\omega RC}\right) = \exp\left(-\frac{1}{2\Delta f RC}\right) \tag{2.48}$$

where Δf is the bandwidth in the frequency domain. This illustrates that the minimum reflection coefficient obtained by the matching of a lossless network to the load impedance of R and C depends on the desired bandwidth and not on the location of the band on the frequency axis. However, this limit can be overcome if we use non-Foster elements to match the network as we shall study later on.

In the ESA case, we need to match an impedance consisting of frequency dependent very small real part (resistance) and very large negative imaginary part (reactance of a capacitor) to 50 Ω. This could be done traditionally by using an external L-section of lumped elements or it could be done through the use of distributed elements like shorted transmission line sections or stubs. This passive matching network will make the electrically small antenna look large from an electric point of view for the transmitter (source) or we could say that the produced resonance associated with this structure is artificial. The bandwidth of matching a capacitive load to a resistive load (the problem in hand) is limited by the Fano-Bode limit as illustrated by (2.47) and (2.48). The frequency bandwidth associated with such matching is very small. Figure 2.13 shows the performance bandwidth of the various lengths of matched dipoles using an external passive matching network. We see clearly that the useful frequency bandwidth is very small for the cases of the 0.01λ, 0.1λ, 0.25λ dipoles.

Also, the assumption that the matching networks are lossless that we always make is not applicable here. The L-section that we add to match an ESA to a

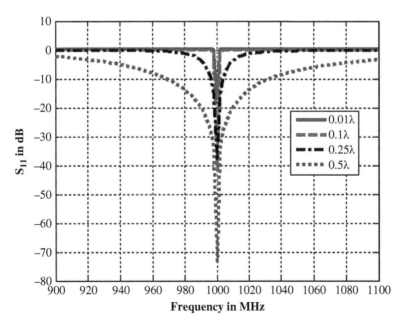

Figure 2.13 S_{11} in dB of various lengths of dipoles matched using an external passive network.

50 Ω system is not lossless and so in general one should compensate for these losses when he/she designs the matching network as these losses will affect the matching process. Furthermore, the resistance associated with the losses that are inherent in the matching networks is generally much higher in value than the radiation resistance of the antenna. This results in loss of most of the supplied input power and this dissipated power is turned into heat and the radiation efficiency of the system becomes low. In other words, if one uses an L-section to match an ESA to 50 Ω then one should compensate for the losses of the inductor/capacitor used in the matching network because in practice any inductor/capacitor has a finite value for the resistance depending on its Q-factor. For example, if we consider a monopole of length 0.025 m, then to match this antenna to a 50 Ω transmission line using a lossy L-section, one has to connect the antenna structure to an inductor of 82.6 nH having a quality factor of 250 in series with the antenna, and then connect a shunt capacitor of 9.29 pF having a quality factor of 1000. The losses associated with the matching network (2.06 Ω from the series inductor only) is comparable to the radiation resistance of 3.19 Ω for the 0.025 m monopole. So, not only will the quality factor of the monopole after matching will be very high but also the efficiency of the antenna will drop dramatically if we assume there are resistive losses associated with the inductor and the capacitor used in our matching network.

Going back to the question: How can we match an ESA? Based on the previous discussion, we cannot achieve a low quality factor or equivalently wide VSWR (voltage standing wave ratio) matched bandwidth for an ESA if we start with a simple z-directed dipole/monopole antenna. One solution is to make the passive matching network automatically tune the antenna based on the desired frequency of operation. This means that the bandwidth of the antenna will change along with the resonant frequency but every time the resonant frequency is changed the bandwidth around it will remain the same if the quality factor remains the same. Another possible way to match the antenna is to use Non-Foster elements in our matching network. Use of Non-Foster elements make it possible to obtain a match over a larger bandwidth determined by the Bode-Fano limit. Non-Foster elements are like using a negative inductance or a negative capacitance represented by − L or − C which involves use of active elements. The use of this kind of elements in our matching network will provide a larger matching bandwidth as it will not be limited by the Fano-Bode limit. The realization of the non-Foster elements could be done using active elements. A recent study [17] have shown a realization of a negative capacitor using Linvill's NIC (negative impedance converter which is a single port operational amplifier acting as a negative load). Another solution is to alter the structure of the ESA so as to increase the radiation resistance and achieve a self-matched environment to 50 Ω with a low quality factor design. Basically, the task is to come up with a design for an ESA such that it is self-resonant at the frequency of interest (in our example it is 1 GHz), self-matched to 50 Ω and which has a low quality factor. In general, there is a fundamental limit (Chu limit) [18] for the lowest quality factor that one can achieve for an ESA [19]. To design an ESA which is self-resonant, self-matched to 50 Ω and with quality factor close to the minimum quality factor (Chu limit), one should maximize the effective volume filled by the ESA structure compared to the volume of the imaginary sphere enclosing the maximum dimension of the ESA antenna (this will assure a low quality factor because this will minimize the non-propagating stored energy) and at the same time make the input impedance of the ESA to be simply 50 Ω at the frequency of interest [11]. One of the techniques to achieve this is to use a multi-arm meander line folded monopole design [11]. We used AWAS [7] to tune the antenna dimensions to achieve our goal which is to get an input impedance close to 50 Ω using the multi-arm meander line folded monopole concept. The input impedance of the final design is (54 + j0.009) Ω, the radius of the imaginary sphere that encloses the maximum dimension of the achieved design and its image is 0.025 m, so that ka = 0.5 approximately at 1 GHz. So there are various other ways to get around the Fano-Bode limit or the Chu limit without using an active device.

Next, we are going to compare the bandwidth of this design to a simple 0.025 m z-directed monopole matched using an L-section. The simulation for the 0.025 m z-directed monopole is performed using 4NEC2 [20] while the simulation for the four-arm meander line folded monopole is done using AWAS. The input S_{11} versus frequency for the both designs are displayed in Figure 2.14 and the fractional VSWR matched bandwidth calculated over which the VSWR is less than 1.5 or equivalently the magnitude of S_{11} is less than 0.2. We found that the fractional bandwidth for the four-arm meander line folded monopole, over which the magnitude of S_{11} is less than 0.2, is 0.016 and for the simple monopole is 0.0047. This implies that a quality factors of 25.5 and 86.85 is obtained for the four-arm meander line folded monopole and the simple monopole, respectively. The Chu limit for the Quality factor in this case is $Q = \dfrac{1}{ka} + \dfrac{1}{(ka)^3} = 10$ for ka = 0.5. Hence, the quality factor of the four-arm meander line folded monopole is approximately 2.5 times the fundamental limit compared to 8.7 times the fundamental limit in the case of the simple monopole.

In summary, designing an ESA is not an easy task because the quality factor of the ESA is high in general and trying to obtain a design for an ESA with low enough quality factor is a difficult task. One can use the multi-arm meander line folded monopole concept [11, 19] to achieve an optimum design of an ESA matched to 50 Ω and still have a quality factor close to the fundamental Chu limit. Also, one can use the non-Foster matching concept to overcome the Fano-Bode limit on the bandwidth of the match. However, the performance of an ESA is not compromised if it is conjugately matched, even though it is applicable in a narrow frequency band. We believe the reintroduction of a tuner that use to exist

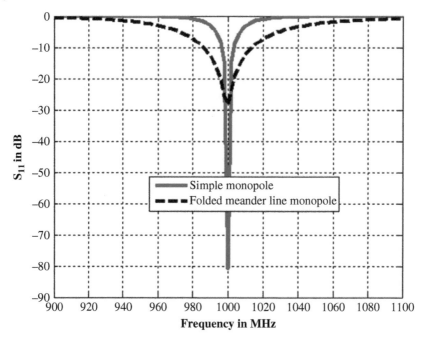

Figure 2.14 S_{11} magnitude versus frequency for the meander line folded monopole and the simple monopole.

in the prehistoric transistor radio receiver sets of the olden days, will result in an improved design particularly the match will be obtained for the antenna in any operating environment. Recent papers illustrate that soon this will be the case in the commercial sector for matching antennas in different bands.

2.9 Near- and Far-Field Properties of Antennas Deployed Over Earth

Another interesting topic is beamforming using antenna arrays. The space division multiple access concept which started the wireless expansion is at the heart of it but so far the original goal has not been accomplished. We now see why this goal has not been reached. An antenna array is a group of antennas in which the relative phases of the respective signals feeding the antennas is reinforced along a defined direction and suppressed along undesired directions. The first point one should think about is: what is meant by the terms *an antenna beam* or *an antenna pattern*? As we have observed it is the field at infinity distance from an antenna and its pattern is independent of the distance *r*.

The antenna pattern is only part of the fields produced by the source which decay linearly with the distance.

So, when we talk about the antenna beam pattern, we always imply the far field pattern. Conversely, in a Maxwellian context one cannot speak of a near field antenna pattern since the radiated power is a function of the distance from the antenna and is a complex quantity. Hence, the angle dependent pattern nulls, are not produced. These subtle issues of an antenna pattern do not come into the picture, if one uses an unrealistic characterization of an antenna as an isotropic point source, neglecting its finite length, as then every region is part of the far field! However, the story is completely different when we are dealing with a finite sized antenna.

As an example, let us consider a half-wave center-fed dipole (L = 15 cm) of radius 1 mm radiating in free space. The far field of the dipole operating, let us say, at 1 GHz (λ = 0.3 m) starts at a distance ($2D^2/\lambda$ = 2 × 0.15 × 0.15/0.3) = 0.15 m from the antenna, where D, the effective span or length of the radiating element, is D = 0.15 m. [Again this value has been assumed for simplicity as a more exact value can be obtained by using the expressions given in Appendix 2A]. At a distance greater than 0.15 m, the field pattern of the dipole is given by the radiation plot as shown in Figure 2.15 [1–4] as a function of θ from 0 to 90° and is independent of the distance from the antenna. [A more detailed accurate quantification of the far fields of antennas for different lengths operating in free space and over Earth is discussed in Appendix 2A]. One fourth of the radiation pattern is shown. The far field pattern has a null along the zenith of the dipole. These principles are well known and is available in any standard textbook on antennas [1–4]. However, the principles presented have very far-reaching implications in a wireless communications environment, as we shall illustrate next.

Figure 2.15 The radiation pattern of a half wave dipole in free space (only one fourth is shown).

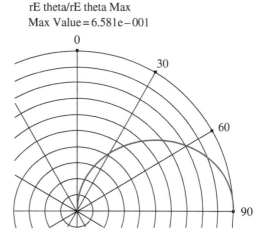

rE theta/rE theta Max
Max Value = 6.581e−001

Now, let us consider the same half wave dipole antenna deployed on top of a 20 m high tower as is conventionally done in a mobile communication environment. Typically the transmitting antennas are placed on top of a tower to serve a cell or a region, where the tower is physically located. Here, we consider the effects of the Earth, first as a perfectly conducting [2] and then as an imperfectly conducting ground plane [7, 21].

First let us assume that the dipole is situated over a perfect ground plane. This is an idealistic situation, but is used as an example to illustrate the physics. For a half wave dipole situated on top of a tower and radiating in free space, the far field will start not at 0.15 m but at a distance of $2 \times 40 \times 40/0.3 = 10.6$ km. As the antenna is situated on top of a 20-m tower over a perfectly conducting ground, the effective transmission aperture D is now 40 m and not 0.15 m. This is because the dipole antenna has an image which is 20 m below the ground, as the source dipole is located 20 m above the ground (Earth) [2, 3]. It can be shown that, if we consider a half wave-length radiating dipole situated 20 m above a perfectly conducting Earth, its far field radiation pattern will be given by Figure 2.16. The image produced by the perfectly conducting ground will be vectorially additive to the original free space field, producing the maximum value of the field strength along the horizon which is exactly doubled as seen from Figures 2.15 and 2.16. It is obvious from a Maxwellian point of view that this deployment of the antenna will produce a large number of maxima and minima and the antenna pattern will look awful in a vertical plane (as shown in the expanded version in Figure 2.17) due to the vectorial interaction between the original field and the field produced by the image. This will also affect the channel capacity as the received power obtained in a receiving antenna will be small because of the integral of the electric field carried out over its length by

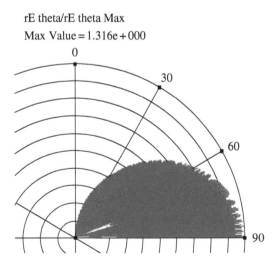

rE theta/rE theta Max

Max Value = 1.316e + 000

Figure 2.16 Radiation pattern of a half wave-length dipole situated on top of a 20-m tower located over a perfectly conducting ground (only one fourth is shown).

(a)

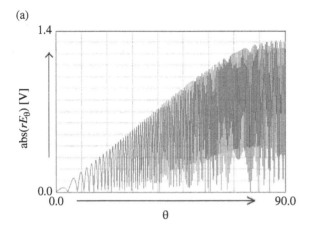

(b)

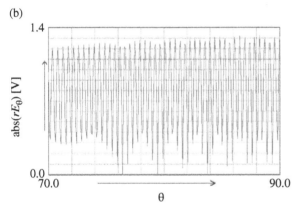

Figure 2.17 Radiation pattern of a dipole (a) normal, and (b) expended, situated on top of a 20 m tower located over a perfectly conducting ground.

the receiving antenna will yield a small value. This is because as part of this radiation pattern (in the form of the complex vector **E**, the electric field) when integrated over the length of the receiving antenna induces a voltage ($V = \int \mathbf{E} \cdot d\mathbf{l}$) at the feed point of the receiver. However, the radiation pattern is omni-directional along a horizontal plane. In any case, the far field pattern is not pertinent for a wireless system since a mobile user will invariably be in the near field for all practical purposes under current deployment situations over a high tower. Hence, one needs to carefully look at the near field of this antenna, which can be computed accurately using the Maxwell's equations and taking into account the effects of the environment [7, 20, 21] and will be discussed in details in Chapter 3.

For the sake of completeness, we consider the same half wave dipole antenna now placed on top of the tower but situated over a real Earth which is considered to be an imperfectly conducting ground. Thereby, we can characterize the real Earth, let us say, by sandy soil, for an assumed relative permittivity of $\varepsilon_r = 10$, and a conductivity of $\sigma = 2\times10^{-3}$ mhos/m [2, p. 893]. The radiation pattern for this elevated dipole is shown in Figure 2.18. (The field pattern is calculated using the commercial software package AWAS [7], which uses the Sommerfeld formulation as explained in Chapter 3). Observe that near the ground, the pattern is dramatically different from the perfectly conducting case. The plots are similar to Figure 2.15, except that the fields are zero along the Earth in contrast to the perfectly ground case. This is due to the losses in the earth and the fields at infinity becomes zero. But near the zenith the field plots are different as the strength of the field from the image is weaker due to the assumption of an imperfectly conducting ground. The expanded version of the pattern along the ground is shown in Figure 2.19. The interference pattern in Figure 2.18 is produced by the interaction of the free space field due to the dipole located in free space and the ground wave produced by the imperfectly conducting Earth [7]. The transmitting antenna launches the wave propagating along the air-earth interface which attenuates exponentially as we go away from the antenna but otherwise decays slower than the space wave. This will be discussed at length later in Chapter 3.

It is quite clear from Figures 2.16 to 2.19 that placing an antenna on top of a tower produces a highly undesirable radiation pattern. More will be discussed in Chapter 3 and a detailed analysis will be given.

It is quite pertinent to ask at this point how with such a severely distorted radiation pattern one is going to perform beam steering or even beam forming!

rE theta/rE theta Max
Max Value = 1.264e + 000

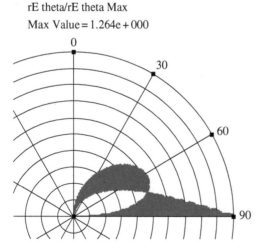

Figure 2.18 Radiation pattern of a half wave-length dipole situated on top of a 20-m tower located over an imperfectly conducting ground (only one fourth shown).

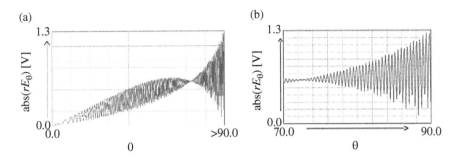

Figure 2.19 Radiation pattern of a dipole over a real ground. (a) normal, and (b) expended, situated on top of a 20 m tower located over an imperfectly conducting ground.

Beam forming or beam steering is out of the question in the near field as a beam or field pattern is not defined as seen from the Figure A2.1 presented in the Appendix. In the far field pattern with so many nulls it is difficult to see what beam steering will imply? Moreover, since the far field for a dipole situated 20 m on top of a ground plane starts at 10.6 km from the tower, then the cell, which this antenna is going to serve, will be definitely be in the near field. If that is the case, then what does beam forming mean in a near field environment? This point must be addressed from a Maxwellian point of view, based on actual physics, if we want to have an effective and properly working operating system! Assumption of a point source for the antenna bypasses all these subtleties as a point source has no near fields. But as shown before, such an assumption is not correct from a physics point of view!

In view of these subtleties in the analysis of the vector antenna problem, if we consider a set of transmitting antennas directing energy to a receiving antenna in a communication system, then, what does it mean to direct a beam if the receiving antenna is in the near field of the transmitting antenna? For example, what do we mean by directing a beam formed by using an array of transmitting antennas to a designated receiving antenna in a cell or a microcell? Another important issue, is that in the plots of Figures 2.16–2.19 the antenna is operating in a free space environment and there are no other scatterers nearby. It is often assumed that if there are many scatterers present near the tower then these nulls in the pattern will fill up! However, for a vector electromagnetic scattering problem, it is not clear why that will be the case. In fact, the problem will get much more complicated if the presence of other near field scatterers is taken into account in the computation of the beam pattern. In the near field, the antenna beam can only be defined at a specific distance as it varies from point to point and moreover there cannot be any pattern nulls independent of the distance. Therefore, we need to know precisely the spatial locations of the transmitters and the receivers, if one is willing to direct the signal-of-interest (SOI) to the desired receiver. In addition, we must also know the near field environment, i.e., the location and shape of any possible scatterers in order to compute a distance dependent antenna beam. If such is the case then it is difficult to see how the simplistic approach used to advocate Multiple-Input-Multiple-Output (MIMO) that many communication papers present will really work in practice, with such a severely distorted element pattern. A disbeliever of the Maxwellian principles may say, MIMO does not deal with beams but direct energies, or employs the principle of superposition to increase the probability of communication. However, the principles of MIMO must be demonstrated using electromagnetics principles and not from some heuristic erroneous nonphysical statistical reasoning, as we illustrate this fallacy later in this chapter! In fact, the message we are trying to convey is that all these presently deployed electromagnetic systems can be analyzed with great numerical accuracy using currently available numerical electromagnetics code, for example [9], and most physical principles thus can be verified and a system can be designed with high accuracy and no statistics-based heuristic conjectures are necessary!

Secondly, in a microcell and in a picocell and with a deployment of the antennas at a large distance from the ground, we are always operating in the near field. Hence under these conditions, this complicated element fields can be analyzed accurately using Maxwell's equations as is done in [9]. Many researchers often use ray tracing to characterize the multipaths in this propagation MIMO environment in order to characterize the channel. We need to initiate a discourse about how meaningful is near field ray tracing from a Maxwellian point of view, and even in the far field when the antenna element pattern has a large number of maxima and minima!

2.10   Use of Spatial Antenna Diversity

To correct for signal cancellation in a complex environment, as illustrated, even with a simple example presented in the previous section (an antenna located on top of a tower), it is often suggested that deployment of more transmitting antennas by increasing spatial transmit diversity is going to mitigate the signal loss due to multipaths at the receiving antenna. Again, it is difficult to accept such a nonscientific conjecture under the present scenario when we observe the radiation patterns in Figures 2.15–2.19. The electric field is a vector quantity and obviously adds according to the specific rules that apply for these types of vector quantities. In Figures 2.15–2.19, the interference of the signal produced by the antenna and its image due to the ground over which the antenna is placed occur in free space in the absence of any obstacles. The presence of obstacles will further complicate the scenario. Given the highly distorted element pattern of the half wave dipole located on top of a tower, it is impossible to see how adding more antenna elements in the transmit mode is going to eliminate some of the pattern nulls. To the contrary, addition of more antenna elements will deteriorate the scenario further as an accurate solution of Maxwell's equations demonstrate. For example, we know from a simplified point source antenna array theory that the resultant radiation pattern is the product of the single antenna element radiation pattern and the array factor, which is due to the presence of multiple antenna elements in the array [2–4]. If we now consider an array of isotropic omni-directional point radiators (radiators as mentioned in section 2.2, by the way, do not exist in reality), then the array pattern will no longer be isotropic, as the array factor by itself produces pattern nulls. Thus, if an antenna element pattern has M number of nulls, an array of such elements will definitely have more than M nulls under all conditions! Therefore, one needs to initiate a dialogue to see how diversity with a highly distorted element pattern makes any scientific sense in improving the mode of communication.

However, we may reach some bizarre conclusions if we apply probability theory. For example, because the antenna element pattern has so many maxima and minima then the antenna pattern can be treated hypothetically as a random variable, and using the *central limit theorem* [22] one can reach the erroneous conclusion that the addition of more antenna elements is going to improve the scenario and the field strength at a spatial location will improve due to ensemble averaging. Often when dealing with the analysis of very large communication system problems, we introduce the concepts of probability theory either to introduce uncertainty into the model or supplement our incomplete knowledge about the system [22]. In addition when a problem is too large or too complex to solve by any other method we take recourse to using probability theory. ***One point to be made is that a probabilistic model cannot be used to supplement insufficient or inadequate knowledge about a system*** [22]. **Through the introduction of a probabilistic model we cannot obtain *more* knowledge about a system.** We merely make some assumptions about the underlying process (through the use of the ensemble) so that we can apply available analytical tools to solve a problem without giving further thought as to whether the assumptions made for the system are indeed relevant or even correct! A probabilistic description of a random process is not adequate since additional information about the system may be required, with such information generally unavailable. This is why assumption of a Gaussian or a Markov process is often made, as for such probability distributions all the higher order probability density functions can easily be determined from the lower order density functions. In short, the analysis is often simplified through the use of wide sense stationarity or the assumption of an ergodic system. Again, it should be pointed out, the scalar probability theory does not handle correctly the various vector interactions of the different components of the electric fields. Also, the introduction of the concept of a random process in the analysis of a system presupposes that one will obtain a solution, which will be a possible outcome, in an ensemble. However, the probabilistic solution may not be an appropriate one. Sometimes the application of probability theory may provide some very interesting counterintuitive result [23].

Taking a physics-based approach not only tells us which system does not perform as we would like, but also we could predict what can be a viable solution and correct the short comings. Therefore, to understand basic properties of an antenna and its fundamental limits in performance, it is imperative to apply Maxwell's theory to obtain the design guidelines rather than using statistical modeling. Unless backed by the Maxwellian physics the statistical modeling will lead to erroneous conclusions as illustrated in [24]. For example, if we use a different deployment of the transmitting antennas instead of using the concept of spatial diversity, then one might get a better solution to the signal transmission problem instead of applying some statistical procedures. The signal cancellation problem can be largely diminished if the antenna is placed close to the Earth and not on top of a tower as the following example will illustrate. We have seen that the gain of a vertically oriented half wave dipole located in free space is given by the plot marked '1' of Figure 2.20 and this gain is minimum only along the zenith. We now place the same antenna almost touching an infinitely conducting ground plane. The gain pattern of such an

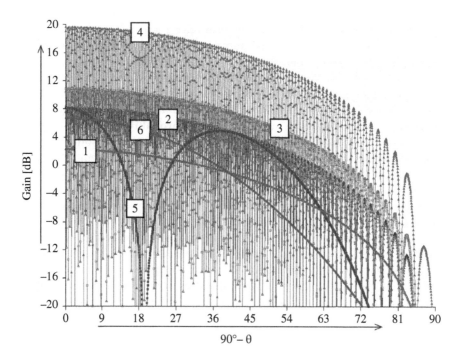

Figure 2.20 Radiation pattern of a half wave dipole elements located in various environments.

antenna system is given by the plot marked '6' in Figure 2.20. For the same voltage excitation, the gain of an antenna very close to the ground along the horizontal plane is increased by 6 dB over the same antenna when it is located in free space. This is the positive aspect due to the image in the ground.

Next, we consider the same half wave dipole antenna on top of a 20-m tower, but also operating over a perfect ground plane. It is seen in the plot marked '2' of Figure 2.20 that the gain pattern has too many nulls due to the interference between the direct ray and the rays reflected from the ground, and the interesting part is that it does not get any better whether we use 3 antennas instead of one (plot marked '3' of Figure 2.20) or 21 antennas instead of one (plot marked '4' of Figure 2.20) dipole element. For the cases of multiple antennas, the separation distance between them is half a wavelength and they are all located on top of the tower. The only difference is that the radiated fields get stronger along certain directions and nulls are produced in other directions. The y-axis provides the value of the gain in dB and the x-axis is the elevation angle. If the single antenna is now brought close to the ground and placed 15 cm on top of it, the gain is given by the plot marked '5' of Figure 2.20.

Also, observe that the gain along the horizontal plane of the plots '2' and '5' are very similar along the azimuth directions. Hence, it is quite illustrative that diversity for antennas located on top of a tower has very little real physics. However, by treating the plot marked '2' in Figure 2.20 as a scalar random variable one can essentially play any theoretical game one likes but it is not going to change the reality that this antenna will have an awful pattern and putting in more antennas in reality will not improve the situation! The only way to make things work is to place the antenna close to the ground and generate similar gain patterns marked as '5' in Figure 2.20, which does not contain a large number of antenna pattern nulls.

Bringing the transmitting antenna closer to the ground may at first sound paradoxical because most broadcast antennas in TV or mobile communications are placed on top of a tower. This is because in broadcast the transmitting antenna serve a different purpose as they cover a very wide area which could be of the order of tens of miles. Even though these antennas operate at frequencies, where propagation is primarily line-of-sight (LOS), we all know that the TV and mobile communication signals are often received inside a house or a car. For example, one can see the objects inside the closet in the basement of a house where the direct LOS of the sun can never reach! So first, we must realize that the propagation is not strictly LOS, but can also be due to diffraction, refraction, scattering and diffusion. Secondly, for TV, often we transmit or receive the signals with a high gain Yagi-Uda antenna [It is a directive antenna and for details refer to the encyclopedia in reference 25] and thus we discriminate between the source and its image, thereby eliminating the patterns with multiple nulls. In mobile communication, since we typically do not use such highly directive receiving antennas it makes sense to place the antenna closer to the ground, reduce the transmitting power at least by a factor of ten, and use more of the transmitting antennas, if desired, strategically located to increase the signal coverage rather than using the concept of spatial diversity by deploying more antennas on top of a tall tower as seen in the gain patterns marked by '5' of Figure 2.20. Addition of more antennas under the guise of diversity clearly indicates that there is no improvement in the transmit radiation pattern. The scientific basis of such a deployment will be discussed in Chapter 3 from both a theoretical point of view and using experimental data.

However, in a receiver one can use multiple antennas and then use a switch to couple to the signal of largest amplitude in any of the receiving antennas to improve the received signal to noise ratio. But then, we are not vectorially adding all of the electromagnetic field components incident on the receiving antenna simultaneously, but selecting an individual candidate. A better simple solution is to reintroduce the prehistoric tuner concept in a receiver which can dynamically match any system operating in any environment.

A disbeliever of the Maxwellian principles may argue baselessly that if there are buildings or other near field scatterers then the pattern nulls shown in Figure 2.20 may fill up due to the multiple scattering of the patterns and that perhaps the pattern may have no nulls! Yet no evidence is provided, to demonstrate that such a situation may even remotely be true for the vector electromagnetic problem. In the next chapter we will see why such conjectures do not hold!

In summary, deploying transmitting antennas close to the ground will also be a better solution as demonstrated next, when we try to communicate with a receiver in the near field. The physics and the mathematics of this concept will be presented in Chapter 3. Here we illustrate the problem with some numerical examples.

2.11 Performance of Antennas Operating Over Ground

As one observes that typically the base station antennas are generally deployed on top of a high tower typically in the range from 30 m to 120 m. The question that needs to be addressed now: is it really necessary to deploy the antennas on top of a high tower? Because we have seen from the previous section that such a deployment over ground produces an antenna pattern which is full of maxima and minima and this makes it useless for them to do any adaptive processing. The antennas for a microwave relay link are generally deployed on top of tall towers as the goal there is get a communication distance of 50 to 100 miles between the towers. For such a long distance of transmission without any repeaters it is then necessary to account for the curvature of the earth as this is a line-of-sight communication. This goes also for TV transmission as a radiating antenna typically serves 20–50 miles. Hence microwave relay links are placed on top of tall towers so as to overcome the curvature of the earth even though they have line of sight transmission. Typically in a microwave relay link a rule of thumb is to relate the height of the tower H to the active distance of communication by the effective radio line of sight. This is computed from the following: If R is the radius of the Earth, h is the height of the antenna above the ground and d is the distance between the two antennas, then $(R+h)^2 = R^2 + d^2$. By assuming a mean radius of the earth as 6380 km, one obtains $d(km) = 3.57 \times \sqrt{1.33 \times H}$, where the factor 1.33 is to account for the refraction in the atmosphere. This rule of thumb applies for effective line of sight communication in order to avoid the curvature of the earth. If one were to use this rule of thumb in cellular wireless communication as carried out in microwave relay links then for the cell size of 5 km the antenna height should **ONLY** be approximately half a meter above the ground!! As we shall see later on that this is a good choice for a variety of reasons. In addition, the cost of a tower scales approximately as the square of the height of the tower. In addition, the microwave relay links are made taller so that no obstruction occurs in the first Fresnel zone of operation. However, it is important to note that microwave relay links generally operate at 10 GHz where buildings trees have more pronounced effects than in mobile communication which works around 1 GHz. It is important to note that due to diffraction, a principle that implies that light rays bend around obstacles, there will be still transmitted energy even if the LOS is blocked. An example of this is when one is walking in a dense environment of trees, where one can still see objects on the ground even when the line of sight is blocked. Or if one goes to the basement closet of their home and opens it, it is still possible to see inside the closet due to the diffused sunlight even though there is no line-of-sight link with the sun. This principle is illustrated by Figure 3.3 of Chapter 3.

However, in a cellular wireless communication the radius of the cell is between 3–5 km. Hence, the curvature of the earth is not an issue and so the antennas can be placed at a much lower height even upto a height of 0.5 m as the above formula predicts. If such a deployment is made, then immediately, one would hear questions like: how does the buildings and the trees affect the propagation? Should we not try to avoid them! The answer to these questions as we will see in the next section is that, they produce second order effects as the primary path loss is due to free space propagation of the signal is much greater!

Another point of argument that is always put forward is: We need diversity due to fading. The problem with this statement is that indeed in microwave relay links there can be fading due to multipaths created by the reflecting Earth. Hence, over a large distance there can be fading due to multipaths which may be generated from the point reflections from the reflective ground. In addition, for multipaths to be generated one needs to operate in the far field of the antennas where the rays can be traced from the antenna. In the near field, an antenna has no pattern nulls and it is difficult if not impossible to introduce ray concepts to characterize propagation. So, if any antenna is placed on top of a 15 m tower, the far field (according to Appendix A) for an antenna operating at 1 GHz will start approximately at $\dfrac{2D^2}{\lambda} = \dfrac{2 \times 30 \times 30}{0.3} = 6km$ from the radiating element which in most cases will be outside the operating cell. Hence use of diversity is not at all necessary to deal with multipath fading. We will see in the next chapter as to what is the real cause of slow fading. In addition, the principle of fast fading does not exist as illustrated in Chapter 1. In any case, for a cellular communication the radius of the cell is generally only a few kilometers. In that case, the height of the tower could be less than a 1 m from the ground. And now, the transmitted power can be reduced at least by a factor of ten for acceptable quality of service as illustrated in the next chapter. That is a good way to deploy the base station antenna closer to the ground. As will be illustrated later, this type of deployment can increase the receive signal strength several times and so the transmitting power can easily be reduced by a factor of ten reducing the concern of health hazards irrespective of whether it is a valid concern or not!

As an example consider the transfer of power from a half wave base station transmitting dipole antenna located at various heights above the ground and excited by 1 W of input power operating at 1 GHz to a half wave receiving dipole located at 2 m above Earth. Both the antennas are conjugately matched. We would like to know what is the power received in the load of the receiving antenna. We consider both a perfectly electrically conducting ground and an urban ground at different separation distances of 100 m and 1000 m between the transmitting and the receiving antennas. Different deployment heights are considered for the base station antenna for different orientations as one is tilted to towards the ground presumably to improve the performance. Table 2.5 illustrates that in the near field, bringing the antenna closer to the ground actually enhances the signal strength in the receiver whereas in the far field there is the usual conventional height gain, namely, the field strength increases as the height of the base station antenna is increased as illustrated in Chapter 3. Also, as will be demonstrated in the next chapter the propagation path loss in the near field is 30 dB per decade of distance and whereas in the far field it is 40 dB/decade. In addition, the fading disappears in the far field and when the antenna is brought closer to the ground. This will be treated quantitatively from a physics point of view in Chapter 3.

Finally it is shown that the use of a high gain antenna like a horn instead of a dipole antenna for the base station may increase the radiation efficiency even when the LOS to the receiver is blocked by a dielectric wall. Here, the transmitting dipole is replaced by a standard 20 dB gain horn. The direct LOS between the transmitting and receiving antennas is interrupted by a finite dielectric wall that is 85 m from the transmitter. The wall has a dielectric constant of $\varepsilon_r = 2.0$ and

Table 2.5 Transfer of power between a transmitting and a receiving antenna for a fixed input power for various heights and separation distances.

Type: $\lambda/2$ base station antenna	PEC ground $\rho = 100$ m Received power (μW)	PEC ground $\rho = 1000$ m Received power (μW)	Urban ground $\rho = 100$ m Received power (μW)	Urban ground $\rho = 1000$ m Received power (pW)
20 m, vertical	0.0337	0.0014	0.11	1564.1
20 m, tilted down 11°	0.04	0.0013	0.124	1506.2
10 m, vertical	0.08	0.003	0.139	490.9
2 m, vertical	0.14	0.0031	0.144	21.5
1 m, vertical	0.26	0.0031	0.046	5.5
0.5 m, vertical	0.297	0.0031	0.013	1.42

is 5 m high, 3 m wide, and 6 cm thick. The rest of the parameters remain the same as before and the ground is considered to be perfectly conducting for simplifying the analysis. For example, even when the transmitting horn is 20 m above the ground and the LOS is blocked by the dielectric wall as seen in Figure 2.21. In this example, both the transmitter and the receiver are conjugately matched so as to transmit and receive the maximum power. The radius of the receiving dipole is 0.0033λ. The received voltages and the received power at the receiving half wave dipole for different nature of the deployment is shown in Table 2.6.

So even the presence of an obstruction, like a wall, to the LOS mode of propagation does not mean that the received signal strength will be less. This non-intuitive result arises from the vector nature of electromagnetic wave propagation, where the diffracted, scattered, diffused and refracted fields may in fact add to provide a higher signal strength in some particular deployments. Of course, this isolated example does not imply that such signal enhancements will occur all the time. For the particular problems under consideration, the nature of the signal strength is determined by the physical dimensions and electrical parameters of the system, and the results will be dictated solely by Maxwell's equations.

2.12 Fields Inside a Dielectric Room and a Conducting Box

Primary objective of this group of simulations is to gain insight into the impact of placing the receiving antenna in a completely enclosed structure like a room inside building, or a metal container and the like. It illustrates two points, first,

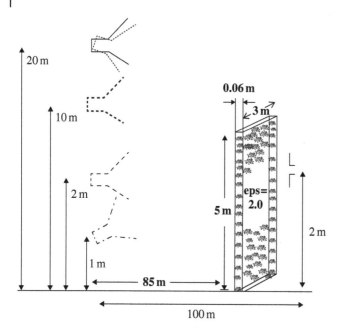

Figure 2.21 Different transmitting horn configurations at a height of 20 m, 20 m tilted downwards at an angle of 11°, 10 m, 2 m, and 1 m tilted upwards by an angle of 1° above the ground plane. The receiving dipole is located at 2 m above the ground plane and at a horizontal distance of 100 m from the transmitter. The LOS component between the transmitter and the receiver is blocked by a dielectric wall.

Table 2.6 Transfer of power between a transmitting horn and a receiving dipole antenna for a fixed input power for various heights and separation distances.

Location of the transmitting horn over ground	Received voltage (mV)	Received power (µW)
20 m above ground	12.24	1.26
20 m above ground and tilted down 11°	25.0	5.27
2 m above ground	44.4	16.7
1 m above ground tilted 0.58° towards the sky	62.0	32.3

it is shown that encapsulating the receiving antenna inside a room does not necessarily impair the receive signal strength. Secondly, often a rich multipath environment for MIMO is simulated by placing the transmit/receive antennas inside a metal chamber. It is illustrated that such an exercise is meaningless and it is really ridiculous that some manufacturers of equipment are in the process of constructing systems to emulate this scenario. Unfortunately an antenna does not radiate at all when it is placed inside a metallic container, even though it may produce some fields! Thirdly, in a finite difference numerical methodology like in finite element method or in finite difference time domain, often an absorbing boundary are placed near the radiating system to truncate the mesh so as to reduce the size of the solution space. This methodology yields acceptable results for certain thickness of the walls and choices of the material parameters of this boundary. However, the nature of this absorption boundary condition is highly dependent on the angle of incidence of the wave onto this boundary. When the thickness of the terminating mesh is small, some numerical instabilities can be anticipated. This is where the efficiency and accuracy of an integral equation based methodology becomes quite apparent as then the solution procedure does not influence the radiating properties of the antenna as the radiation condition is automatically built in the solution procedure and no artificial methodology needs to be introduced. The simulations also illustrate that one can get more received power when the antenna is located closer to the ground for the near field scenario.

Consequently, the basic question is: Which receiving dipole will have a larger induced voltage, hence larger received power from the transmitting antenna? – a receiving dipole operating in free space with the direct LOS link with the transmitter or a receiving dipole encapsulated by the dielectric box with no LOS link? Since researchers in communication theory are more involved with space-time coding and intricate statistical details, they have neither addressed nor solved this system-related problem. Electromagnetic practitioners, who generally lack the computational resources to solve generic problems like the one presented above, may predict that the received power in the dielectric encapsulated dipole will be much less than when the receiving dipole is operating in free space in the presence of the direct LOS link to the transmitting antenna. The possibility that the received power in the encapsulated dipole with no LOS can be larger is often ignored as a possible solution in the contemporary wireless propagation literature. However, if one looks at a standard graduate text book on basic electromagnetic theory, it is quite clear that some situations do exist [3, p. 370] where the transmission coefficient of a system can actually exceed unity. This phenomenon generally occurs when the characteristic impedance of a medium is effectively greater than that of free space, which may occur near a parallel resonance. Under this condition, the radiated fields coming from the currents induced on the other side of the plate due to the field penetration through the aperture may add with the incident field going through the aperture [3]. This may result in a higher value of the transmitted field, which does not imply that the transmitted power will be greater!

As an example, consider two parallel, half-wave, center-fed, conjugately matched, vertically oriented, transmit-receive dipoles operating in free space over a perfectly conducting ground plane. The dipoles have a 1-mm radius and

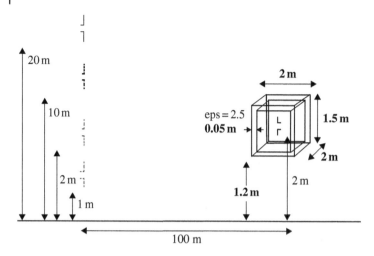

Figure 2.22 Different transmitting dipole configurations at a height of 20 m, 20 m tilted downwards at an angle of 11°, 10 m, 2 m, and 1 m tilted upwards by an angle of 1° above the ground plane. The receiving dipole is located at 2 m above the ground plane and at a horizontal distance of 100 m from the transmitter. The receiving dipole is enclosed by a dielectric shell.

operate at 1 GHz. Initially, the bottoms of the transmitting and receiving dipoles are located at heights of 20 m and 2 m, respectively, above the ground. The basic differences between this example and the previous ones dealing with dipole transmitting antenna are: the feed of the transmit and receive dipoles are displaced vertically upward by 15 cm, the equivalent of a half wavelength; and the receive antenna is encapsulated by a dielectric box. The outer dimensions of the dielectric box are: 1.5 m high and 2 m × 2 m in the cross-sectional plane. The thickness of the dielectric is 5 cm, and the bottom of the box is located 1.2 m above the ground as seen in Figure 2.22. The dielectric constant of the box is assumed to be $\varepsilon_r = 2.5$ and the value of the loss tangent is zero for simplicity (i.e., a low loss dielectric material). For all the simulations, the input power is restricted to 1 W at the transmitting antenna. The simulated environment is displayed in Figure 2.22. The computed results are tabulated in Table 2.7 where the induced received voltages and the power are tabulated.

The reason the signal levels are higher at the receiving antenna when it is operating inside a dielectric box than in free space is because the room has a signal enhancement effect. It appears that the room is acting as a dielectric lens. It is important to note that even if the room is very large the fields inside the room are due to the near field effects of the antenna. This is due to the six images formed by the four walls, the ceiling and the floor along with multiple copies that are generated by the various walls of the dielectric room. First of all due to the principle of images the dipoles operating inside the room will

Table 2.7 Transfer of power between a transmitting dipole and a receiving dipole antenna for a fixed input power and for two different scenarios (one inside a box and the other operating in free space).

Orientation of the transmitting antenna		Receiving antenna operating in FREE space		Receiving antenna operating inside a DIELECTRIC box	
Height	Type of deployment	Received voltage	Induced power	Received voltage	Induced power
20 m	Tilted down towards Earth by 5°	1.99 mV	0.036 µW	2.17 mV	0.043 µW
20 m	Vertical	1.93 mV	0.034 µW	2.12 mV	0.041 µW
10 m	Vertical	2.97 mV	0.08 µW	4.15 mV	0.16 µW
2 m	Vertical	3.92 mV	0.14 µW	6.22 mV	0.25 µW
1 m	Vertical	5.35 mV	0.26 µW	8.57 mV	0.66 µW

generate six images reflected from the six walls of the room. Hence the fields inside the room will create a near field scenario where there will be interference between the fields due to the images and will generate an interference pattern. Depending on where the dipole is located inside the box the fields will be higher or lower as illustrated by the following plots.

The next set of examples illustrates that, the power induced at the receiving dipole operating in free space is 0.11 µW for a transmit antenna located centered on top of a 20 m tower and fed with an input power of 1 W. When the receiving dipole is encapsulated by the dielectric box, the received power at the dipole is 0.128 µW for the same 1 W transmitter input power. The received power is considerably larger than before. It is also noteworthy that a shift of a half wavelength in the positions of the transmit and receive antennas also has an impact on the received power, even when the transmit powers are constrained to be the same. This clearly indicates the possibility that the presence of buildings may actually enhance the signals inside them in some cases. Since the voltage induced in the resistive load is 3.9 mV when the dipole is operating inside a dielectric box, as opposed to an induced voltage of 3.64 mV when the dipole is operating in free space with a direct LOS, the case of a larger induced voltage in the receiver is substantiated.

Next let the receiving antennas and their environments remain the same and we lower the bottom of the transmitting dipole centered to 10 m above the ground as shown in Figure 2.22. As discussed earlier [26], the induced power will be more than when the end of the transmitting antenna is located centered 20 m above the ground. Specifically, the induced powers at the receiving dipole in free space and when encapsulated by the dielectric box respectively are is 36.11 nW and 102.76 nW. As this simulation clearly indicates, situations exist for which the non-LOS power induced on a receiving antenna exceeds (significantly in this case) the received power achieved for a free-space transmit-receive configuration.

Finally, the transmitting dipole is lowered so that its bottom is 2 m above the earth, while keeping the receiving dipoles at the same locations as seen in Figure 2.22. For an input power of 1 W to the transmitting antenna, the received power (0.14 μW) in the free-space LOS mode is less than half the received power (0.25 μW) of the encased dipole. Similarly, the induced voltage increases from 3.92 mV to 6.22 mV as tabulated in Table 2.7. As expected, the interference between the fields from the actual source and its ground-plane image again causes the received signal voltage to change as the height of the transmit dipole increases while keeping the receiving dipole in a dielectric box. However, what is most intriguing is that the signal strength inside the room is higher than when there is a direct LOS link between the transmit/receive system! This phenomenon is not observed by a standard application of probability theory. Thus, it is clear that simple application of probability theory, without including the Maxwellian principles of basic electromagnetics, does not provide a complete and accurate solution to the problem.

Another problem that is often treated extensively in the Antennas and Propagation literature is the use of ray tracing for an indoor mobile communication environment [27]. Typically, to find the fields inside a room, a hallway, or for roof top propagation, one generally uses a ray tracing technique to compute the field distribution in the neighboring environment. Such a methodology should be highly questionable from a scientific view point because of the existence of the floors and walls in the model which will produce multiple images of complex amplitudes. Hence, it is not sufficient to use just the antenna element pattern in ray tracings but one should take into account the existence of the various images for the antenna element and moreover one will be dealing with a near field prediction problem which precludes the use of ray tracing as the latter is more suitable for far field predictions.

As an example, consider a center fed half wave $\hat{z}$-directed dipole located at the center (given by the origin of the coordinate system) of a $7\,\lambda \times 7\,\lambda \times 7\,\lambda$ room which has a wall of thickness 0.03 λ. The relative dielectric constant of the wall all around this cube is considered to be $\varepsilon_r = 2.5$. Next, we plot the various components of the fields inside this dielectric cube produced by the center fed half wave dipole as computed by the computer code of [9]. The fields inside the room are shown in Figures 2.23–2.26 at various locations and compared with the field that would exist if the dipole were to be operating in free space.

As seen from Figures 2.23–2.26, it is clear that the effects of the images of the initial dipole source generated by the dielectric walls, floor and the ceiling of the room will clearly distort the fields in such a way that simple ray tracing

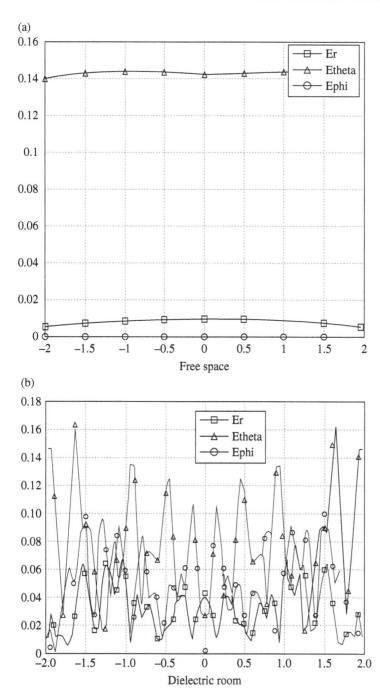

Figure 2.23 The three components of the fields E_r, E_θ, E_φ in (a) free space and (b) inside the room at $x = -3.75\lambda$, $z = -3.75\lambda$, as a function of y.

(a)

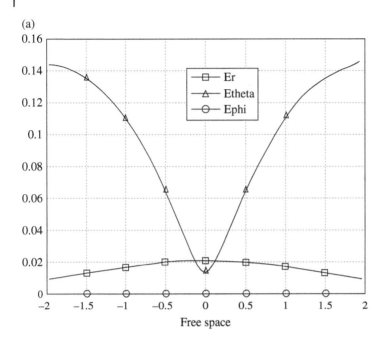

Free space

(b)

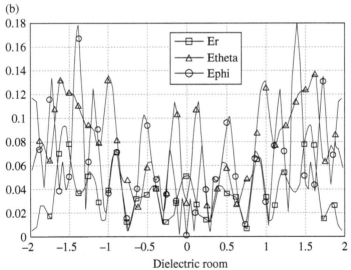

Dielectric room

Figure 2.24 The three components of the fields E_r, E_θ, E_φ (a) free space and (b) inside the room at $x = -0.0198\lambda$, $z = -3.75$, as a function of y.

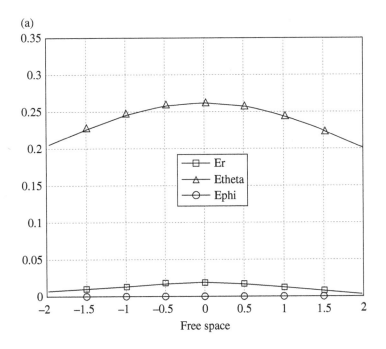

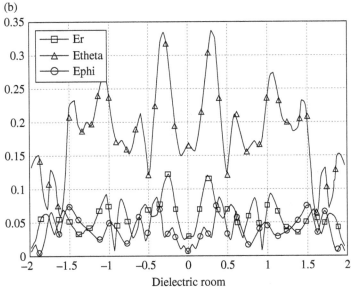

Figure 2.25 The three components of the fields E_r, E_θ, E_φ (a) free space and (b) inside the room at $x = -3.75\lambda$, $z = -0.0198\lambda$ as a function of y.

(a)

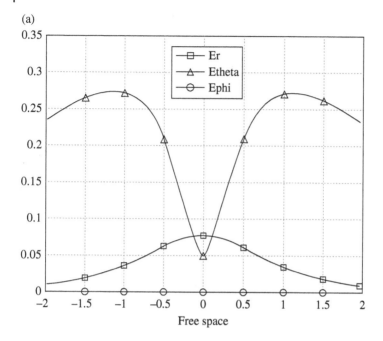

Free space

(b)

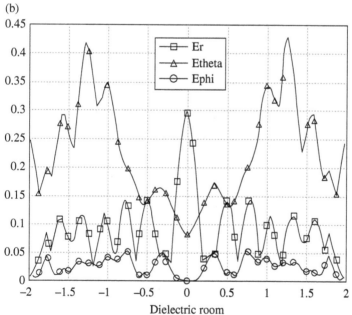

Dielectric room

Figure 2.26 The three components of the fields E_r, E_θ, E_φ in (a) free space and (b) inside the room at $x = -0.0198\lambda$, $z = -0.0198\lambda$ as a function of y.

using the antenna element pattern of free space is not sufficient to predict the correct fields. Another important reason why ray tracing is not an efficient methodology to compute the fields inside a room as one is operating in a near-field scenario where many rays need to be considered to obtain a meaningful solution. Because of the generations of the various images, computation of the fields inside the dielectric cube is a near field problem. Hence the applicability of ray tracing itself may be questionable to compute fields in an indoor environment. This clearly demonstrates that in the computation of the fields inside a room, a hall way or for a roof top propagation, the effect of the images produced by the sources must be taken into account to produce any meaningful simulation results for the fields. Instead of using ray tracing, one can solve the entire problem accurately using a numerical electromagnetics code [9] providing stable and accurate results.

In summary, the value of the electric field inside the big dielectric cube representing a simple room will depend on the physical characteristics of the room, the frequency of the incident wave and the physical location of the transmitting dipole. It is seen that the values of the field may be enhanced at some locations due to a focusing lens effect associated with a dielectric structure or can be reduced at other locations! It is impossible to predict a priori where such construction and destructive interference of the fields will occur. That is why computational electromagnetic tools provide valuable information in the design of a real system. The actual value depends on the spatial location of the transmitting antenna as the vectorial addition of the fields from various images can cause severe fluctuations to the electric field strength.

Also it is interesting to speculate what will happen to the value of the fields if this dielectric box is transformed to a perfectly conducting box. Typically an antenna is placed within a metal box to characterize its radiation efficiency through the Wheeler-Cap method [28]. The basic principle involved is that an antenna radiates when it is operating in free space. So it has a complex input impedance, the real part of which consist of the loss (ohmic) resistance and the radiation resistance. However, when it is made to operate inside a metal cavity an antenna cannot radiate and so its radiation resistance will be zero. Now if one measures the real part of the input impedance one will obtain only the ohmic resistance for the antenna. By knowing the loss resistance and the radiation resistance it is possible to evaluate the radiation efficiency of the antenna. In recent times this basic principle has been forgotten and one is measuring characteristics of antenna systems in a conducting box with a mode stirrer to simulate the rich multipath MIMO environment as if one is preparing a soup!!! It is rather unfortunate that these researchers have overlooked some fundamental properties of radiation related to antennas. It is quite straightforward to numerically simulate an environment where an antenna is operating inside a box made of metal using a frequency domain integral equation code as the radiation conditions are built-in [9]. An analysis of an antenna inside a metal

(a)   (b)

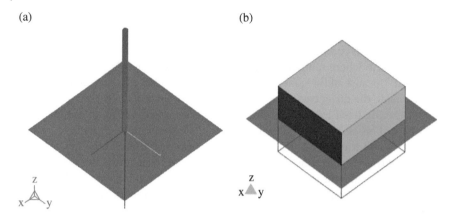

Figure 2.27 A half-wave dipole antenna radiating in (a) in free space and (b) operating inside a conducting cube.

box using a time domain code will not produce any meaningful results as the waves inside the box will bounce back and forth from the metallic walls and the solution WILL NOT CONVERGE! So it is not clear what type of results are being generated by various researchers and does it really contain any meaningful information when they use a time domain code to generate the performance of the antenna operating inside a metal can? However, if the antenna is not radiating at all then it is not clear what actually is being measured and what some of the commercial enterprises are doing in claiming that they really can simulate such a rich multipath environment for an antenna radiating inside a conducting box. To illustrate the absurdity and the foolishness of such a way of thinking we now simulate the operation of a quarter wave monopole first operating in free space and then radiating inside a conducting box of different dimensions as shown in Figure 2.27b. The computed results are given in Table 2.8.

Consider a monopole of length 0.25λ operating over an infinite perfectly conducting ground plane and radiating in free space. Let the radius of the quarter wave monopole be 5 mm and it is operating at 300 MHz, just for illustration purposes. Actual value of the frequency is not important as the results are scaled in terms of the wavelength. The monopole is displayed in Figure 2.27a. The simulated input impedance of this quarter wave monopole is given by the first line of the Table 2.8 and the computer code provides a good solution for the input impedance of the quarter wave monopole antenna. Now consider the same monopole antenna to be completely encapsulated by a perfectly conducting metallic box of length = aa m; width = aa m and height = $aa/2$ m as shown in the Figure 2.27b. It is seen in Table 2.8 that in this case, the real part of the antenna impedance is practically zero. This illustrates that the antenna is not

Table 2.8 Input impedance of the half wave dipole antenna when operating inside a perfectly conducting box of different sizes.

Model	Input impedance of the antenna
Radiating in Free space	$Z_{in} = 46.7 + j\ 20.2$
Covered by a metal box of side $aa = 1\ \lambda$	$Z_{in} = 0.009 - j\ 4.2$
Covered by a metal box of side $aa = 5\ \lambda$	$Z_{in} = 0.0086 - j\ 7.8$
Covered by a metal box of side $aa = 10\ \lambda$	$Z_{in} = 0.0091 - j\ 42.2$
Covered by a metal box of side $aa = 20\ \lambda$	$Z_{in} = 0.048 - j\ 205.3$
Covered by a metal box of side $aa = 50\ \lambda$	$Z_{in} = 0.046 + j\ 127.3$

radiating at all. This is really the basis of the Wheeler-Cap method to measure the radiation efficiency of an antenna, where an antenna is placed inside a conducting structure to make the radiation resistance to be zero. The last five lines of Table 2.8 provide the value of the input impedance of the monopole when it is operating inside a conducting box of different sizes. The conclusion is that the antenna is not radiating at all when placed inside a box!

However, if the perfectly conducting box is replaced by a dielectric box of different thicknesses and for different values for the complex permittivity and permeability, the results are reasonable. In a differential equation based frequency domain solver one actually places a terminating mesh to simulate the infinite free space. It is interesting to note that in some cases, this actually may provide valid results when terminating a mesh of a differential equation based method. So the goal is here is to see if the presence of a dielectric box actually changes the input impedance to a different value as when it is operating in free space. Tables 2.9 and 2.10 provide the input impedance for the half wave dipole when operating inside a dielectric cube of a finite thickness but of different material parameters. Table 2.9 provides the input impedance for the half wave

Table 2.9 Input impedance of the half wave dipole antenna when operating inside a dielectric box of thickness of 0.1 λ with different dimensions and having a relative permittivity of $\varepsilon_r = 2 - j\ 2$ and a relative permeability of $\mu_r = 2 - j\ 2$.

Model	Input impedance of the antenna
Radiating in Free space	$Z_{in} = 46.7 + j\ 20.2$
Encapsulated by a dielectric box of material parameters $\varepsilon_r = 2 - j\ 2$ and $\mu_r = 2 - j\ 2$	
Covered by a dielectric box of side $aa = 1\ \lambda$	$Z_{in} = 42.1 + j\ 20.5$
Covered by a dielectric box of side $aa = 3\ \lambda$	$Z_{in} = 46.7 + j\ 20.4$
Covered by a dielectric box of side $aa = 5\ \lambda$	$Z_{in} = 46.7 + j\ 20.4$

Table 2.10 Input impedance of the half wave dipole antenna when operating inside a dielectric box of thickness of 0.1 λ with different dimensions and having a relative permittivity of $\varepsilon_r = 10 - j\,10$ and a relative permeability of $\mu_r = 10 - j\,10$.

Model	Input impedance of the antenna
Radiating in free space	$Z_{in} = 46.7 + j\,20.2$
Encapsulated by a dielectric box of material parameters $\varepsilon_r = 10 - j10$ and $\mu_r = 10 - j10$	
Covered by a dielectric box of side $aa = 1\,\lambda$	$Z_{in} = 42.1 + j\,20.2$
Covered by a dielectric box of side $aa = 3\,\lambda$	$Z_{in} = 46.7 + j\,20.4$
Covered by a dielectric box of side $aa = 5\,\lambda$	$Z_{in} = 46.7 + j\,20.3$

dipole antenna when placed inside a dielectric cube of thickness 0.1 λ but with a relative permittivity of $\varepsilon_r = 2 - j\,2$ and a relative permeability of $\mu_r = 2 - j\,2$. Table 2.10 provides the input impedance for the same half wave dipole antenna enclosed by the same thickness of the dielectric cube but with a relative permittivity of $\varepsilon_r = 10 - j\,10$ and a relative permeability of $\mu_r = 10 - j\,10$. In both cases, results for the input impedance is quite acceptable. So in summary, when an antenna is placed inside a conducting box it does not radiate but operates normally when placed inside a dielectric cube.

Next we study how what exactly an antenna array achieves in practice.

2.13 The Mathematics and Physics of an Antenna Array

When dealing with phased arrays [29, 30], we want to establish one of the mathematical underpinnings of phased array theory, which has to deal with the computation of the direction of arrival (DOA) of the various signals impinging on the array. It is also connected with adaptively enhancing the signal in the presence of interferers and noise. When dealing with sensor elements in acoustics, we define them primarily as isotropic point radiators. This makes perfect sense as a loudspeaker does not distort the input electrical signal when transforming it to acoustics or a microphone does not distort the input acoustic signal when transforming it to electrical ones. A uniform linear array consisting of isotropic point radiators separated by a distance d, is shown in Figure 2.28 (the figure is specialized to an one-dimensional array).

The signal of interest (SOI) impinging on the array is denoted by S, the interferers by J and the clutter, which is diffused reflected-transmitted energy, is characterized by C, respectively. Each of these signals has a complex amplitude

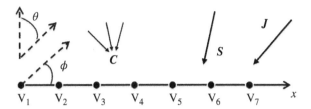

Figure 2.28 A phased array (one-dimensional) of isotropic omnidirectional point radiators as antenna elements.

A_i that impinges on the array from an elevation angle of $(90°-\theta_i)$ and an azimuth angle of φ_i. Then the voltage $V_{n,m}$ induced at the $(n, m)^{\text{th}}$ element of a two-dimensional array oriented along the x and y axes is given by

$$V_{n,m} = \sum_{i=1}^{P} A_i \exp\left[\frac{j2\pi(n-1)d}{\lambda} \cos\varphi_i \sin\theta_i + \frac{j2\pi(m-1)d}{\lambda} \sin\varphi_i \sin\theta_i \right]$$

(2.49)

where λ is the wavelength of transmission and P is the total number of signals impinging on the array. Here, it is assumed that P is much less than the total number of the antenna elements. Therefore, for Figure 2.28, when we examine equation (2.49), it is seen that the DOA $\{\cos\varphi_i \sin\theta_i; \sin\varphi_i \sin\theta_i \}$ of the various signals impinging on the array and the voltages $\{V_{n,m}\}$ induced in the antenna elements in an acoustic phased arrays form a two-dimensional discrete Fourier transform (DFT-implemented through the Fast Fourier Transform, FFT) pair. This is the mathematical basis for an acoustic phased array theory for performing estimation of DOA of the various signals impinging on the antenna array from the individual voltages induced at each of the receiving elements.

Now, let us assume that we are dealing not with the scalar acoustic problem, but with the vector electromagnetic problem. In an electromagnetic problem, isotropic point radiators do not exist. The smallest antenna element in practice is a dipole which may have its pattern omni-directional along certain planes. Also, in electromagnetics one has the added dimension of polarization, which is missing in acoustics. Finally, one has to be careful about the mutual coupling between the antenna elements and the presence of near field scatterers which may distort the basic assumption of an isotropic radiator. Now, if we assume that a plane electromagnetic wave with a linear polarization is impinging on an array consisting of z-directed Hertzian dipoles as shown in Figure 2.29, then the voltages induced in the Hertzian dipoles will be a function of the polarization of the incident field. For the appropriate

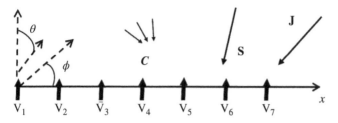

Figure 2.29 A phased array (one-dimensional) of Hertzian dipoles as antenna elements.

polarization, the voltage induced at the $(n, m)^{th}$ Hertzian dipole element could be expressed as

$$V_{n,m} = \sum_{i=1}^{P} A_i \sin\theta_i \exp\left[\frac{j2\pi(n-1)d}{\lambda}\cos\varphi_i\sin\theta_i + \frac{j2\pi(m-1)d}{\lambda}\sin\varphi_i\sin\theta_i\right]$$

(2.50)

Observe that we have now an extra factor of $\sin\theta_i$ in (2.50) when compared with (2.49). By comparing (2.50) with (2.49), it is seen that the DOA $\{\cos\varphi_i \sin\theta_i; \sin\varphi_i \sin\theta_i\}$ and the induced voltages $\{V_{n,m}\}$ do not form a Fourier transform pair, even though they can be when $\theta_i = 90°$, or the incident field is polarized parallel to the Hertzian dipole elements. Moreover, in a real situation, when there are circularly polarized fields or when we have a finite dimensional antenna and not a small Hertzian dipole, it is necessary to modify (2.50) to obtain the DOA of the various signals impinging on an array. This modification is also necessitated due to the existence of mutual coupling effects between the antenna elements or other near field interactions between the antenna and the surroundings or the platform on which they are mounted [29].

In the presence of mutual coupling among the antenna elements and other near field effects, (2.50) actually transforms into a very complicated integro-differential equation and one has to take recourse to Maxwell's equations in addition to signal processing algorithms to solve the complete DOA estimation problem [29, 30]. An example for an antenna, which is not a point source or a small Hertzian dipole, is a finite-sized dipole. Even for a z-directed thin-linear straight wire antenna of finite length, the incident field E^{inc} and the current distribution $I(z')$ on the antenna are related by an integro-differential equation [29]:

$$E_z^{inc} = -\mu_0 \int_{axes} I(z')\frac{e^{-jkR}}{4\pi R}dz' + \frac{1}{\varepsilon_0}\frac{\partial}{\partial z}\int_{axes}\frac{\partial I(z')}{\partial z'}\frac{e^{-jkR}}{4\pi R}dz' \qquad (2.51)$$

where $R = \sqrt{(z - z')^2 + a^2}$, and a is the radius of the wire antenna. The integral is taken along the axis of the antenna. The important point is that one needs to solve this integro-differential equation derived from Maxwell's equations to compute the voltages induced in an actual electromagnetic problem, due to the presence of mutual coupling and the polarization of the incident fields. Thus, we need to refine our conventional scalar acoustic phased array model and its connected theories to deal with the vector electromagnetic case [29, 30]. Other aspects of this issue will be illustrated later. Furthermore, the impulse response of an antenna is different in transmit and receive modes for the vector electromagnetic problem as will be discussed in Chapter 4. Use of point sources in the model instead of using the realistic antenna sizes and shape often ignores the mathematical subtleties introduced by Maxwell's equations. To simulate realistic scenarios it is necessary to characterize antennas by the complex integro-differential equations which correctly encompasses the principles of the vector electromagnetic problem [29, 30].

2.14 Does Use of Multiple Antennas Makes Sense?

The basic purpose of an antenna is to capture the electromagnetic signals that are propagating through space and transform them into an electrical signal which can be physically amplified and can be visualized in practice using an oscilloscope, a vector voltmeter or a network analyzer. An antenna achieves this goal by integrating the incoming electric field that illuminates the antenna and generates a voltage at the terminals of the antenna quantitatively using (1.88). This voltage contains all the information about the incoming electromagnetic wave. The question that arises is: Can this integral of the electric field along the antenna be zero? In that case a single antenna may be insufficient and one needs to deploy multiple antennas. The fact of the matter as we shall see one needs to use multiple antennas if one does not understand the basics of how to deploy a single antenna in the correct fashion. Since electromagnetic wave is a propagating entity it displays the characteristics of a wave. Namely if the functional value of the wave is zero at the current position and/or time then after quarter of an wavelength the wave function will reach either a maximum or a minimum in space or time as seen in Figure 1.3. Now if an antenna is quarter wavelength long then this integral value of (1.88) can never be zero, as one has a maximum and a minimum in this quarter wave section. The only way the integral value of the electric field representing a voltage can be zero is when the antenna is extremely electrically small in size and not matched so that the integral of the electric field will result in a very small value. So for an unmatched small antenna the induced voltage can be too small to be of value. It is important to note that an antenna does not receive power but is sensitive to the incident electric field which it captures and feeds to the first stage of the RF amplifier for further processing. In reality, if the electric field strength is less than 1 µV/m then the first RF stage of the amplifier cannot process the signal. The use of the term power amplifier is a misnomer as there are no power amplifiers in reality, it is either a voltage amplifier or a current amplifier. An amplifier uses the DC power and uses it to amplify the input AC signal and hence the term power amplifier does not carry much sensible meaning! So an unmatched electrically small antenna will not function very well as the received signal will be too small to be of any use and further signal processing cannot create the signal. However, as seen with the older transistor radios that even an electrically small antenna can perform extremely well if it is conjugately matched, as a tuner was used to match the electrically small antenna, which was there at the first stage of all the transistor radios that were designed earlier and so using such a small device one can listen to the electromagnetic fields originating from the other side of the Earth. There were little problems in receiving signals from very far away using a single antenna and the term fading was there in the vocabulary but was not relevant for practical use. So, even though it is true in theory that an electrically small antenna is inefficient, but in practice we have seen in section 2.7 that when the antenna is conjugately matched it behaves as a resonant length antenna. Therefore, the antenna will always produce a finite nonzero value for the voltage.

This is clear from the use of small antennas in transistor radios that were prevalent almost 50 years ago which could receive signals from the other side of the world even though the operating broadcast frequency was only 1 MHz corresponding to a wavelength of 300 m. The antenna inside the transistor radio was no bigger than 10–15 cm. Even though the antenna was very small, they worked flawlessly. This is because if an electrically small antenna is matched it will perform similar to a resonant antenna as seen in Figure 2.12. That is why at the first stage of the ancient radios there was a tuning capacitor so as to tune the antenna to the broadcast frequency of interest. Unfortunately, in the modern designs this key component is missing and this results in a product where how one holds the mobile phone makes a big difference. It is important to note that this key tuning component is going to make a comeback in the future models as some of the companies are now applying for and have received patents for this tuning system!! Just to make this historic saga complete, there was another component in the ancient radio designs called automatic gain control (AGC) This was a closed-loop feedback regulating circuit, the purpose of which was to provide a controlled signal amplitude at its output, despite of significant variation of the amplitude in the input signal from far and near distances, which translates into the near and far power problem in current systems. The average or peak output signal level was used to dynamically adjust the input-to-output gain to a suitable value, enabling the circuit to work satisfactorily with a greater range of input signal levels. And it will not be surprising to see such designs being incorporated in the future systems as this is the right and an intelligent way to design a radio receiver!

In the modern designs, the antennas are not dynamically matched and so the voltages induced in them by an electromagnetic signal is not large enough for the next stage for processing. And if the proper signal is not generated at the antenna terminals then signal processing will not be able to create the signal and this is being realized nowadays! Unless of course one does coding at RF as Shannon and Viterbi advocated and whose implementation would make the systems very complex with high latency. Also development of A/D's in the GHz range is required.

Improper deployment and analysis of an antenna through the assumption of a point source approximation have generated lot of mythologies in design of wireless systems. It is now conjectured that somehow deployment of multiple antennas will provide a better system even though they are deployed incorrectly in a clueless fashion. Systems with multiple antennas at the transmitter and receiver are considered to be MIMO (Multiple-Input-Multiple-Output) systems. First, all the voltage sources are connected in series and never in parallel like connecting multiple batteries to boost the voltages. This is because connecting sources in parallel are not desirable because if they somehow have even a small incompatible voltage between them, it can generate a circulating current which may cause degradation or may even cause explosion of the sources. Also often antenna terminals are unbalanced (that is one of the terminals is grounded to Earth) and so they cannot in general be connected in series. An intelligent design when deploying multiple antennas like in the operation of a phased array is to put a feed network to phase the various antennas in the right sequence so that the voltages add up rather than connecting them in parallel. Again one has to realize that the principle of superposition applies to the voltages and the currents in the circuit, where the principle of convolution is required to compute the total response of the system. The concept of superposition does not apply to power as it is related to the principle of correlation. At least in electrical engineering superposition of power does not apply as it is linear with respect to voltages and currents but not in their product!

As an example, consider two plane waves radiated by two different transmitters of respective power densities 100 and 1 W/m^2. They are allowed to propagate and during that time they are allowed to interact with each other. Even though one of the waves is only 1% in power density of the other, if the two waves interfere constructively or destructively, the resulting variation in the power density received is not $(100 \pm 1) = 101$ or 99 W/m^2 but rather it will be $(\sqrt{100} + \sqrt{1})^2$ or $(\sqrt{100} - \sqrt{1})^2$ (related to $\dfrac{|E|^2}{2\eta}$ W/m^2, where E is the electric field intensity and η is the impedance of free space). This results in a variation of the power densities of 121 or 81 W/m^2 (as electric fields only add or subtract and not power densities) – a 40% change and not 1%, since it is the electric fields related to the voltages in circuit or the magnetic field related to the currents that can be added in the electrical engineering context, i.e., principle of superposition applies to voltages and currents and not to the powers or power densities. Therefore, if one uses two antennas and each one of them radiates 1 W/m^2, the total variation of the received field will be between 4 W/m^2 and 0 W/m^2. So for the result of 4 W/m^2: does this imply there is some sort of power gain generated by using a multiantenna system? Of course not, as power densities cannot be added as superposition only applies to the voltages (electric field) and currents (magnetic field) in a system. This is physics and non-acceptance of this basic fact can lead to design of non-realistic systems. But then if one talks about the upper limit of this interaction between the two of 4 W/m^2 then it is imperative that one must also consider the lower limit which is 0 W/m^2 and in this case there is actually power loss which is never even been mentioned in any discussions!!

Now going back to the topic of deployment of multiple antennas typically one will hear stories like this:

Story 1: A half wave dipole antenna is being excited by a source as seen in Figure 2.30. Then one can measure the radiated field strength coming out of the antenna. The source feeding the antenna is assumed to have an internal resistance of 50 Ω. This gives rise to a value for the channel capacity C_1 computed from the radiated power density as seen in Figure 2.30.

Story 2: Now take two half wave dipole antennas that are being excited by the same single source as seen in Figure 2.31. Then one can measure the radiated field strength coming out from the simultaneous operation of two antennas. The source is assumed to have an internal resistance of 50 Ω as seen earlier. It will be seen that the radiated field strength is much higher in this case than

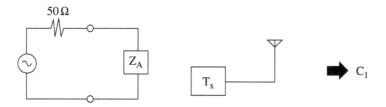

Figure 2.30 A single radiating antenna connected to a transmitter.

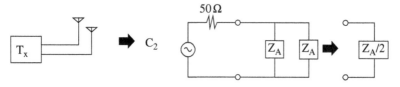

Figure 2.31 Two radiating antennas connected in parallel to a transmitter.

for the single antenna case of Figure 2.30. This gives rise to a value for the channel capacity which is higher from the first case when computed/measured from the radiated power density. So, the conclusion is to use more antennas!?!

The statements made in stories 1 and 2 are both true! But the question is why? It may not be from the principle of deployment of multiple antennas but from the simple straight forward principles of basic electrical engineering, called impedance matching! This is why some basic knowledge of electromagnetic theory is essential otherwise it is not fruitful doing wireless communication in a clueless fashion without grasping the basic physics!

As seen from section 2.12 the input impedance of a thin half wave dipole antenna is approximately 96 Ω. So when this 96 Ω load resistance is connected to a 50 Ω generator approximately one third of the source power is deposited to the antenna. But now when two of these antennas are connected in parallel, their equivalent resistance is now approximately 48 Ω which is a much better match to the 50 Ω generator and hence in the second scenario the two antennas radiate more power densities. Note also that the input power levels for these two systems are different as the antennas are excited by a voltage source and not by a power source. So it is not the number of antennas that radiate larger power densities but it is more of the matching of the antennas which generate a 48 Ω load to the 50 Ω generator. That is why for the second case it radiates a higher electric field. Use of a corporate feed as done by the antenna engineers to connect antennas to a source takes care of the impedance mismatch and so one may get similar types of performance if deployed correctly for both the cases! Now let us pose the following question: Fix the input power to be the same in stories 1 and 2. Then for which case there will be more radiated fields? For the same input power Case 2 can radiate more fields along certain directions due to the two element array gain factor for one of the modes but can be zero for the second mode. Along other directions the field intensities can be less for case 2 when compared to Case 1. **Hence it is not clear that deployment of multiple antennas instead of one will always provide a better solution.**

However, the signal processing approach which is more thoughtful in some special cases than what is presented has some value from a purely theoretical viewpoint. This approach is described next. The philosophical foundation of the signal processing approach has a good foundation as the desire to transmit more information led to implement something like space division multiplexing which is similar to time division multiplexing, frequency division multiplexing and code division multiplexing. Since the latter three are almost saturated then it is a good idea to try to do something in the space domain. Consider a multi-port transmit receive system as shown in Figure 2.32. There are N_T transmit and N_R receive antenna systems in this scenario and the goal is to simultaneously carry out generation of multiple channels in space at the same frequency and using the same antenna systems.

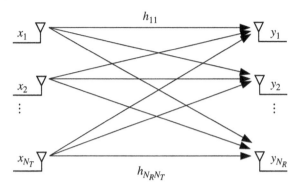

Figure 2.32 A MIMO system model.

A simple input-output relationship for a narrowband MIMO system with N_T transmit and N_R receive antennas can be expressed mathematically through

$$\mathbf{y} = \mathbf{Hx} + \mathbf{n} \tag{2.52}$$

where $\mathbf{y} = [y_1, y_2, ..., y_{N_R}]^T$ is the vector of the received signals and $\mathbf{x} = [x_1, \mathbf{x} = [x_1, x_2, ..., x_{N_T}]^T$ is the vector of the transmitted signals, $\mathbf{n} = [n_1, n_2, ..., n_{N_R}]^T$ is the noise vector where its components are assumed to be additive white Gaussian noise, and $\mathbf{H}$ denotes the $N_R \times N_T$ channel matrix. Here h_{ij} component represents the channel gain from the jth transmit antenna to the ith receive antenna. Here the superscript T denotes the transpose of a matrix. Figure 2.32 shows the MIMO system model. With the knowledge of the system matrix or transfer function $\mathbf{H}$, the MIMO channel can be decomposed into K parallel channels, where $K \leq \min(N_R, N_T)$ as seen in Figure 2.33. A common approach in decomposing the system into the various MIMO channels is via the singular value decomposition (SVD) [30]. The SVD of the channel matrix $\mathbf{H}$ in the complex space C is given by

$$\mathbf{H} = \mathbf{U\Sigma V}^H \tag{2.53}$$

where $\mathbf{U} = [\mathbf{u}_1, \mathbf{u}_2, ..., \mathbf{u}_{N_R}] \in C^{N_R \times N_R}$, $\mathbf{V} = [\mathbf{v}_1, \mathbf{v}_2, ..., \mathbf{v}_{N_T}] \in C^{N_T \times N_T}$, and $\mathbf{\Sigma}$ is a diagonal matrix consisting of the singular values, σ_i of $\mathbf{H}$ arranged in a descending order. $\{\cdot\}_H$ denotes a complex conjugate transpose. $\mathbf{u}$ and $\mathbf{v}$ are column vectors. Both $\mathbf{U}$ and $\mathbf{V}$ are orthogonal matrices and so their inverse are their conjugate transpose, i.e., $\mathbf{U}^H\mathbf{U} = \mathbf{I}_{N_R \times N_R}$ and $\mathbf{V}^H\mathbf{V} = \mathbf{I}_{N_T \times N_T}$, where I is an identity matrix containing unity on the main diagonal and zero elsewhere. By using $\mathbf{U}$ and $\mathbf{V}$ also in encoding the input signal $\mathbf{x}$, it is equivalent to transforming a transmission through MIMO channels into multiple transmissions using weighted parallel single-input-single-output (SISO) channels. This is shown mathematically as follows

$$\tilde{\mathbf{y}} = \mathbf{U}^H \mathbf{y} = \mathbf{U}^H \left(\mathbf{U} \Sigma \mathbf{V}^H \right) \mathbf{x} + \mathbf{U}^H \mathbf{n} = \Sigma \mathbf{V}^H \mathbf{x} + \mathbf{U}^H \mathbf{n} \tilde{\mathbf{y}} = \Sigma \mathbf{V}^H \mathbf{x} + \tilde{\mathbf{n}} = \Sigma \tilde{\mathbf{x}} + \tilde{\mathbf{n}}$$

(2.54)

where $\tilde{\mathbf{y}}$ is the output of the modified systems and $\tilde{\mathbf{n}} = \mathbf{U}^H \mathbf{n}$ is the transformed noise and $\tilde{\mathbf{x}}$ is the input. Note that since $\mathbf{U}$ is a unitary matrix, i.e., $\mathbf{U}^H \mathbf{U} = \mathbf{I}$, and so $\mathbf{U}^H \mathbf{n}$ does not change the distribution of the noise. Thus, $\tilde{\mathbf{n}}$ and $\mathbf{n}$ have the same distribution. However, $\mathbf{V}^H \mathbf{x}$ represents a linear transformation of the input signal $\mathbf{x}$ and transforms it to $\tilde{\mathbf{x}}$. We, therefore, further encode the input signal $\mathbf{x}$ as

$$\mathbf{V}^H \tilde{\mathbf{x}} = \mathbf{x} \text{ or equivalently,} \quad \tilde{\mathbf{x}} = \mathbf{V} \mathbf{x},$$

(2.55)

so that the input signal $\mathbf{x}$ will be transmitted through the uncoupled channels of (2.54) in the form of $\tilde{\mathbf{x}}$. This is sometimes referred to as *MIMO beamforming* in the literature, which should not be confused with the idea of beamforming from an antenna perspective, in which beam patterns are only defined in the far field, whereas here one can be dealing with a near field scenario where beams cannot be defined. The overall channel decomposition (assuming that the multiport channel characteristics of $\mathbf{H}$ are known) can be illustrated through the following formulation and is intuitively explained by Figure 2.33. Thus,

$$\tilde{\mathbf{y}} = \Sigma \tilde{\mathbf{x}} + \tilde{\mathbf{n}}$$

(2.56)

Equation (2.56) clearly demonstrates that a M-transmit and M-receive antennas is equivalent to a parallel combination of M-SISO channels weighted by the appropriate singular values of the channel matrix for each of the spatial modes. In a near field environment, where antenna patterns are not defined this singular value decomposition can still be carried out and so one has transmission of simultaneously M-SISO channels using the same physical parameters of a $M \times M$ MIMO system. Hence, space division multiplexing looks

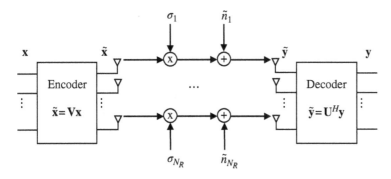

Figure 2.33 Uncoupled parallel decomposition of a MIMO channel.

feasible, at least in theory, from this scenario. However, this is where physics takes over the mathematics which even though is elegant cannot result in the realization of a real system.

Based on this SVD decomposition, multiple signals can be transmitted simultaneously, which increases the data rate of the system without changing the bandwidth of the systems. **However, the performance of this transmission system strictly depends on the singular values of the channel matrix since they represent gains of the decomposed channels, over the SISO channels.** It is also possible that the matrix $\mathbf{H}$ may not be of full rank. In that case, the decomposed channel will be useful for transmission of data only by the modes with singular values larger than a threshold. This results in a multiplexing gain of K, where $K \leq \min(N_R, N_T)$. When $\mathbf{H}$ is full rank and well-conditioned, $K = \min(N_R, N_T)$. At least, in principle, this is where we gain benefits at least in theory, from the use of multiple antenna systems. To implement such a concept, different assumptions are made about the knowledge of the channel matrix. These assumptions lead to different schemes for MIMO systems.

The problem is that in real physical system one generally have **ONLY** one dominant singular value of the $\mathbf{H}$- matrix in (2.53) and rest of the singular values are quite small resulting in useless channels. That is why the antenna engineers use only the dominant singular value in their design of the phased arrays. The reason why the remaining singular values except the first one is small is because in order to make these type of decomposition it is necessary to decompose the signal into the various orthogonal eigenvectors which is quite possible from a theoretical point of view but does not result in a good physically realizable system. For example, let us take a 2×2 MIMO system. The two orthogonal modes that can carry the signal from the input to the output are the two eigenvectors of $\{+1, +1\}$ and $\{+1, -1\}$. It is seen that two antennas when excited in phase unison of the same voltage corresponding to the first singular/eigenvector will generate a strong field pattern at the receiver and it is this mode that is generally used in a phased array. Now consider the second mode where the two excitations of two antennas are occurring with opposing excitations. In this case there will be little radiated field at the receiver and so no practical system will use such excitations except may be in theory. Therefore, the second mode is useless from a practical point of view and only one mode is useful. Use of multiple antennas generate the same problem with orthogonal decompositions as we will see in the remainder of this section.

In summary, a moment of reflection at this point will reveal that the concept of the best region of operation of a MIMO is an environment where all the singular values of $\mathbf{H}$ are large and different from zero. Because if the singular value is very small or close to zero then that channel is useless from a practical point of view.

This theory so far has been developed from a purely scalar point of view. So from a philosophical view point, it is possible to have some pathological scenarios where **H** may be of full rank containing large singular values so that K indeed can be min(N_R, N_T). And here lies another fallacy. The ideal situation in the SVD decomposition in (2.53) will be the case when all the singular values will be equal and close to each other. This implies that Σ will be a diagonal matrix with all entries as unity. In that case **H** of (2.53) will degenerate into an identity matrix as **U** and **V** will reduce to similar orthogonal matrices and one will have K parallel SISO (Single-Input-Single-Output) channels!

It is extremely doubtful that for the vector electromagnetic problem, such a scenario can indeed occur where **H** will be full rank with similar large singular values as will be illustrated later. This goes back to the case that if one has a collection of vector electric fields representing the various signal transmission mechanisms then it is not clear at all that by combining all the various vector signals without paying any attention to their relative phases will provide a better solution! Alternately, one can view a MIMO system as an over-moded waveguide where simultaneous operation is carried out utilizing the various orthogonal modes which can be generated inside a waveguide. The difficulty in such a system is how to simultaneously excite all the orthogonal modes and then how to extract the signal from each individual mode at the receiver? That is why waveguides are generally operated in a single mode regime.

So far all the discussions about MIMO has been based on the scalar statistical methodology, which illustrates that use of multiple antennas for transmit and receive have superior performance over a single transmit receive system. In this section, we investigate how the introduction of the vector Maxwellian principles impacts the previous discussions. So, the first departure from the scalar theory lies in using real antennas instead of using point sources, which really does not exist in practice. There are additional hidden subtleties which are missed in the point source model. For example, the entire field radiated by a point source is akin to a far-field electromagnetic radiation and which is real and the power can be computed from either the electric or the magnetic field. But for a finite length dipole there are both near and far fields. In the near field the approximations of point sources for realistic antennas done in the signal processing literature does not hold. Moreover, there is mutual coupling between the antennas, which can completely alter the nature of the conclusions if their effects are taken into account. Finally, we limit the total input power to the systems so that we can make meaningful performance comparisons between various operating systems.

Research topics in MIMO systems under an electromagnetic point of view have been recently studied [30]. Also for an antenna the space and time variables are not independent but they are related as an antenna is both a temporal filter and also a spatial filter and that is why one uses Maxwell's equations to study antennas as these equations capture the fundamental real life physics which is missing in a scalar statistical analysis.

In this section, we illustrate the vector nature of the MIMO electromagnetic system through a few simulated numerical examples. The actual parameters of the example are not important in the sense that the examples presented will demonstrate that MIMO does not always perform better than a SISO system, from a real system standpoint but may provide numerical values using a statistical methodology which seem to indicate that the Shannon channel capacity performance is better. Hence, it is important to make a distinction between conclusions based on statistical aberrations as opposed to basic physics. Also, due to the existence of complicated multipath environments, which are non-existent in a near-field scenario, the discussion in this section will be based only on numerical simulations, using an accurate numerical electromagnetic analysis computer code [9]. Our goal is to provide some numerical examples to illustrate the basic principles when using a system standpoint.

2.14.1 Is MIMO Really Better than SISO?

As an example, consider two dipole antennas of 15 cm in length and of radius 1 mm separated in free space by 1.5 m. The center-fed dipoles are conjugately matched at 1 GHz by connecting the load of $90.7 - j\,42.7\,\Omega$ at their feed points. For a 1 V excitation of the transmitting dipole we have an excitation current of $5.5 + j\,0.0015$ mA in the transmitter antenna which induces a current of $0.045 + j\,0.14$ mA at the load located at the feed point of the receiving dipole. Therefore for a total input power of 1 W in the transmitting antenna, there will be a received power of 0.36 mW in the load of the receiving antenna. This describes a typical SISO system.

Next we consider two conjugately matched center-fed dipoles as transmitters and two additional conjugately matched center-fed dipoles as receivers replacing the single antenna system to represent a MIMO system. We consider the two half wave dipole antennas at both the transmitter and the receiver to be separated by half a wavelength. The four center-fed dipoles have the same length, radius and loading as before. They are also separated by the same distance of 5 wavelengths or 1.5 m. The basic philosophy is that since there are two antennas each for the transmit and receive systems, one can communicate with two spatial orthogonal modes of the system. The two orthogonal modes of the excitation of the antennas will be of 1 V each fed to the transmitting antennas so that they operate in phase. The other orthogonal mode will have a +1 V and −1 V excitations to each of the transmitting antennas so that the excitations are orthogonal. The basic principle of MIMO is to simultaneously use both of these spatial modes for transmission. For a co-phase 1 V excitation of the dipoles we have an excitation current of $6.0 + j\,1.2$ mA in each of the transmitting antennas which induces a current of $0.014 - j\,0.33$ mA in each of the receiving dipoles. Therefore for a total input of 1 W of power to the two transmitting antennas, they will produce a total received power of 1.6 mW in the loads of the receiving antennas. For an anti-phase excitation of +1 V and −1 V in each of the transmitting antennas representing the second orthogonal mode we have an excitation current of $4.7 - j\,0.74$ mA in each of the transmitting antennas which induces a current of $12.9 - j\,13.3$ µA in each of the receiving dipoles. Therefore for a total of 1 W of input power to the two transmitting antennas will result in a received power of only 6.5 µW in the loads of the receiving antennas. Even though there are two spatial modes, the first mode has a higher radiation efficiency than the second mode, by a factor of 246 approximately. Electromagnetically this second mode will never be used in practice because of its poor radiation efficiency. Communication using the first mode in the antenna literature is called a phased array. Using the first spatial mode in this two antenna system it is possible to get a gain of 4.44 over the SISO system for the same total power input. It is important to note that this value is greater than the number 4 which is the limit obtained from a scalar statistical analysis!

The discussion should generally stop here. However, since a different metric called the channel capacity other than the received power is used to compare the performance between systems, we need to explore what is the system performance under this new metric. The channel capacity is a formula that has been derived from the concept of entropy as illustrated in Chapter 1 which is purely philosophical in nature and not connected with the basic physics. The Shannon channel capacities for the SISO and the MIMO systems for bandwidth B will then be given by [30]

$$C_{SISO} = B\log_2\left(1 + \frac{0.00036}{P_N}\right) \tag{2.57}$$

$$C_{MIMO} = B\log_2\left(1 + \frac{0.0016}{2 \times P_N}\right) + B\log_2\left(1 + \frac{0.0000065}{2 \times P_N}\right) \tag{2.58}$$

where B is the bandwidth of the system as per (1.117). P_N is the thermal noise power. The factor of 2 appearing in the denominator of (2.58) is due to the fact that for the same input power of 1 W for the two spatial modes, we may be feeding 0.5 W to each of the spatial modes so that the total input power remains constant. Now if the denominator of both (2.57) and (2.58) is the thermal noise power, as per Shannon theory, then even the minuscule useless power received at the receiver for the second MIMO mode will contribute to the formula for the channel capacity, even though it may be useless from a practical standpoint. One may even argue that by some appropriate coding this mode can be put to use. But then wireless communication does not use coding at the radio frequency transmission stage. The physics disappear at this point and the logarithm of the radiated to the thermal noise power appears really very attractive even though its contribution is dismal in a real system! This topical discussion is often clouded by introducing the additional factor of unequal power distribution over the various modes to increase the presumed capacity as per Shannon's theory.

It is important to reemphasize again that Shannon was addressing an important problem of how to recover the signal of interest when it is below the background noise level. And Shannon showed that by introducing redundancy in the form of a code at the transmitter, signals below the thermal noise level can be recovered at the receiver. However, in cellular wireless communication there is no coding at RF and therefore Shannon theory is not applicable. What Shannon made possible was satellite communication where the signal levels can be below even 100 dB the thermal noise level. Another system that uses coding at RF is the global positioning system (GPS). The metric of channel capacity is then not a good one to use in wireless communication as there is no coding performed at the RF stage like done in the satellite communication and in GPS.

In summary, a MIMO antenna system is thought of as simultaneously transmitting multiple orthogonal modes in a multi-moded wave guiding system. This type of multi-moded transmission is seldom used in real life because of the dispersion in the system and the logistics involved in simultaneously exciting all the orthogonal modes and combining them in the same waveguide for transmission and then separating the various modes at the receiver, is quite complex. In the capacity formula the linearity of the addition of the separate channels overwhelms any gain achievable through the hardware of the antenna systems. The thinking appears to be that the logarithmic increase in power is not as relevant as the multiple channels even though they may be unacceptable vehicles for transmission from a hardware point of view! However, if one uses the expression for the Hartley capacity given by (1.126) instead of the Shannon capacity (1.117) then due to the discretization of the induced voltages, one would get a more realistic value for the capacity of the system as the second mode may be weeded out as it may yield an induced voltage comparable to the first quantization level and hence below the noise threshold.

So far so good, and one can relate the physics with mathematics for any system. However, the next step really becomes bizarre if we now pose the problem as follows: In the MMO system that is just described, the direct line of sight creates the more efficient channel and if we take the direct line of sight out then the linearity of the two terms will predominate in (2.58). In actual practice there may seldom be a direct line of sight communication. There could be many multipaths and the second orthogonal mode whose performance is really dismal in the line of sight operation perhaps may be a viable mode of propagation. Unfortunately, this way of thinking clearly misses the vector nature of the wireless communication problem and is mostly guided by the scalar channel capacity theorem. Let us illustrate the fallacy of such a concept by another example.

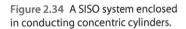

Figure 2.34 A SISO system enclosed in conducting concentric cylinders.

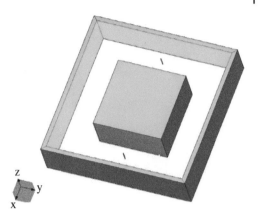

Let us now introduce a concentric region inside the transmit-receive antennas described in the previous example so that there will be no direct line of sight of communication. The spacing between the antennas, dimensions and the load remains the same as before. Let us place the SISO system described earlier in a similar environment characterized by two conducting structures as shown in Figure 2.34. The closed inner conducting box has a dimension of 1 m × 1 m × 0.5 m. The outer conducting shell has an inner dimension of 2 m × 2 m × 0.5 m. The thickness of the conducting walls is 6 cm. In this case there is no line-of-sight of communication as the inner conducting box prevents such a mode of operation. Also, the conducting structure will guide the signals more to the receiver and therefore if we place multiple antennas, one of them may pick up more signal. Unfortunately, such naive simplistic reasoning does not hold for the vector electromagnetic problem as we will observe next.

For a 1 V excitation of the transmitting dipole in the SISO system in Figure 2.34 will produce an excitation current of $5.2 + j\,3.4$ mA in the transmit antenna. It will also induce a current of $0.99 + j\,0.4$ mA in the receiving dipole. Therefore for an input power of 1 W in the transmitting antenna, it will produce a received power of 16.64 mW in the load of the receiving antenna. An increase in the received power over the earlier open-air SISO example is expected as the signals are directed in this case to the receiving antenna by the concentric guiding structure.

Next we consider a 2 × 2 MIMO antenna systems where the two transmitting and the receiving antennas as discussed before are encapsulated by the concentric cylinders. The situation is depicted in Figure 2.35. The two dipoles as transmitters and two dipoles as receivers are now replacing the single antenna systems to represent a MIMO system. We consider the same configuration of two half wave dipole antennas instead of the single one but separate them by half a wavelength. All the four antennas will have the same length, radius, and loading as before. The transmit receive sets are also separated by

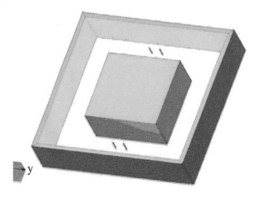

Figure 2.35 A 2 × 2 MIMO system enclosed in a conducting concentric box.

the same distance of 5 wavelengths. The two antenna transmit receive systems can communicate using two spatial orthogonal modes. One of the orthogonal modes of the antenna systems will be an excitation of 1 V to each to the transmitting antennas so that they will be operating in phase. There are no direct paths linking the transmitting and the receiving antennas. For a co-phase 1 V excitation of the dipoles we have an excitation current of 4.2 + j 1.8 mA induced in each of the transmitting antennas will induce a current of 0.075 + j 0.86 mA in each of the receiving dipoles. Therefore, for a total input power of 1 W to the two transmitting antennas, there will be a total received power of 14.79 mW in the loads of the receiving antennas. For the second spatial orthogonal mode, an anti-phase excitation of +1 V and −1 V in each of the transmitting antennas will produce an excitation current of 4.7 + j 0.22 mA in the first transmitting antenna and a current of 0.53 + j 0.3 mA in the first receiving dipole.

Therefore for the second orthogonal spatial mode, for a total input power of 1 W to the two transmitting antennas there will be a received power of only 7.15 mW in the loads of the receiving antennas. Even though there are two spatial modes, the first mode is more efficient in transmitting power than the second mode, by a factor of only 2.07 approximately in this case, as the line of sight has been eliminated completely. Electromagnetically it appears that both the two spatial orthogonal modes in this case have poorer radiation efficiency than the SISO case. Yet, if one writes the capacity in this case for the SISO and the MIMO system, one obtains the following two expressions:

$$C_{SISO} = B \log_2\left(1 + \frac{0.01664}{P_N}\right) \tag{2.59}$$

$$C_{MIMO} = B \log_2\left(1 + \frac{0.01479}{2 \times P_N}\right) + B \log_2\left(1 + \frac{0.00715}{2 \times P_N}\right) \tag{2.60}$$

Assuming a background thermal noise floor of about 2 pW, one evaluates the two capacities as

$$C_{SISO} = B \times 32.95 \tag{2.61}$$

$$C_{MIMO} = B \times 31.78 + B \times 30.74 = B \times 62.52 \tag{2.62}$$

Here is the dichotomy. The total power received by the MIMO antenna systems using the two spatial modes is less than the total power received by the SISO system. So, from an electromagnetic system point of view, we have two inferior modes of propagation than over a single antenna system, yet if one were to claim that (2.62) is a better system than (2.61), then it must be based on statistical aberrations and not from a sound physical system point of view. The other point is quite clear, that in this case two independent SISO systems will always be better than a 2 × 2 MIMO system under all conditions and keeping the total amount of input power to the system fixed. The problem is if we use the Shannon channel capacity where the denominator P_N represents thermal noise, then any signal above thermal noise will yield an acceptable value for the channel capacity even if that mode is useless from a practical standpoint. But form a systems point of view the limiting factor in this case is not thermal noise but the smallest signal that the first stage of the RF amplifier in the system can discern and gainfully amplify without introducing too much noise. Using this criterion the second mode will never work in practice. Finally, equations (2.61) and (2.62) indicate that depending on the value of P_N, the values for C_{SISO}, or C_{MISO} will be larger. Hence, there is no guarantee for a general situation that a SISO will be inferior in performance than a MIMO unless the actual system parameters are exactly specified.

The situation is actually much worse as if one considers only the superior single MIMO mode and put all the power in that mode and discard the second mode we still obtain

$$C_{SISO} = B \log_2\left(1 + \frac{0.01664}{P_N}\right) > C_{MIMO-the\ domainant\ mode\ only} = B \log_2\left(1 + \frac{0.01479}{P_N}\right)$$

which clearly demonstrates that MIMO in reality is a statistical aberration!

Finally, the statistical analysis of Shannon which is responsible for the derivation of the channel capacity in this case (it is important to note that Shannon never applied this principle to the wireless case but for wired transmission) does not support basic physics. Examples of such non-intuitive results based on application of probability theory are available in the literature [24]. The objective of MIMO is to provide spatial diversity through the use of multiple transmit and receive antennas. So, if there is N transmit and N receives antennas, then one can generate N spatially orthogonal modes to communicate between these transmit-receive systems. Even though the MIMO principle is

located on receiver R_1 will also induce a voltage at the antenna located on receiver R_2, which is ignored because it is superfluous to this discussion. This does not mean that this induced voltage is small, but it does not enter into the theory! Now exciting antenna on transmitter T_1 with 1 V will induce currents $I_{R_1}^{T_1}$ and $I_{R_2}^{T_1}$, respectively, at the loads located at the feed points of the two receiving antennas. The superscript T_1 indicates that in this case only transmitter T_1 is active. Moreover, the 1 V excitation of transmitter T_1 also induces a current at transmitter T_2 that is not germane to this development. Now if one applies the principle of reciprocity between the excitation voltages and currents flowing at the two feed ports corresponding to antennas located on transmitter T_1 and receiver R_1, one observes that the respective currents are related by

$$I_{T_1}^{R_1} = I_{R_1}^{T_1},\tag{2.63}$$

for 1 V excitations. Similarly, if the reciprocity principle is applied to the feed ports of antennas on transmitter T_1 and receiver R_2, then

$$I_{T_1}^{R_2} = I_{R_2}^{T_1}.\tag{2.64}$$

Therefore, exciting the antenna on transmitter T_1 with voltage W^{T_1} in (2.63) and (2.64) induces currents equal to $W^{T_1} I_{T_1}^{R_1}$ at antenna on receiver R_1 and $W^{T_1} I_{T_1}^{R_2}$ at antenna on receiver R_2, by reciprocity applied to the respective ports of the transmitting and receiving antennas.

If we now excite the antenna on receiver R_2 with 1V, then currents $I_{T_1}^{R_2}$ and $I_{T_2}^{R_2}$ are induced at the antennas on transmitters T_1 and T_2, respectively, and the induced current in the antenna on receiver R_1 is ignored, as it is not germane to the present discussions. Recall that the superscript R_2 on the currents implies that antenna on receiver R_2 is transmitting. Now exciting antenna on transmitter T_2 with 1 V induces currents $I_{R_1}^{T_2}$ and $I_{R_2}^{T_2}$, respectively, at the loads located at the feed points of the two receiving antennas, as well as an inconsequential current at the inactive transmitting antenna T_1. If one now applies the principle of reciprocity between the excitation voltages and currents at the two feed ports corresponding to antennas on transmitter T_2 and receiver R_1, then the respective currents are related by

$$I_{T_2}^{R_1} = I_{R_1}^{T_2},\tag{2.65}$$

when antennas on transmitter T_2 and receiver R_1 are excited with 1 V. Similarly, if we apply the same principle of reciprocity between the feed ports of antennas located on transmitter T_2 and receiver R_2, then we will obtain

$$I_{T_2}^{R_2} = I_{R_2}^{T_2}.\tag{2.66}$$

Therefore, exciting antenna on transmitter T_2 with voltage W^{T_2} and applying the principle of reciprocity to the respective ports of the transmitting and the receiving antennas will induce currents equal to $W^{T_2}_{R_1} I^{R_1}_{T_2}$ at the antenna on receiver R_1 and $W^{T_2}_{R_1} I^{R_2}_{T_2}$ at the antenna on receiver R_2.

Next, the principle of superposition is applied to modify some of these induced currents. Suppose antennas on transmitters T_1 and T_2 are excited with voltages $W^{T_1}_{R_1}$ and $W^{T_2}_{R_1}$, respectively. The subscript R_1 symbolically specifies the goal of maximizing the induced current in antenna on receiver R_1 while ensuring the current in antenna on receiver R_2 is practically zero. Assume that this maximum current is 1 A. Under these conditions, the desired total currents induced in antennas on receivers R_1 and R_2, respectively, are

$$W^{T_1}_{R_1} I^{R_1}_{T_1} + W^{T_2}_{R_1} I^{R_1}_{T_2} = 1, \tag{2.67}$$

$$W^{T_1}_{R_1} I^{R_2}_{T_1} + W^{T_2}_{R_1} I^{R_2}_{T_2} = 0. \tag{2.68}$$

Similarly, it is possible to choose a set of excitations, $W^{T_1}_{R_2}$ and $W^{T_2}_{R_2}$, to be applied to the antennas on transmitters T_1 and T_2, such that no current is induced in antenna on receiver R_1 and the current induced at antenna on receiver R_2 is 1 A. Under these conditions, the total current induced in antenna on receivers R_1 and R_2 will be given by

$$W^{T_1}_{R_2} I^{R_1}_{T_1} + W^{T_2}_{R_2} I^{R_1}_{T_2} = 0, \tag{2.69}$$

$$W^{T_1}_{R_2} I^{R_2}_{T_1} + W^{T_2}_{R_2} I^{R_2}_{T_2} = 1. \tag{2.70}$$

Equations (2.67)–(2.70) can be written more compactly in matrix form as

$$\begin{pmatrix} I^{R_1}_{T_1} & I^{R_1}_{T_2} \\ I^{R_2}_{T_1} & I^{R_2}_{T_2} \end{pmatrix} \times \begin{pmatrix} W^{T_1}_{R_1} & W^{T_1}_{R_2} \\ W^{T_2}_{R_1} & W^{T_2}_{R_2} \end{pmatrix} = \begin{pmatrix} 1 & 0 \\ 0 & 1 \end{pmatrix} \tag{2.71}$$

By using (2.71), one can solve for an a priori set of excitations that will direct the signal to a pre-selected receiver by vectorially combining the signal from the two transmitting receivers. The excitations are obtained by inverting the current matrix in (2.71) to yield

$$\begin{pmatrix} W^{T_1}_{R_1} & W^{T_1}_{R_2} \\ W^{T_2}_{R_1} & W^{T_2}_{R_2} \end{pmatrix} = \begin{pmatrix} I^{R_1}_{T_1} & I^{R_1}_{T_2} \\ I^{R_2}_{T_1} & I^{R_2}_{T_2} \end{pmatrix}^{-1} \begin{pmatrix} 1 & 0 \\ 0 & 1 \end{pmatrix} \tag{2.72}$$

The caveat here is if antennas on transmitters T_1 and T_2 are excited by $W^{T_1}_{R_1}$ and $W^{T_2}_{R_1}$, respectively, then the total induced current due to all the electromagnetic signals will be vectorially additive at the load, which is located at the feed point of antenna on receiver R_1, and would be destructive at the feed point of antenna on receiver R_2. In contrast, if we apply $W^{T_1}_{R_2}$ and $W^{T_2}_{R_2}$ to antennas on transmitters

T_1 and T_2, then the received electromagnetic signal will be vectorially destructive at the load, which is located at the feed point of antenna on receiver R_1, and will be vectorially additive at the feed point of antenna on receiver R_2, generating a large value for the induced load current.

In short, by knowing the voltages that are induced in each of the transmitting antennas by every receiver, it is possible to select a set of weights based on reciprocity that will induce large currents at a specific receiving antenna. This relationship, based on the principles of reciprocity and superposition, can be applied only at the terminals of the transmitting and receiving antennas. This also assumes that there exists a two-way link between the transmitter and the receiver. Furthermore, this principle of directing the signal energy to a pre-selected receiver is independent of the sizes and shapes of the receiving antennas and the near field environments.

As an example, we place six antennas inside two concentric cylinders, as shown in Figure 2.37. The inner cylinder is composed of a dielectric and the outer one is made of conductor. The dimension of the outer conducting cylinder is 1 m × 1 m × 0.5 m. The transmitting and receiving antenna sets are separated by 0.8 m and they are spaced 0.25 m from each other. The inner conducting cylinder has a dielectric of relative permittivity $\varepsilon_r = 4$, as shown in Figure 2.37. In this case, one can choose a set of weights W so that the energy is directed to a particular receiver. For the three sets of weights we see that the energy can be directed to a preselected receiver as shown by the results of Table 2.11. The three different rows in the Table 2.11 correspond to three different excitations when applied to each of the transmitters will direct the energy to the terminals of a particular receiver while simultaneously inducing no currents at the remaining receivers.

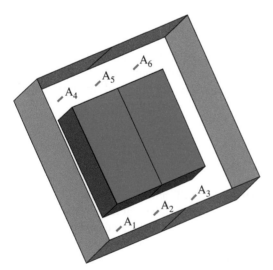

Figure 2.37 Simulation situation when the conducting inner cylinder is replaced by a dielectric cube.

Table 2.11 Magnitude of the currents measured at the three receiving antennas 1, 2, and 3 for three different choices of excitations.

$\left|I_{R_1}\right|$	$\left|I_{R_2}\right|$	$\left|I_{R_3}\right|$
0.99	0.002	0.001
0.003	1.0	0.0008
0.001	0.004	1.0

For the next example, consider three transmitting helical antennas A_1, A_2, and A_3, located at a base station. Each helical antenna has a circumference C = 0.3 m. The parameter of each of the antenna is: diameter of the helix D = C/π = 0.0955 m, pitch angle α = 13°, the spacing between turns S = C × tan (α), length of one turn L = $\sqrt{C^2 + S^2}$ = 0.3079 m, number of turns n = 10, and the axial length A = n*S = 0.6926 m. The operating frequency is 1 GHz. Next the three receiving antennas marked as A_4, A_5, and A_6 in Figure 2.38 are considered. Dimension of the receiving antennas are the same and they are separated from the transmitters by a distance of 6 m. All of the six antennas are loaded with 140 Ω at the feed point.

First, consider maximizing the currents induced in the antenna on receiver A_4. The antenna on receiver A_4 is excited with 1 V, which induces currents in antennas of receivers A_5 and A_6 and antennas on transmitters A_1, A_2, and A_3. These currents are computed by an electromagnetic analysis code. In turn, these induced currents generate voltages across the loads of the other five loaded helices. The induced currents at the antennas of transmitters A_1, A_2,

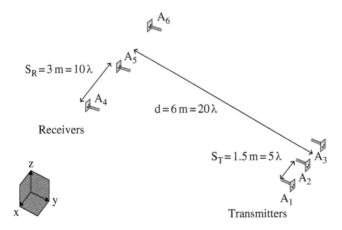

Figure 2.38 A six helical antenna transmit/receive system.

and A_3 are $I_{A_1}^{A_4}$, $I_{A_2}^{A_4}$, and $I_{A_3}^{A_4}$, respectively. As noted earlier, the currents induced in antennas of A_5 and A_6 are not considered, as they are not relevant in the present discussions. The induced currents have been obtained using the electromagnetic analysis code *HOBBIES* [9].

Next, when antenna on receiver A_5 is excited with 1 V, it induces the currents $I_{A_1}^{A_5}$, $I_{A_2}^{A_5}$, and $I_{A_3}^{A_5}$ on antennas located on transmitters A_1, A_2, and A_3, respectively. Similarly, exciting antenna of receiver A_6 with 1 V induces currents $I_{A_1}^{A_6}$, $I_{A_2}^{A_6}$, and $I_{A_3}^{A_6}$ in antennas of transmitters A_1, A_2, and A_3. Based on the available information, the claim is that one can choose a set of complex voltages $\{W_{A_i}^{A_1}$, $W_{A_i}^{A_2}$ and $W_{A_i}^{A_3}\}$, for $i = 4, 5$, or 6, which when exciting the three antennas located on each of the transmitters will result in an additive vectorial combination of the electromagnetic fields at antenna on receiver A_i while inducing zero currents at the antennas on the other two receivers. The currents in the antennas located on the receivers then would be

$$I^{A_4} = W_{A_4}^{A_1} I_{A_1}^{A_4} + W_{A_4}^{A_2} I_{A_2}^{A_4} + W_{A_4}^{A_3} I_{A_3}^{A_4} \quad \left(\text{at receiver } A_4\right)$$

$$I^{A_5} = W_{A_4}^{A_1} I_{A_1}^{A_5} + W_{A_4}^{A_2} I_{A_2}^{A_5} + W_{A_4}^{A_3} I_{A_3}^{A_5} \quad \left(\text{at receiver } A_5\right)$$

$$I^{A_6} = W_{A_4}^{A_1} I_{A_1}^{A_6} + W_{A_4}^{A_2} I_{A_2}^{A_6} + W_{A_4}^{A_3} I_{A_3}^{A_6} \quad \left(\text{at receiver } A_6\right)$$

The objective now is to select the excitations $W_{A_j}^{A_i}$ for each j of $\{1, 2$ or $3\}$ in such a fashion that the received currents are maximal at antenna of receiver A_j and zero at the other antennas of the other receivers.

To determine the weight vectors that should induce the maximum current to antenna of only one receiver A_i ($i = 4, 5$ and 6), one can solve

$$\begin{pmatrix} I_{A_1}^{A_4} & I_{A_2}^{A_4} & I_{A_3}^{A_4} \\ I_{A_1}^{A_5} & I_{A_2}^{A_5} & I_{A_3}^{A_5} \\ I_{A_1}^{A_6} & I_{A_2}^{A_6} & I_{A_3}^{A_6} \end{pmatrix} \begin{pmatrix} W_{A_4}^{A_1} & W_{A_5}^{A_1} & W_{A_6}^{A_1} \\ W_{A_4}^{A_2} & W_{A_5}^{A_2} & W_{A_6}^{A_2} \\ W_{A_4}^{A_3} & W_{A_5}^{A_3} & W_{A_6}^{A_3} \end{pmatrix} = \begin{pmatrix} 1 & 0 & 0 \\ 0 & 1 & 0 \\ 0 & 0 & 1 \end{pmatrix} \quad (2.73)$$

to obtain

$$\begin{aligned} \left[\mathbf{W}_4, \mathbf{W}_5, \mathbf{W}_6\right] &= \begin{pmatrix} W_{A_4}^{A_1} & W_{A_5}^{A_1} & W_{A_6}^{A_1} \\ W_{A_4}^{A_2} & W_{A_5}^{A_2} & W_{A_6}^{A_2} \\ W_{A_4}^{A_3} & W_{A_5}^{A_3} & W_{A_6}^{A_3} \end{pmatrix} \\ &= \begin{pmatrix} 0.323 + 4.636j & -0.368 + 0.509j & 0.318 - 0.631j \\ 2.217 + 1.405j & -2.302 - 1.432j & 2.188 + 1.390j \\ 0.322 - 0.634j & 0.360 + 0.504 & 0.309 + 4.613j \end{pmatrix} \end{aligned}$$

$$(2.74)$$

To demonstrate the feasibility of this methodology, one can use these voltages of $\mathbf{W}_4$ as excitation inputs in the electromagnetic analysis code [9] to compute the induced currents on the antennas located on receivers A_4, A_5, and A_6 as

$$I^{A_4} = 1 - 0.001j, \quad I^{A_5} = -0.001 + 0.001j, \quad I^{A_6} = 0.001 - 0.001j. \qquad (2.75)$$

where $R_{A_1} = R_{A_2} = R_{A_3} = 140\Omega$ and all the currents are multiplied by 10^{-3}A. Clearly, all the electromagnetic signals are vectorially additive at the antenna on receiver A_4, and the currents are practically zero at the feeds of the antennas of receivers A_5 and A_6. Similarly, to direct the signals to the antenna of receiver A_5, one can use the computed voltages $\mathbf{W}_5$ in the electromagnetic analysis code to find the currents induced at the feed point of the antennas located on the various receivers as

$$I^{A_4} = 0, \quad I^{A_5} = 1.0, \quad I^{A_6} = 0, \qquad (2.76)$$

which clearly shows that the induced energy can be directed to antenna of receiver A_5 while producing no appreciable induced currents at the other two antennas located on receivers A_4 and A_6. Finally, to direct the signal from the transmitting antennas to the antenna located on receiver A_6, one can use the computed voltages $\mathbf{W}_6$ in the electromagnetic analysis code to find the feed currents at the antennas located on the receivers as

$$I^{A_4} = -0.001, \quad I^{A_5} = -j0.001, \quad I^{A_6} = 1.0 \qquad (2.77)$$

This clearly demonstrates that by appropriately choosing the complex values of the excitations at the different transmitting antennas it is possible to direct the signal so that it vectorially adds up at the antenna of a pre-selected receiver. No electromagnetic characterization of the environment is necessary.

Next, the behavior of the magnitude of the currents on antennas located on receivers A_4, A_5, and A_6 as a function of frequency is analyzed to observe what the useful bandwidth of the proposed methodology is. The performance in bandwidth is relevant as the uplink operates at a different frequency than the downlink. So when the system is characterized at the frequency of uplink, the question is will it work at the frequency of the downlink. So one fixes the weights at the transmitting antennas which have been evaluated at 1 GHz and then one can use the same weights at other frequencies to observe how well this methodology works. The induced currents at each of the receivers are simulated in Figures 2.39–2.41 over the 12.0% bandwidth from 0.94 GHz to 1.06 GHz when using the three set of frequency independent voltages $\{\mathbf{W}_4, \mathbf{W}_5, \mathbf{W}_6\}$ obtained for 1 GHz. For mobile communication, in the GSM band the uplink is at 890-915 MHz and the downlink is at 935-960 MHz and so a 12 % bandwidth is sufficient. So calibration of the system either at uplink or the

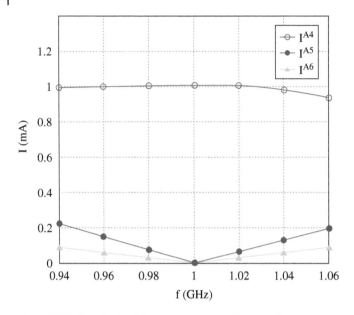

Figure 2.39 Magnitude of the measured currents at the three receivers with **W₄**.

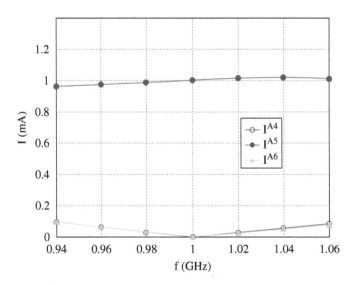

Figure 2.40 Magnitude of the measured currents at the three receivers with **W₅**.

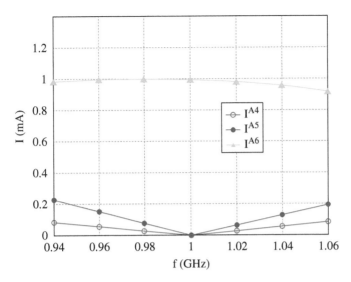

Figure 2.41 Magnitude of the measured currents at the three receivers with $\mathbf{W}_6$.

down link will work. As indicated in Figure 2.39 for $\mathbf{W}_4$, the induced currents at the antennas of the other receivers are down by a factor of 4.3 at the lower frequency and by a factor of 4.8 at the upper frequency. For the middle receiver, the induced currents at the antennas of the other receivers are down by a factor of 10 at the higher frequency end and by a factor of 12 at the lower frequency end points as shown in Figure 2.40. Finally the result in Figure 2.41 is just the reverse of Figure 2.39 as the six helical antennas have an axis of symmetry.

Now, the six helical antennas are placed inside a concentric perfectly conducting cylinder as shown in Figure 2.42. The dimension of the conducting cylinder is 7.8 m × 4.8 m × 1.2 m. The transmitting and the receiving antenna sets are separated by 6 m and the inter-element spacing within each set is 1.5 m. All the six antennas are loaded by 140 Ω.

Along the same line as outlined before, one can also choose a set of excitation voltages as follows:

$$
\left[\mathbf{W}_4,\mathbf{W}_5,\mathbf{W}_6\right] =
\begin{pmatrix}
W_{A_4}^{A_1} & W_{A_5}^{A_1} & W_{A_6}^{A_1} \\
W_{A_4}^{A_2} & W_{A_5}^{A_2} & W_{A_6}^{A_2} \\
W_{A_4}^{A_3} & W_{A_5}^{A_3} & W_{A_6}^{A_3}
\end{pmatrix}
$$

$$
=
\begin{pmatrix}
-1.359-0.995j & -0.803+0.326j & 0.333-0.394j \\
-0.805+0.328j & -0.989-0.256j & -0.831+0.335j \\
0.336-0.397j & -0.830+0.341j & -1.473-0.994j
\end{pmatrix}. \tag{2.78}
$$

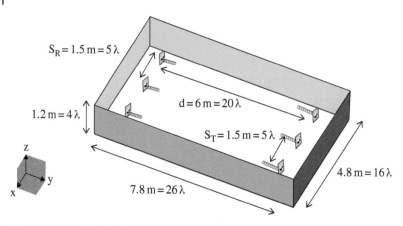

Figure 2.42 A six helical antenna transmit/receive system inside a conducting rectangular cylinder.

Table 2.12 Complex values of the currents measured at the three antennas located on the three receivers marked as 4, 5, and 6 for three different choices of excitations given by (2.78).

By the excitation voltages	I^{A_4}	I^{A_5}	I^{A_6}
$\mathbf{W}_4$	$1.0013 - j0.0008$	$-0.0023 - j0.0010$	$0.0004 + j0.0009$
$\mathbf{W}_5$	$-0.0004 + j0.0017$	$1.0016 + j0.0001$	$-0.0009 - j0.0029$
$\mathbf{W}_6$	$-0.0022 - j0.0010$	$0.0031 + j0.0042$	$0.9970 + j0.0007$

so that the signals can be directed to a preselected receiver. For the three sets of excitation voltages Table 2.12 shows that the signal can be directed to each of the receivers placing nulls along the undesired ones.

In summary, using the principle of reciprocity, it is possible to direct a signal along the direction of the desired user while simultaneously placing a null along the undesired ones without characterizing the channel. This can be achieved when simultaneously both the transmitter and the receiver are sending reference signals so as to calibrate the system.

2.16 Conclusion

In this chapter, a frequency domain analysis of some simple antennas is carried out. Concept of the near field and far field properties of the antennas is extremely essential to carry out system design and modeling of the electromagnetic environment in which the antennas are deployed. Several examples have been presented to illustrate the subtleties involved in the near and far fields electromagnetic modeling and correct prediction of the fields emanating from the structures. Even though a realistic environment may be quite complex, numerical tools that solve Maxwell's equations can provide a very good approximation to the actual results even for a very realistic environment.

The concepts of maximum power transfer as opposed to increase operational efficiency has been demonstrated. It is shown that for antenna problems radiation efficiency is the key parameter instead of the maximum power transfer. The concepts of an Electrically Small Antennas (ESAs) has been discussed and shown that contrary to popular belief, an ESA can perform well compared to larger antennas under matched conditions. The difficulties in achieving a good design of an ESA matched to a system has also been discussed.

The concepts of near and far field properties of an antenna have been presented when deploying them in free space and over Earth. The fields of a radiating antenna located inside a room is illustrated and demonstrated that it is operating in a near field environment irrespective of the size of the room as multiple images are created by the various walls. It is shown that the antenna does not radiate when placed inside in a conducting room. The mathematics and physics of an antenna array is also discussed.

The concepts and the philosophy of the use of Multi-Input-Multi-Output (MIMO) antenna principles have been described. Even though from a theoretical point of view: the concept of using the same spatial channel for simultaneous transmission of multiple signals is quite interesting, there are some major practical problems as we have seen arises when it comes to a physical realization! It is also illustrated how the principle of reciprocity can be used to direct a signal to the receiving antenna of interest and simultaneously placing null along the undesired receivers without even characterizing the channel. This is equivalent to adaptivity on transmit. In the signal processing literature this is sometimes been categorized as the zero-forcing algorithm.

Finally, the Appendix 2A describes where the far field of an antenna starts when it is operating in free space and/or over an Earth.

Appendix 2A   Where Does the Far Field of an Antenna Really Starts Under Different Environments?

Summary

The far field of an antenna is generally considered to be the region where the outgoing wavefront is planar and the antenna radiation pattern has a polar variation and is independent of the distance from the antenna. Hence, to generate a locally plane wave in the far field the radial component of the electric field must be negligible compared to the transverse component. Also, the ratio of the electric and the magnetic far fields should equal the intrinsic impedance of the medium. These two requirements: namely that the radial component of the field should be negligible when compared with the transverse component and the ratio of the electric and the magnetic fields equal the intrinsic impedance of the medium must hold in all angular directions from the antenna. So to determine the starting distance for the far field we need to examine the simultaneous satisfaction of these two properties for all θ and φ angular directions, where θ is the angle measured from the z-axis and φ is the angle measured from the x-axis. It is widely stated in the antenna literature that the far field of an antenna operating in free space, where all the above properties must hold, starts from a distance of $2D^2/\lambda$, where D is the maximum dimension of the antenna and λ is the operating wavelength.

The objective of this appendix can be summarized in three points: *First*, we intend to illustrate that the distance of the far field as $2D^2/\lambda$ formula is not universally valid, and we include the derivation of this formula to show that it is generally valid for antennas where $D \gg \lambda$. *Second*, we intend to compute a more specific constraint and so instead of $D \gg \lambda$ we compute a threshold for D after which the $2D^2/\lambda$ formula applies. *Third*, we intend to properly interpret D in the formula $2D^2/\lambda$ when the antenna is operating over an imperfect ground plane. In this case D is defined in terms of the height of the antenna located over the ground, which is related to the effective aperture size and not necessarily to the dimension of the antenna. In addition, since we focus our attention on the prediction of the fields primarily near the Earth-air interface, a modification of this general formula is possible. Particularly, the far fields start at a distance dictated by the height of the transmitting and the receiving antennas and also depends on the ratio of the radial to the azimuth component of the fields. This is because the higher the antenna is located over the ground the greater is the radial component of the field near the ground and so the radiating fields do not form a locally plane wave at close distances to the antenna even though the ratio of the electric and the magnetic fields settle down to the intrinsic impedance of the medium rather quickly as one moves away from the antenna.

Numerical results are presented to illustrate these points. This is an important topic as the antenna radiation pattern is only defined in the far field and the antenna has no pattern nulls in the near field irrespective of the distance from the antenna and therefore cannot carry out space division multiple access (SDMA).

2A.1 Introduction

In order to characterize the far field of an antenna, it is necessary to understand what properties characterize the far field. In fact there are several conditions and if all these conditions are met simultaneously then only it is safe to say that one is located in the far field. The relevant properties for an antenna operating in free space have been presented in section 2.2.

In summary, the outward directed complex power over a sphere of radius r is complex. The time averaged power radiated is the real part of this power and is related to the radiation resistance of the antenna. A knowledge of the reactive power, which cannot be obtained from the radiation zone fields, is needed to evaluate the input reactance.

Determining the starting distance of the far field is crucial in any wireless communication system since it tells the designer where the radiation pattern starts to evolve. Beyond the starting distance of the far field, the designer can use the pattern properties like gain, sidelobes, beamwidth, nulls .... etc. to achieve various goals like for example to be able to perform SDMA (Space Division Multiple Access) and introduce adaptive or smart antenna concepts.

Let us consider an antenna array operating in free space consisting of 11 z-oriented half–wave dipoles aligned along the x-axis, and the spacing between the elements is half wavelength. All the elements are center fed with an excitation of 1 V located at each dipole. The frequency of operation is 1 GHz. Using the electromagnetic simulator HOBBIES [9], we plotted the total field as a function of the azimuthal angle in the xy-plane at different radial distances to observe how the field pattern will evolve. Figures 2A.1(a)–(c) plot the magnitude of the total electric field as a function of the azimuthal angle in the xy-plane for $r = 0.85$ m, 3 m, and 100 m, respectively, where r is the radial distance from the center of the antenna array. Please note the scale along the y-axis is linear. Clearly we see that the far field pattern does not evolve till we go quite far away from the antenna array. Also, from Figure 2A.1a we see that at $r = 0.85$ m, the field is almost uniform and omnidirectional and (no nulls and no peaks) so applying Space Division Multiple Access (SDMA) or introduction of smart antenna concepts at this distance will not make any sense as no nulls can be generated in the near field of an antenna.

Some antenna text books discuss the point about the far field criteria and provide a starting distance without any condition and typically refer to a particular angular direction along which this field is being observed, for example [32, section 1.4]. Care should be exercised when defining the distance for the far field so that the established criterion hold universally. Analysis become more involved when the antenna is operating over an imperfect ground, i.e., radiating over Earth. Hence, we should be careful in analyzing under what conditions deployment of SDMA is possible in cellular networks.

It is interesting to note that the widely used formula $2D^2/\lambda$ to characterize the far field of an antenna operating in free space applies generally for $D \gg \lambda$. The problem that we are trying to emphasize here is that researchers do not pay attention to this condition under which $2D^2/\lambda$ applies. Predicting the far field of any antenna operating in free space using this formula is not useful in all situations [33]. The question arises, does the far field start at $2D^2/\lambda$? If not then what is the proper distance? This appendix intends to answer these questions.

The appendix is arranged as follows: in the second section, we discuss the derivation of the $2D^2/\lambda$ formula and we show that it is not universally valid for characterizing the far field. In the third section, we provide some analytical results to illustrate that the $2D^2/\lambda$ formula is not applicable for antennas where D is comparable to λ. We also compute a specific threshold for D after which

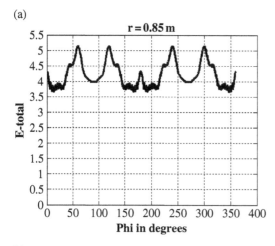

(a)

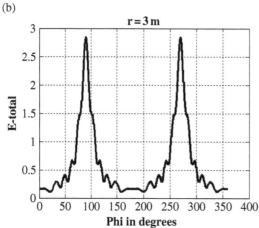

(b)

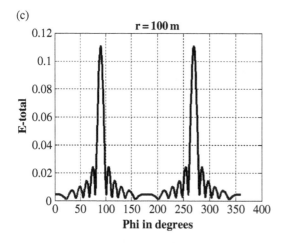

(c)

Figure 2A.1 The total electric field magnitude as a function of the azimuthal angle φ at (a) r = 0.85 m, (b) r = 3 m, (c) r = 100 m.

the $2D^2/\lambda$ formula applies. In the fourth section, we provide numerical results from which we introduce a new formula to determine the starting distance of the far field for antennas radiating over imperfect ground. In the fifth section, some conclusion is given.

2A.2 Derivation of the Formula $2D^2/\lambda$

For any antenna, one can find the magnetic vector potential using

$$A = \frac{\mu}{4\pi} \int_{v'} \frac{J(r')e^{-jkR}}{R} dv' \tag{2A.1}$$

where $J(r')$ is the volume current density on the radiating volume v' and R is the distance between the field point r and the source point r' so that $R = r - r'$, $R = |r - r'|$, μ is the permeability of the medium, and k is the wave number. Even though a succinct proof is given in [34, sections 1.7.2–1.7.3], we present the outline for the derivation here as it is quite involved. The bold letters denote the quantities are vectors.

For the case of a z-oriented linear antenna, (2A.1) becomes:

$$A = \hat{a}_z \frac{\mu}{4\pi} \int_l \frac{I(z')e^{-jkR}}{R} dz' \tag{2A.2}$$

where $r' = z'\,\hat{a}_z$, and $r = x\hat{a}_x + y\hat{a}_y + z\hat{a}_z$. Therefore,

$$R = \sqrt{x^2 + y^2 + (z-z')^2} = r\sqrt{1 - \left(2\cos\theta \frac{z'}{r} - \left(\frac{z'}{r}\right)^2\right)} \tag{2A.3}$$

Using the Taylor series representation one can write

$$
\begin{aligned}
R &= r\sqrt{1 - \left(2\cos\theta \frac{z'}{r} - \left(\frac{z'}{r}\right)^2\right)} \\
&= r\left[1 - \frac{1}{2}\left\{2\cos\theta \frac{z'}{r} - \left(\frac{z'}{r}\right)^2\right\} - \frac{1}{8}\left\{2\cos\theta \frac{z'}{r} - \left(\frac{z'}{r}\right)^2\right\}^2 \right. \\
&\quad \left. - \frac{1}{16}\left\{2\cos\theta \frac{z'}{r} - \left(\frac{z'}{r}\right)^2\right\}^3 + \ldots\ldots\right] \\
&= r - z'\cos\theta + \frac{1}{2}\frac{(z')^2}{r}\sin^2\theta + \frac{1}{2}\cos\theta\sin^2\theta \frac{(z')^3}{r^2} + \text{higher order terms}
\end{aligned}
\tag{2A.4}
$$

where ω is the angular frequency, and ε is the permittivity of the medium. From (2A.11), the ratio of $|E_\theta/E_r|$ is of order of kr, if $kr \gg 1$ for which it suffices that $r \gg \lambda$ and then the radial component is small compared to the transverse component and can be neglected. In the far field, the radial component must be negligible compared to the transverse component so the first term of (2A.11), i.e., the radial component, does not contribute to the electric far field. Therefore, requiring that $kr \gg 1$ is important to determine the far field. Finally, the electric far field is given by:

$$E = j\frac{\omega\mu}{4\pi}\frac{e^{-jkr}}{r}\sin\theta \int_l I(z')e^{+jkz'\cos\theta}dz'\,\hat{a}_\theta \qquad (2A.12)$$

Hence, it is imperative that requiring $r \geq \dfrac{2D^2}{\lambda}$ is not enough for the field point to be located in the far field of the antenna. For cases where $D \gg \lambda$, the condition $r \geq \dfrac{2D^2}{\lambda}$ assures that $r \gg \lambda$ and $r \gg z'$ at the same time, so $r \geq \dfrac{2D^2}{\lambda}$ is enough to guarantee that all three conditions for the far field are satisfied only when $D \gg \lambda$. For all other cases, $r \geq \dfrac{2D^2}{\lambda}$ condition is not enough to satisfy all the three required conditions for the far field. It is important to mention that many references stated that the far field's starting distance is $2D^2/\lambda$ as long as $D \gg \lambda$, for example [2, 34, 36]. In this section, we emphasize on this fact and show why it is important to be careful in using the formula $2D^2/\lambda$. In the next section, we try to compute a more specific constraint so instead of $D \gg \lambda$ we compute a threshold for D after which the $2D^2/\lambda$ formula applies. An interesting point, when $D \gg \lambda$ then $r = 2D^2/\lambda$ determines the starting location of the far field. So, when $D \gg \lambda$ then the near field is located for $r \geq 2D^2/\lambda$. The near field region is divided into two sub regions, one is called the "reactive near field" region which is closest to the antenna where the reactive field dominates the radiative field. The second sub region is called "radiating near field" in which the radiation field dominates but the angular distribution depends on the distance from the antenna so it is not in the far field because in the far field the pattern is independent of r. We can calculate the distance that separates the two sub near field regions by equating the maximum path length deviation introduced by neglecting the fourth term in (2A.4) to $\lambda/16$ [34]. The fourth term of (2A.4) is $\dfrac{1}{2}\cos\theta\sin^2\theta\dfrac{(z')^3}{r^2}$, and one needs to consider the maximum of this term which occurs at $z' = D/2$ and $\theta = \tan^{-1}(\sqrt{2}) = 54.7°$. Under this circumstance, r is evaluated as

$$\frac{1}{2}\cos\left(54.7°\right)\sin^2\left(54.7°\right)\frac{(D/2)^3}{r^2} = \frac{\lambda}{16} \Rightarrow \frac{1}{2}(0.3849)\frac{D^3}{8r^2}$$

$$= \frac{\lambda}{16} \Rightarrow r = 0.62\sqrt{D^3/\lambda}$$

The conclusion is that when $D \gg \lambda$, the reactive near field region is from 0 to $0.62\sqrt{D^3/\lambda}$, and the radiating near field region is from $0.62\sqrt{D^3/\lambda}$ to $2D^2/\lambda$, and then the far field starts after $2D^2/\lambda$. This is true for $D \gg \lambda$ only.

Observing the results given in (2A.10) and (2A.12), we see that the far E and H fields decrease as the inverse of the distance for an antenna radiating in free space. Furthermore, the electric and the magnetic far fields are in space quadrature but they are coherent in time. The fields form a locally plane wave, and the ratio $\left| E_\theta / H_\phi \right| = \dfrac{\omega\mu}{k} = \dfrac{\omega\mu}{\omega\sqrt{\mu\varepsilon}} = \sqrt{\dfrac{\mu}{\varepsilon}} = \eta$; where η is the intrinsic impedance of the medium. This concludes the discussion for the derivation of the various subtleties in the near and far fields of an antenna.

It is interesting to observe again the patterns of the array given by Figures in 2A.1. For a distance $r < 0.62\sqrt{D^3/\lambda}$ the antenna pattern is close to an omni directional antenna as seen in Figure 2A.1a, when r is between $0.62\sqrt{D^3/\lambda}$ and $2D^2/\lambda$ the antenna pattern is beginning to take shape as seen in Figure 2A.1b, and finally when $r > \dfrac{2D^2}{\lambda}$, one is in the far field and the antenna radiation pattern is given by Figure 2A.1c.

Next we observe the angular variation for the criteria to be in the far field.

2A.3 Dipole Antennas Operating in Free Space

Here, we use the analytical expressions given in [37, equations 10-74, 10-75, and 10-76] to compute the fields for a z-oriented dipole antenna of various lengths operating in free space. For each specific length of the dipole antenna we provide two plots, one is for the ratio of $\left| E_\theta / H_\phi \right|$ as a function of the radial distance in meters for various angular directions θ (for $\theta = 5°$, $10°$, $15°$, $30°$, $45°$, $60°$, and $90°$), and the second plot is for the ratio $\left| E_r / E_\theta \right|$ in dB as a function of the radial distance in meters for various angular directions θ as before.

The first z-oriented dipole antenna we consider is 0.1λ long, λ is assumed to be 1 m or equivalently the operating frequency is 300 MHz for all the examples in this section. The value of $2D^2/\lambda$ in this case becomes $0.02\ \lambda$ which is 0.02 m. Figure 2A.2a displays that the ratio of $\left| E_\theta / H_\phi \right|$ settles down to a value of 377 after about 2 m for all θ directions. Therefore in this case the rule of thumb $2D^2/\lambda$ is not applicable to characterize the far field. The intrinsic impedance of the medium is reached after $2\ \lambda$ and not after $0.02\ \lambda$ as the formula predicts. Figure 2A.2b plots the ratio $\left| E_r / E_\theta \right|$ as a function of the distance from the antenna as a function of the angular rotation. In Figure 2A.2b, the plot for $\theta = 90°$ does not appear because the radial component is zero along that specific direction. It is seen that as the angular variation changes the radial component of the field becomes more dominant near the antenna and it does not become smaller in value till one goes away like 80 m or 80 λ to achieve a ratio of $\left| E_r / E_\theta \right|$ less than -20 dB for all θ directions.

(a)

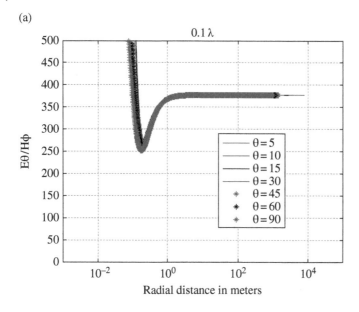

(b)

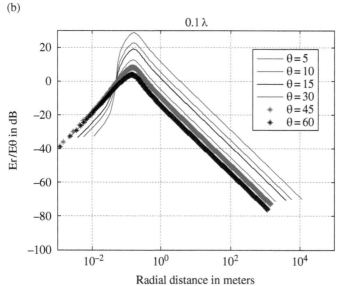

Figure 2A.2 (a) The ratio of $|E_\theta/H_\phi|$ for a 0.1λ z-oriented dipole antenna as a function of the radial distance in meters for various azimuthal orientations in θ. (b) The ratio $|E_r/E_\theta|$ in dB for a 0.1 λ z-oriented dipole antenna as a function of the radial distance in meters for various directions in θ.

In this case, 80 m is the appropriate starting distance for the far field and it is nowhere near the value of 0.02 λ. In addition, the ratio of the electric and magnetic fields settles down to the free space intrinsic impedance very close to the antenna satisfying this property for the far field. However, a locally plane wave does not exist at such a distance because there is a strong radial component of the electric field. Hence, it is necessary to go 80 λ away from the antenna to be in the far field for all angular directions. In this case the expression $2D^2/\lambda$ does not provide the correct result for the location of the far field.

The second z-oriented dipole antenna we consider is 0.5 λ long. If we calculate $2D^2/\lambda$ for this case, it gives 0.5 λ which is 0.5 m. Figure 2A.3a shows that the ratio of $|E_\theta/H_\phi|$ is 377 after 2 m for all θ directions. Figure 2A.3b plots the ratio $|E_r/E_\theta|$ and it is observed that to have the ratio of $|E_r/E_\theta|$ less than −20 dB for all θ directions one has to be away 80 m or 80 λ. In Figure 2A.3b, the plot for θ = 90° does not appear because the radial component is zero along that specific direction. The conclusion from these results is that when making a far-field pattern measurement of a half-wave dipole, one has to be at least 80-wavelengths away from it for certain θ directions to be in its far field. This means that when making a far-field pattern measurement of a half-wave dipole covering all possible θ directions, one has to be at least 80-wavelegths away from it. It is interesting to mention that if we plot the ratio $|E_r/E_\theta|$ of a Hertzian vertical dipole at θ = 1° we see that the ratio is 0 dB at a distance of 18-wavelengths away from it.

The last example deals with a dipole antenna that is 5 λ long. If we calculate $2D^2/\lambda$ it results in 50 λ which is 50 m. Figure 2A.4a shows that the ratio of $|E_\theta/H_\phi|$ is 377 after 20 m for all θ directions. Figure 2A.4b shows that the ratio $|E_r/E_\theta|$ takes considerable more distance to reach a negligible value. However the small value of the ratio is reached for most angles after 50 λ which is the same distance predicted by the $2D^2/\lambda$ formula, this means that the $2D^2/\lambda$ formula provides the correct result for the location of the far field in this case because $D \gg \lambda$. This again proves that the $2D^2/\lambda$ formula applies only if simultaneously $D \gg \lambda$ condition is satisfied.

In [33], it is mentioned that if $D \ge 2.5\lambda$ then it is safe to use $2D^2/\lambda$ to locate the starting distance of the far field. However, the author of [33] referred to [34] which does not include a derivation for the limits from which the conclusion in [33] was drawn. In our work, we actually provided some analytical results to show that we need $D \ge 5\lambda$ to assure that the far field criteria is satisfied at a distance given by $2D^2/\lambda$ so that it approximately forms a plane wave as the strength of the radial component is significantly decreased.

An alternate methodology and a different expression has been recently given in [38].

(a)

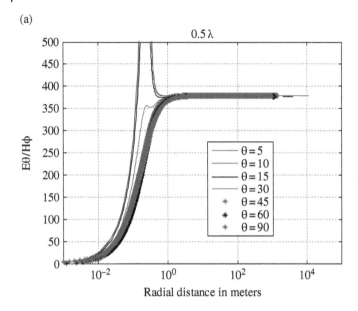

(b)

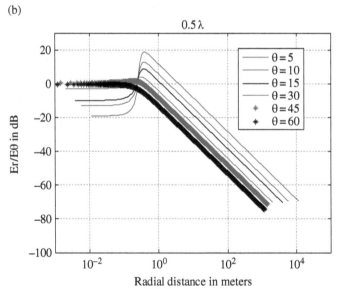

Figure 2A.3 (a) The ratio of $|E_\theta/H_\phi|$ for a 0.5λ z-oriented dipole antenna as a function of the radial distance in meters for various directions in θ. (b) The ratio of $|E_r/E_\theta|$ for a 0.5λ z-oriented dipole antenna as a function of the radial distance in meters for various directions in θ.

(a)

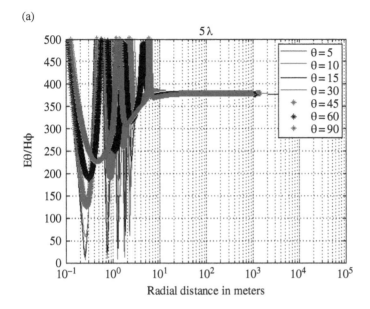

(b)

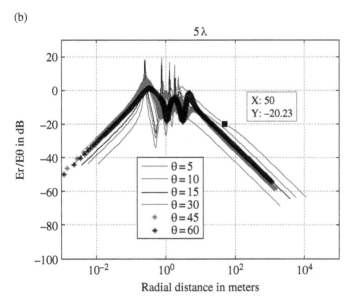

Figure 2A.4 (a) The ratio of $|E_\theta/H_\phi|$ for a 5 λ z-oriented dipole antenna as a function of the radial distance in meters for various directions in θ. (b) The ratio $|E_r/E_\theta|$ in dB for a 5 λ z-oriented dipole antenna as a function of the radial distance in meters for various directions in θ.

2A.4   Dipole Antennas Radiating Over an Imperfect Ground

For the characterization of an antenna operating over Earth, generally one is interested not only in the prediction of the fields near the interface but also as a function of the transmitting and receiving antenna's height over the air-Earth interface. In this section, we do not use $2D^2/\lambda$ to predict the starting distance of the far field for antennas operating over an imperfect ground. However, if someone tends to use $2D^2/\lambda$ expression for the start of the far field for antennas operating over an imperfect ground then D has to be interpreted correctly to be the effective radiating aperture and not the maximum dimension of the antenna. To predict the starting distance of the far field for antennas operating over an imperfect ground we use the formula given in [39] where it has been shown that for the case of antenna operating over PEC (Perfect Electric Conductor) ground, the far field starts at a distance d, given by

$$2\pi \frac{H_{Tx}^2 H_{Rx}}{\lambda d^2} = \frac{1}{\Psi}; \Rightarrow d = H_{Tx}\sqrt{\frac{2\pi H_{Rx} \Psi}{\lambda}} \qquad (2A.13)$$

where Ψ is the ratio of the transverse component of the electric field with respect to the radial component. H_{Tx} and H_{Rx} are the height of the transmitting and the receiving antennas over the Earth, respectively. So in the case of an antenna operating over ground, the starting distance for the far field depends on the measurement set up and more specifically on the antenna height over ground for the measuring device.

Here we use the electromagnetic analysis code, Analysis of Wire Antennas and Scatterers (AWAS) [7], to simulate a z-oriented dipole antenna of length $10\,\lambda$ radiating over an urban ground with relative permittivity $\varepsilon r = 4$, conductivity $\sigma = 2 \times 10^{-4}$ Siemens/m, the height of the transmitter antenna is 60 m above the ground (the height is defined with respect to the center feed point of the antenna), the height of the field point is 10 m from the ground. The frequency of operation is 300 MHz. Figure 2A.5a plots the ratio of $|E_\theta/H_\phi|$ as a function of the horizontal distance in meters, the ratio of $|E_\theta/H_\phi|$ is 377 after 700 m. Figure 2A.5b plots the ratio $|E_r/E_\theta|$ in dB as a function of the horizontal distance in meters, the ratio $|E_r/E_\theta|$ is less than −35 dB after a distance of about 2000 m so the fields represent a locally plane wave after a distance of about 2000 m. Notice that we only consider the direction of $\theta \approx 90°$, so we expect the 2000 m distance to increase if we check the behavior of the fields along other directions just like what we noticed in the free space scenario in section 2A.3. Using the expression in (2A13) and assuming that the radial component of the field has to be 35 dB less than the transverse component to represent an approximate locally plane wave which means that

(a)

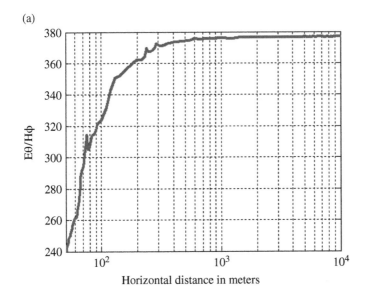

(b)

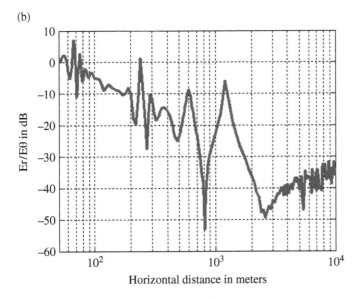

Figure 2A.5 (a) The ratio of $|E_\theta/H_\phi|$ for a 10 λ z-oriented dipole antenna located at a height of 60 m above an urban ground. (b) The ratio $|E_r/E_\theta|$ in dB for a 10 λ z-oriented dipole antenna located at a height of 60 m above an urban ground.

$\Psi = \sqrt{10^{3.5}} = 56.23$, one would expect the far field to start approximately at

$$d = 18.8 H_{Tx} \sqrt{\frac{H_{Rx}}{\lambda}} = 18.8 \times 60 \times \sqrt{10} = 3567 \, \text{m}.$$

As a second example, consider a z-oriented dipole antenna of 5 λ length radiating over an urban ground located at a height of 30 m. All other simulation parameters were the same as the first example. Figure 2A.6a shows that the ratio of $|E_\theta/H_\phi|$ is 377 after 700 m. Figure 2A.6b shows that ratio $|E_r/E_\theta|$ represents a locally plane wave after about 950 m. We can calculate the predicted value as to where the far field will start if the ground was a PEC using (2A.13). This results in $d = 18.8 H_{Tx} \sqrt{\frac{H_{Rx}}{\lambda}} = 18.8 \times 30 \times \sqrt{10} \approx 1783 \, \text{m}.$

As the last example, consider a z-oriented dipole antenna of $\lambda/2$ length radiating over an urban ground located at a height of 30 m. All other simulation parameters were the same as the previous example. Figure 2A.7a shows that the ratio of $|E_\theta/H_\phi|$ is 377 after 700 m. Also, Figure 2A.7b shows that ratio $|E_r/E_\theta|$ is approximately a locally plane wave after about 950 m. We can calculate the predicted value as to where the far field will start if the ground was a

PEC using (2A.13) from $d = 18.8 H_{Tx} \sqrt{\frac{H_{Rx}}{\lambda}} = 18.8 \times 30 \times \sqrt{10} \approx 1783 \, \text{m}.$

One can observe that the ratio $|E_r/E_\theta|$ becomes almost constant as we go away from the antenna in the case of an imperfect ground while the ratio $|E_r/E_\theta|$ goes down as we go away from the antenna in the case of antennas radiating in free space as shown in the examples of section 2A.3. The reason for this is that the radial component of the field at very large distances from the antenna decays as $1/r^2$ in the case of a vertical dipole radiating over an imperfect ground while it decays as $1/r^2$ or higher in the case of free space and it decays as $1/r^3$ in the case of a PEC ground [39], whereas the transverse component decays as $1/r^2$ in the case of vertical dipole radiating over an imperfect ground while it decays as $1/r$ in the case of free space and it decays as $1/r$ in the case of PEC ground [39].

2A.5 Epilogue

Predicting the far field of an antenna operating in free space using $2D^2/\lambda$ is not useful when the maximum dimension of the antenna is smaller or comparable to λ. From the derivation of $2D^2/\lambda$, it is shown that this distance guarantees that the maximum phase error introduced from neglecting the third term of (2A.4) will not exceed $\pi/8$. However, it does not guarantee that the radial component is negligible compared to the transverse component, i.e., an important criteria for the far field. We must pay attention to the constraint under which it is safe to use $2D^2/\lambda$ to predict the starting distance of the far field. If $D \geq 5\lambda$ then it is

(a)

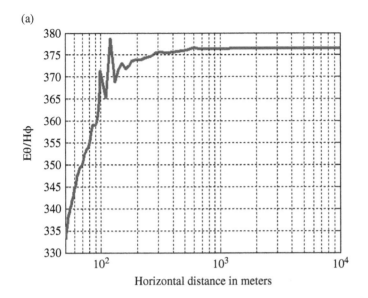

(b)

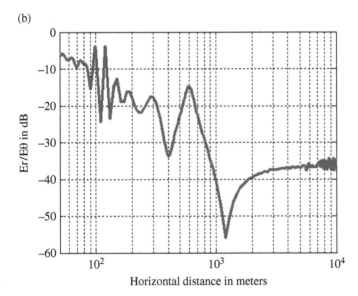

Figure 2A.6 (a) The ratio of $|E_\theta/H_\phi|$ for a 5 λ z-oriented dipole antenna at height of 30 m above an urban ground. (b) The ratio $|E_r/E_\theta|$ in dB for a 5 λ z-oriented dipole antenna at height of 30 m above an urban ground.

(a)

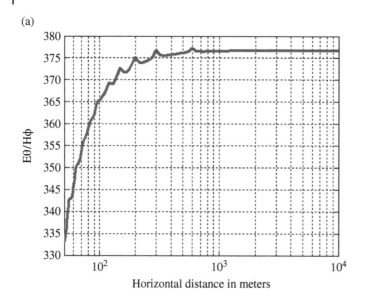

(b)

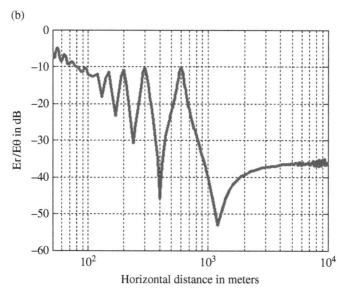

Figure 2A.7 (a) The ratio of $|E_\theta/H_\phi|$ for a $\lambda/2$ z-oriented dipole antenna at height of 30 m above an urban ground. (b) The ratio $|E_r/E_\theta|$ in dB for a $\lambda/2$ z-oriented dipole antenna at height of 30 m above an urban ground.

safe to use $2D^2/\lambda$ to predict the starting distance of the far field for an antenna radiating in free space. Locating the starting distance of the far field using $d = H_{Tx}\sqrt{\dfrac{2\pi H_{Rx}\,\Psi}{\lambda}}$ is sufficient for antennas operating on top of an imperfect ground.

References

1 R. G. Brown, R. A. Sharpe, W. L. Hughes, and R. E. Post, *Lines, Waves, and Antennas*, John Wiley & Sons, Inc., New York, Second Edition, 1973.

2 J. D. Kraus and R. J. Marhefka, *Antennas*, McGraw Hill, New York, Third Edition, 2002.

3 R. F. Harrington, *Time-Harmonic Electromagnetic Fields*, McGraw-Hill, New York, 1961.

4 C. A. Balanis, *Antenna Theory: Analysis and Design*, Harper and Row Publishers, New York, 1982.

5 S. P. Thompson, *Dynamo-Electric Machinery: A Manual for Students of Electrotechnics*, E. & F. N. SPON, London/New York, 1884.

6 Electric generator. http://edison.rutgers.edu/generator.htm. Accessed 17 November 2017.

7 A. R. Djordjevic, M. B. Bazdar, T. K. Sarkar, and R. F. Harrington, *AWAS Version 2.0: Analysis of Wire Antennas and Scatterers, Software and User's Manual*, Artech House, Norwood, MA, 2002.

8 S. Llorente-Romano, A. Garcia-Lampérez, T. K. Sarkar, and M. Salazar-Palma, "An Exposition on the Choice of the Proper S Parameters in Characterizing Devices Including Transmission Lines with Complex Reference Impedances and a General Methodology for Computing Them," *IEEE Antennas and Propagation Magazine*, Vol. 55, No. 4, pp. 94–112, 2013.

9 Y. Zhang, T. K. Sarkar, X. Zhao, D. Garcia-Donoro, W. Zhao, M. Salazar-Palma, and S. Ting, *Higher Order Basis Based Integral Equation Solver (HOBBIES)*, John Wiley & Sons, Inc., Hoboken, NJ, 2012.

10 T. K. Sarkar, E. L. Mokole, and M. Salazar-Palma, "An Expose on Internal Resonance, External Resonance, and Characteristic Modes," *IEEE Transactions on Antennas and Propagation*, Vol. 64, No. 11, pp. 4695–4702, 2016.

11 S. R. Best, "The Performance Properties of Electrically Small Resonant Multiple-Arm Folded Wire Antennas," *IEEE Antennas and Propagation Magazine*, Vol. 47, No. 4, pp. 13–27, Aug. 2005.

12 H. A. Wheeler, "Fundamental Limitations of Small Antennas," *Proceedings of the IRE*, Vol. 35, No. 12, pp. 1479–1484, Dec. 1947.

13 J. S. McLean, "A Re-examination of the Fundamental Limits on the Radiation Q of Electrically Small Antennas," *IEEE Transactions on Antennas and Propagation*, Vol. 44, No. 5, pp. 672, May 1996.

14 H. Bode, *Network Analysis and Feedback Amplifier Design*, Van Nostrand, New York, 1947, p. 367.

15 R. M. Fano, "Theoretical Limitations on the Broadband Matching of Arbitrary Impedances," *Journal of the Franklin Institute*, Vol. 249, pp. 57–83, Jan. 1950, and pp. 139–155, Feb. 1950.

16 D. C. Youla, "A New Theory of Broadband Matching," *IEEE Transactions on Circuit Theory*, Vol. CT-11, pp. 30–50, Mar. 1964.

17 S. E. Sussman-Fort and R. M. Rudish, "Non-Foster Impedance Matching of Electrically-Small Antennas," *IEEE Transactions on Antennas and Propagation*, Vol. 57, No. 8, pp. 2230–2241, Aug. 2009.

18 L. J. Chu, "Physical Limitations of Omni Directional Antennas," *Journal of the Applied Physics*, Vol. 19, pp. 1163–1175, 1948.

19 A. R. Lopez, "Fundamental Limitations of Small Antennas: Validation of Wheeler's Formulas," *IEEE Antennas and Propagation Magazine*, Vol. 48, No. 4, pp. 28–36, Aug. 2006.

20 R. Lewallen, "EZNEC/4 Antenna Modeling Software," http:www.eznec.com. Accessed 21 November 2017.

21 T. K. Sarkar, "Analysis of Arbitrarily Oriented Thin Wire Antennas over a Plane Imperfect Ground," *AEU*, Band 31, Heft 11, pp. 449–457, 1977.

22 T. K. Sarkar, H. Schwarzlander, S. Choi, M. Salazar-Palma, and M. C. Wicks, "Stochastic versus Deterministic Models in the Analysis of Communication Systems," *IEEE Antennas and Propagation Magazine*, Vol. 44, No. 4, pp. 40–50, Aug. 2002.

23 S. Kay, "Can Delectability Be Improved by Adding Noise?," *IEEE Signal Processing Letters*, Vol. 7, No. 1, pp. 8–10, Jan. 2000.

24 T. K. Sarkar, E. L. Mokole, and M. Salazar-Palma, "Relevance of Electromagnetics in Wireless Systems Design," *IEEE Aerospace and Electronic Systems Magazine*, Vol. 31, No. 10, pp. 8–19, 2016.

25 T. K. Sarkar, R. S. Adve, and M. Salazar-Palma, "Phased Array Antennas," in *Wiley Encyclopedia of Electrical and Electronic Engineering*, J. G. Webster, editor, John Wiley & Sons, Inc., New York, 2001.

26 T. K. Sarkar, Z. Ji, K. Kim, A. Medouri, and M. Salazar-Palma, "A Survey of Various Propagation Models for Mobile Communication," *IEEE Antennas and Propagation Magazine*, Vol. 45, No. 3, pp. 51–82, June 2003.

27 T. K. Sarkar, S. Burintramart, N. Yilmazer, S. Hwang, Y. Zhang, A. De, and M. Salazar-Palma, "A Discussion about Some of the Principles/Practices of Wireless Communication under a Maxwellian Framework," *IEEE Transactions on Antennas and Propagation*, Vol. 54, No. 12, pp. 3727–3745, Dec. 2006.

28 H. A. Wheeler, "The Radiansphere around a Small Antenna," *Proceedings of the IRE*, Vol. 47, No. 8, pp. 1325–1331, 1959.

29 T. K. Sarkar, M. C. Wicks, M. Salazar-Palma, and R. Bonneau, *Smart Antennas*, John Wiley & Sons, Inc., New York, 2003.

30 T. K. Sarkar, M. Salazar-Palma, and E. L. Mokole, *Physics of Multiantenna Systems and Broadband Processing*, John Wiley & Sons, Inc., Hoboken, NJ, 2008.

31 S. Hwang, A. Medouri, and T. K. Sarkar, "Signal Enhancement in a Near-Field MIMO Environment through Adaptivity on Transmit," *IEEE Transactions on Antennas and Propagation*, Vol. 53, No. 2, pp. 685–693, 2005.

32 R. H. Clarke and J. Brown, *Diffraction Theory and Antennas*, John Wiley & Sons, Inc., Hoboken, NJ, 1980.

33 R. Bansal, "The Far-Field: How Far Is Far Enough," *Applied Microwave and Wireless*, Vol. 11, pp. 58–60, 1999.

34 W. L. Stutzman and G. A. Thiele, *Antenna Theory and Design*, John Wiley & Sons, Inc., Hoboken, NJ, Third Edition, 1998.

35 A. W. Rudge, K. Milne, A. D. Olver, and P. Knight, *The Handbook of Antenna Design*, Vols. 1 and 2, Peter Peregrinus Ltd., London, U.K., 1986.

36 "IEEE Standard Definitions of Terms for Antennas," IEEE Std. 145-1983.

37 E. C. Jordan and K. G. Balmain, *Electromagnetic Waves and Radiating Systems*, Prentice Hall Inc., New Delhi, India, Second Edition, 1968.

38 J. Asvestas, "How Far Is the Far Field?," *IEEE Antennas and Propagation Magazine*, Submitted for publication.

39 A. De, T. K. Sarkar, and M. Salazar-Palma, "Characterization of the Far Field Environment of Antennas Located over a Ground Plane and Implications for Cellular Communication Systems," *IEEE Antennas and Propagation Magazine*, AP-52, 6, pp. 19–40, Dec. 2010.

3

Mechanism of Wireless Propagation: Physics, Mathematics, and Realization

Summary

This chapter starts with presenting all the various experimental results available in the literature on the propagation path loss for cellular wireless propagation systems. From the experimental data it is quite clear that trees, buildings, and other man made obstacles contribute second order effects to the propagation path loss as the dominant component is the free space propagation of the signal and the effect of the Earth over which the signal is propagating. In addition, the propagation path loss of 30 dB per decade of distance inside a cell is independent of the material parameters of the Earth and this holds even when propagation is over water. This path loss is also independent of frequency. Next it is shown that an electromagnetic macro model can accurately predict the dominant component of the propagation path loss in a cellular wireless communication system which is first 30 dB per decade of distance and later on, usually outside the cell, it is 40 dB per decade of distance between the transmitter and the receiver irrespective of their heights from the ground. This implies that the electric field decays first at a rate of $\rho^{-1.5}$ inside the cell and later on, usually outside the cell, as ρ^{-2}, where ρ stands for the distance between the transmitter and the receiver. It is also illustrated that the so called slow fading is due to the interference between the direct wave and the ground wave as introduced by Sommerfeld over a hundred years ago. All these statements can be derived from the numerical integration of the Sommerfeld integrals using a modified path for the steepest descent method and also using an accurate purely numerical methodology. An optical analog model is described based on the image theory developed by Van der Pol to illustrate the physical mechanism of radio wave propagation in a cellular wireless communication system where the path loss is 30 dB per decade or the field decays as $\rho^{-1.5}$. This is due to the image of the transmitting antenna generated by the lossy imperfectly conducting Earth. The macro model developed in this chapter is used to refine the

The Physics and Mathematics of Electromagnetic Wave Propagation in Cellular Wireless Communication, First Edition. Tapan K. Sarkar, Magdalena Salazar Palma, and Mohammad Najib Abdallah.
© 2018 John Wiley & Sons, Inc. Published 2018 by John Wiley & Sons, Inc.

experimental data collection system for the propagation path loss and it is also illustrated how the antenna tilt both mechanical and electrical can be incorporated in the macro model to predict the propagation path loss. Finally, an observation is made on how to further improve the propagation mechanism by moving the transmitting antenna closer to the ground and tilting it slightly upwards towards the sky – a very non intuitive solution which can improve the path loss in the system. It is seen that within the cell the field strength significantly increases as the antenna is brought closer to the ground and thus one may do away with the most expensive part of the deployment which is the tower and since the electric field strength increases significantly when the antenna is placed closer to the ground implies that for the same quality of service the transmitted power both in the base station and in the mobile can be reduced at least by a factor of ten! Also, it will be quite profitable if the path loss can be tailored to 20 dB per decade which is the lowest free space path loss possible over the usual 30 dB per decade and the 40 dB per decade achieved for the vertical oriented transmitting antenna radiating over Earth. This is illustrated by providing a slight tilt to the vertically oriented antenna and simulation shows that such a deployment can generate a horizontal component of the field which approaches a path loss of 20 dB per decade and indeed this field can exceed in strength over the vertical component of the field over long distances. This phenomenon becomes more enhanced as the antenna height is reduced and it is deployed closer to the ground. This has an additional effect of minimizing the effect of slow fading which is due to the effect of the earth. Finally, it is illustrated that the communication in cellular systems takes place through the elusive Zenneck wave and not a surface wave. The four appendices at the end of the chapter provides the mathematical details for the derivation of the macro model described earlier.

3.1 Introduction

In [1] D. Gabor states: "*The wireless communication systems are due to the generation, reception and transmission of electro-magnetic signals. Therefore all wireless systems are subject to the general laws of radiation. Communication theory has up to now been developed mainly along mathematical lines, taking for granted the physical significance of the quantities which are fundamental in its formalism. But communication is the transmission of physical effects from one system to another. Hence communication theory should be considered as a branch of physics.*" This illustrates that to visualize the physics of the propagation mechanism in cellular wireless communication systems it is necessary to apply the basic principles of electromagnetic theory as a statistical model is incapable of capturing the fundamentals of the propagation of the fields that travel with a decay of $\rho^{-1.5}$ with horizontal distance ρ inside the cell.

To illustrate this point, we first present results for various experiments that have been carried out over several years in the past. Then an analytical model [2–5] is derived from the classical Sommerfeld integrals to reproduce the experimental data of Okumura et al. [6] as this is one of the few papers which clearly state the nature of the transmitting and receiving antennas used for measurements and in what particular environment they were operating thus making it possible for an interested reader to completely duplicate not only their experiments but also can perform an accurate numerical simulation of their experimental setup using an appropriate electromagnetic code, thus validating their measurements. Comparison is also made with other experimental data to illustrate the accuracy of the prediction carried out using an electromagnetic macro model based on available computational electromagnetic codes. The reason why a macro model can be used to predict the propagation path loss is because of the fact that most of the loss is due to propagation over an imperfect ground characterizing the earth, and the buildings, trees, and the like introduce second-order effects. Thus, it is shown that these electromagnetic analysis codes can provide path loss models which are as good as carrying out real time path loss measurements using a van and other accompanying equipment which are both very expensive and time consuming to perform an actual measurement campaign. An optical analog model is provided to visualize the actual propagation mode for the nature of the signal transmission in a cellular environment [2–5]. Finally, it is illustrated how to deploy antennas in a different way in a cellular system to improve their performance.

3.2 Description and Analysis of Measured Data on Propagation Available in the Literature

In Figure 12 of Okumura et al.'s classic paper [6], which is presented here as Figure 3.1, it is seen that for most practical heights of the base station antenna it is quite clear that the decay of the electric field strength is at a rate of $\rho^{-1.5}$ where ρ stands for the horizontal distance from the transmitting antenna. As described by Okumura et al. in Figure 3.1, it is seen that the decay of the electric field strength of the propagating fields inside the cell in an urban environment is at a rate of $\rho^{-1.5}$ or equivalently the power in the wave decays as ρ^{-3}. This is equivalent to the power in the wave decaying as $10\mathrm{Log}_{10}$ $(\rho^{-3}) = -30$ dB/decade of distance.

In Figure 3.2, we present the prediction data from Ericsson for an in-building path loss model. This was shown in Chapter 3 of *Mobile Antenna Systems Handbook* [7]. Figure 3.2 shows that for a distance of 1 m to 10 m the path loss is typically 30 dB, the same value stated earlier by Okumura et al. but for measurements done in an urban environment. Similar experimental results for the propagation path loss have been obtained in measurements carried out in an

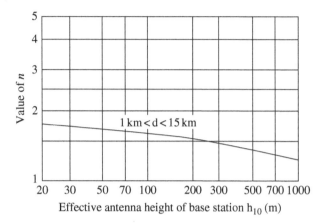

Figure 3.1 Distance dependence of median field strength in urban area ($E \propto d^{-n}$)

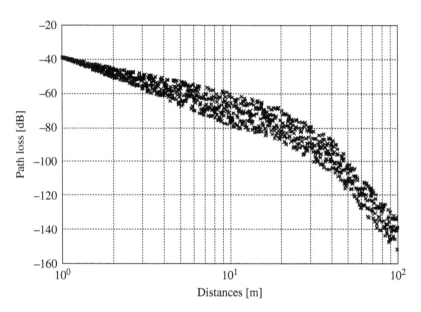

Figure 3.2 Prediction from Ericsson in-building path loss model. (Simon R. Saunders, Advances in mobile propagation prediction methods, Chapter 3 of *Mobile Antenna Systems Handbook*, Ed. Kyohei Fujimoto, Artech House, 2008.

indoor environment at 945 MHz [8]. However, if measurements are carried out in the near fields of the transmitter (the concept of the near and far fields have been explained in Chapter 2) then a lower value for the path loss exponent will be obtained. In Figure 3.3 an empirical model for macrocell propagation at 900 MHz is presented [7]. The dots are measurements taken in a suburban

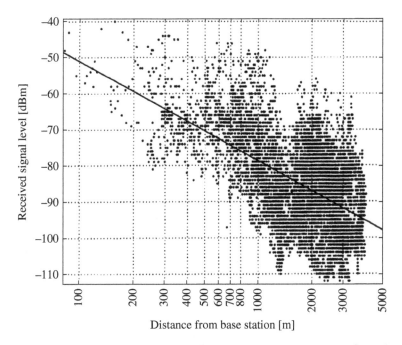

Figure 3.3 Empirical model of macro cell propagation measurements performed at 900 MHz. The dots are actual experimental data points taken in a suburban area, whereas the solid line represents the best fit empirical model.

area, whereas the solid line represents a best fit empirical model (the solid line was not drawn by us but by the authors of that chapter). If one goes from 100 m to 1000 m it is seen that the path loss is again 30 dB for the best fit line as presented in [7] even though the parameters of the experiment are different from that of Figure 3.2.

In 1937, C. R. Burrows from Bell Labs [9] carried out measurements over Seneca Lake, New York, USA to find the nature of electromagnetic wave propagation over water. The transmitting and receiving antennas were placed on two different boats as seen in Figure 3.4a. The measured propagation path loss over water is illustrated in Figure 3.4b. Again, it is seen that as one moves from 10 m to 100 m the path loss is 30 dB and from 100 m to 1000 m, away from the transmitter, the path loss is 40 dB as mentioned before. So the propagation path loss is still the same when measured over water!

Next results are presented displaying actual measurements carried out at different frequencies for three base stations located in a dense-urban, urban and suburban environments. The special feature of the data taken in an urban environment is that the houses were not uniformly spaced. First, the results are presented for data taken at the carrier signal frequency of 1800 MHz GSM base

(a)

(b)

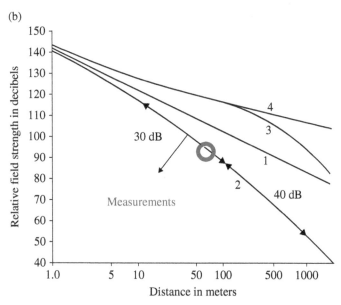

Figure 3.4 (a) Experimental arrangement for the measurements of the received field strength over water carried out by Burrows in Seneca Lake located in New York State. (Reproduced with permission of IEEE.) (b) Measurements of the relative field strength with distance over water.

station transmitters located in the dense-urban/urban environments of New Delhi area belonging to *Idea Cellular Network* which have been monitored with Nokia GSM receiver (model 6150) generally used as a drive-in tool for planning cellular networks along with a GPS receiver to know the latitude and longitude of the mobile. The sensitivity of the receiver is −102 dBm. The transmitting power of all the base stations is +43 dBm. The receiving equipment was installed in a vehicle along with the data acquisition system. The vehicle moved on a normal road at a permissible speed in the traffic and the downlink signal strength was monitored by the receiver. The observed signal levels were converted into path loss values based on the received signal level, transmitter power, and the transmitting and receiving antenna gains for further analysis. The gain of the transmitting antenna was 18 dB. The estimated measurement r.m.s. (root mean square) error is around 1.5 dB. The height of the mobile receiving antenna was 1.5 m from the ground.

For the case of the 900 MHz results, the experiment is carried out with the help of *Aircom International Limited*, a UK company based in India. The transmitting power of all the base stations used in this study is 43.8 dBm and the transmitting antenna gain is 2dBi for all the base stations. The gain of the receiving antenna is 0 dB and the height from the ground is 1.5 m. The receiver is a standard Nokia equipment used in drive-in tools used for field trials. For all the measurements, the position of the mobile is determined from the GPS receiver and this information with the coordinates of the base station was utilized to deduce the distance traveled by the mobile from the base station. The signal strength information recorded in dBm was converted into path loss values utilizing the gains of the antennas. The data was recorded with 512 samples in one second on a laptop computer and the number of samples collected for each site varied from 1×10^5 to 2×10^5. Measured r.m.s. error is around 1.5 dB. Data was averaged over a conventional range of 40λ.

The first base station operating at 1800 MHz called OM-1 was located in a dense urban environment in the Indian capital city of Delhi and the height of the transmitting base station antenna was 24 m above ground. The variation of the path loss is shown in Figure 3.5. It is seen that inside the cell the path loss has settled down approximately to a value of 3 at a distance of $4H_{TX}H_{RX}/\lambda = 4 \times 24 \times 1.5/(1/6) = 864$ m as predicted by the theory [5]. Here, H_{TX} and H_{RX} represent the height of the transmitting and the receiving antenna from the ground and λ is the wavelength of the operating frequency.

Next, the data at 900 MHz for the base station UA located in a medium urban environment of the Indian capital city of Delhi are shown. The base station antenna is located at a height of 24 m. From Figure 3.6, it appears that the path loss exponent factor in the cell has settled down to a value of 3 at around

$$\frac{4H_{TX}H_{RX}}{\lambda} = \frac{4 \times 24 \times 1.5}{1/3} = 432\,\text{m}$$ as predicted by the theory [5].

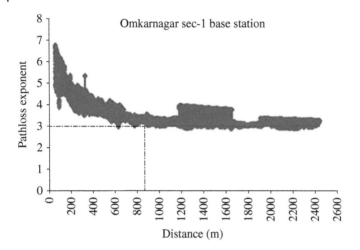

Figure 3.5 Variation of path loss exponent with distance for OM-1 base station (1800 MHz) located in a dense urban environment. Base station antenna height is 24 m.

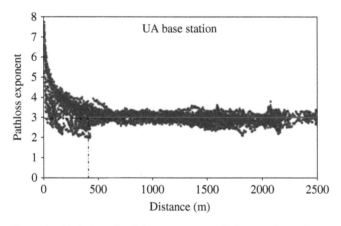

Figure 3.6 Variation of path loss exponent with distance for UA base station (900 MHz) in a medium urban environment. Base station antenna height is 24 m.

Figure 3.7 plots the variation of the path loss exponent from the base station AURNIA 19. This station is located in an industrial area in the city of Aurangabad, in the Maharashtra state of western India. The transmitter height is 50 m above ground and it operates at 900 MHz. The path loss exponent is expected to start at 3 at a distance of approximately $(4 H_{TX} H_{RX}/\lambda) = \dfrac{4 \times 50 \times 1.5}{1/3} = 900$ m [5] for the operating frequency 900 MHz. This value starts where the slow fading region ends as illustrated in [5]. After some distance,

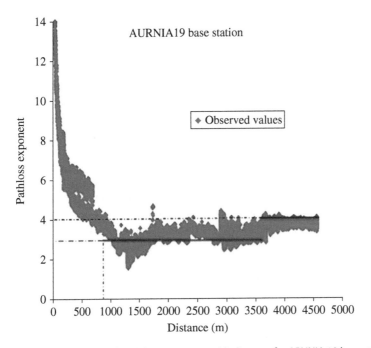

Figure 3.7 Variation of path loss exponent with distance for ARUNIA 19 base station (900 MHz) in an industrial area. Base station antenna height is 50 m.

the path loss exponent gradually increases to 4 in the far field region, as predicted by theory [5].

All three measurements show that in an intermediate region from the base station antenna the propagation path loss exponent is 3 and that region is close to the far field region. In the near fields, the path loss exponent change a lot because of the presence of the various field components other than radiation. Outside the cell and far away from the transmitter the path loss exponent is 4. This implies that the electric field varies as $\rho^{-1.5}$ inside the cell and at the far field region, usually outside the cell as ρ^{-2}.

Another important observation to be made is that at 1000 m from the transmitter the path loss due to the decay of the free space fields propagating over Earth is $10 \log_{10} \rho^3$ which is equal to 90 dB. This is quite a large value. In addition, there is variability of the measurement data due to buildings, trees and so on at the site, which shows an additional large variation which is of the order of 30–40 dB in the Figures 3.5–3.7. This secondary variation in the results due to buildings, trees and so on provide a dimensionality to this experimental plot, as seen by the width of the line plotting the results. This is also illustrated in the variation of the path loss exponent. Clearly, a variation of 30–40 dB due to the environment of trees, buildings and so on is quite a large number but compared to the primary propagation path loss of 90 dB to traverse a distance of 1 km which is due to the propagation over the Earth, the former variation is quite small. Hence, the path loss due to propagation over the imperfect ground namely Earth, is of prime importance and the effects of the buildings, trees and the nature of the terrain are secondary as illustrated by the variability of the plots. Therefore an electromagnetic macro model should be able to predict the primary propagation path loss. Also, the path loss exponent seems to be insensitive to the nature of the ground be it an urban, a suburban ground or for propagation over water. It is also insensitive to the height of the transmitting antenna. However, the height of the transmitting antenna affects the starting distance at which the decay of the fields will start at −30 dB/decade.

From the above discussions, it is now asserted that a physics based electromagnetic macro model can provide accurate predictions for the path loss exponent in a cellular network using electromagnetic simulation tools that depend only on some physical parameters of the macro model, primarily those related to the antennas and the electrical parameters of the ground. In a macro model, one needs to include only the electrical parameters of the environment without including clutter effects of buildings, trees, and so on.

Furthermore as illustrated in [10] and displayed in Figure 3.8, the reason that the 0.8 to 2.1 GHz cellular band was chosen for mobile broadband is that the reflection from buildings is negligible, and yet the signals can penetrate buildings and do not significantly bounce inside the rooms. In addition, in Figure 3.9, the one way attenuation through common building materials illustrates that at the cellular band, signals can penetrate buildings very easily [11]. Hence, the effects of clutter generated by buildings or trees are considered to be

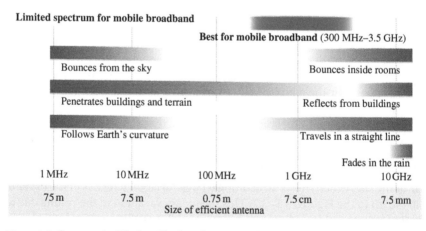

Figure 3.8 Opportunity Window: The best frequencies for mobile broadband are high enough so that the antenna can be made conveniently compact yet not so high that signals will fail to penetrate buildings [9]. That leaves a narrow range of frequencies reliable for use (red band). (Reproduced with permission of IEEE.)

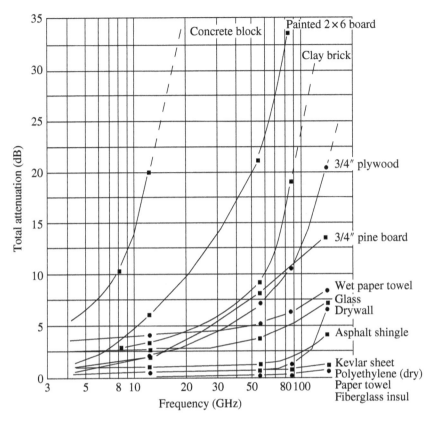

Figure 3.9 One way attenuation through common building materials as a function of frequency. Observe the curve does not even start till 4.5 GHz.

second order effects, even though they are large. The primary factor influencing propagation is the effect of the imperfectly conducting ground which however is seldom accounted for in any propagation model. Informations on Figures 3.8 and 3.9 make it clear, that the use of an electromagnetic macro model for the analysis of the propagation path loss in mobile communication using physical parameters like height, gain, and tilt of an antenna over an imperfect ground, can accurately duplicate experimental data taken in a real environment. The non-planar nature of the real ground is deemed not to be a show stopper for the propagation analysis using a macro model as illustrated in [2–5]. It is primarily because of these preponderance of physical evidence that wireless signals penetrate through buildings and do not get reflected significantly from them, that an electromagnetic macro model has been proposed based on fundamental physics replacing the statistical models which do not address the underlying electromagnetics, namely the effect of the propagation over Earth!

Typically, in the current state of the art, the drive test data is fit in a least squares fashion to a statistical model known as the log-normal model. Sometimes, empirical-based models are given based on the path loss exponent values calculated using the log-normal model. The fact of the matter is that current propagation modeling tools cannot capture the fundamental physics, namely inside a cell the path loss is typically 30 dB/decade [2–4] or equivalently the electric field strength varies as $(1/\rho^{1.5})$ with the horizontal distance from the antenna ρ, and outside the cell the propagation path loss is approximately 40 dB/decade or equivalently the electric field strength varies as $(1/\rho^2)$. Thus statistical based models do not capture the basic physics of the propagation mechanism. Furthermore, it does not make sense to have the path loss exponent tied to a value of a reference distance in the current state of the art ad hoc models which is quite meaningless from a scientific point of view as the propagation effects should be dependent on the physical parameters of the system. It is imperative that this reference distance in these models be located in the far field of the transmitting antenna, which is seldom considered in the implementation.

As a final example consider two parallel vertical center-fed half wave dipoles constituting the transmitting and receiving antennas operating in free space over a perfectly conducting ground. The dipoles has a 1 mm radius and is operating at 1 GHz. The dipoles are separated by 100 m and the receiving dipole is located 2 m above the ground. The height of the base station antenna can be adjusted from 1 m to 10 m above the ground. We consider two scenarios. In the first case the receiving antenna is in direct line-of-sight with the transmitting antenna both being located in free space, In the second scenario, we consider the receiving dipole to be encapsulated by a dielectric box of 1.5 m in height and in the cross sectional plane of 2 m × 2 m. The thickness of the dielectric is 5 cm and is located 1.2 m from the ground. The dielectric constant of the box is 2.5 and it is assumed to be lossless for simplicity. The scenario is illustrated in Figure 3.10. It is seen that the received power computed using the automatized parallel electromagnetic simulation code HOBBIES [12] is more when the receiving dipole is encapsulated by the dielectric cube than when it is located in the line of sight (LOS) configuration. This is reflected through Figure 3.11 – a plot of the Shannon Channel capacity as per the definition in section 1.10. The reason for that is there may be resonances due to the dielectric cube which for this specific configuration may enhance the received signal strength. Hence, it is not universally true that locating a receiver inside a building will impair the signal level received.

In summary, all the measured data show that within the cell a typical 3.5 km radius the path loss is 30 dB/decade whereas outside the cell it is 40 dB/decade irrespective of the nature of the ground be it dense urban, medium urban, industrial or even over water. This clearly illustrates that to go from 1 m to 1000 m or 1 km the signal undergoes a 90 dB attenuation in comparison to buildings, trees and the like which introduce a mere 40 dB variation. Also, it is not true that buildings, trees and the like introduce large insertion loss and

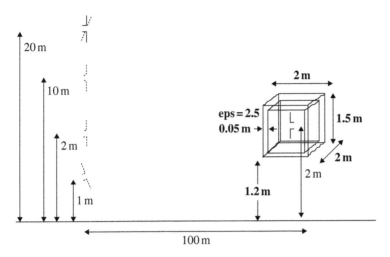

Figure 3.10 Different transmitting and receiving configurations for LOS and non-LOS scenarios.

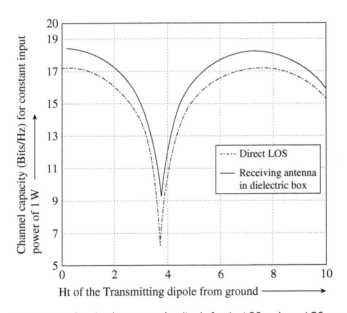

Figure 3.11 Received power at the dipole for the LOS and non-LOS scenario.

reflections. This is established through the standards that the specific band of Figure 3.8 was chosen for wireless communication was simply because the signals at those frequencies can penetrate through buildings and the antenna can be made compact. The path loss is independent of the height of the base station antenna and the frequency of operation at the cellular frequency.

Because the effect of the man-made structures are a second-order effect compared to the free-space path loss, we envision that a physics based solution methodology will be able to predict the path loss in a cellular wireless communication as predicted by Gabor.

Next, the electromagnetic macro model that is going to predict the path loss exponents is presented.

3.3 Electromagnetic Analysis of Propagation Path Loss Using a Macro Model

To simulate the electromagnetic propagation path loss in a cellular environment accurately using the physical parameters related to the environment, first we present an analytical solution obtained by Arnold Sommerfeld over a hundred years ago in terms of some semi-infinite integrals. The objective is to demonstrate that an exact analytical formulation based on integrating the semi-infinite integrals by a modified method of steepest descent one can reach the same conclusion as illustrated in the earlier sections using different measurements.

The complete analytical formulation [13–17] is presented in details in Appendix 3A for completeness. In the Sommerfeld formulation, one considers an elementary dipole of length dz located at (x', y', z') and carries a current of strength I, thus having a dipole moment Idz oriented along the z-direction. The dipole is situated over an imperfect ground plane characterized by a complex relative dielectric constant ε as seen in Figure 3.12. The complex relative dielectric constant is given by $\varepsilon = \varepsilon_r - \dfrac{j\sigma}{\omega\varepsilon_0}$ where ε_r represents the relative permittivity of the medium, ε_0 is the permittivity of vacuum, σ is the conductivity of the medium, ω stands for the angular frequency, and j is the imaginary unit, i.e., $j = \sqrt{-1}$.

It is possible to formulate a solution to the problem of radiation from the dipole operating in the presence of the imperfect ground in terms of a single Hertzian vector potential $\hat{u}_z\Pi_z$ of the electric type. It is seen from the presentation in Appendix 3A the fields radiated by the dipole can be deduced from the Hertz potentials which can be written as

$$\Pi_{1z} = P\left[\frac{\exp(-jk_1R_1)}{R_1}\right.$$

$$\left. +\int_0^\infty \frac{J_0(\xi\rho)}{\sqrt{\xi^2 - k_1^2}} \frac{\varepsilon\sqrt{\xi^2 - k_1^2} - \sqrt{\xi^2 - k_2^2}}{\varepsilon\sqrt{\xi^2 - k_1^2} + \sqrt{\xi^2 - k_2^2}} \exp\left(-\sqrt{\xi^2 - k_1^2}\left(z + z'\right)\right)\xi\, d\xi\right]$$

$$(3.1)$$

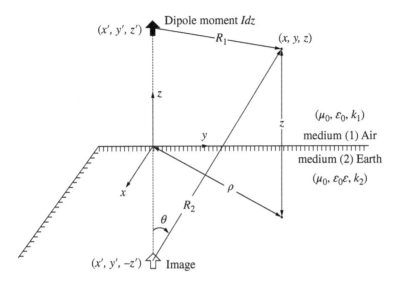

Figure 3.12 A vertical dipole over a horizontal imperfect ground plane.

and

$$\Pi_{2z} = 2P \int_0^\infty \frac{J_0(\xi\rho)\exp\left(\sqrt{\xi^2 - k_2^2}\,z - \sqrt{\xi^2 - k_1^2}\,z'\right)}{\varepsilon\sqrt{\xi^2 - k_1^2} + \sqrt{\xi^2 - k_2^2}}\,\xi d\xi \tag{3.2}$$

where

$$k_1^2 = \omega^2 \mu_0 \varepsilon_0 \tag{3.3}$$

$$k_2^2 = \omega^2 \mu_0 \varepsilon_0 \varepsilon \tag{3.4}$$

$$P = \frac{I\,dz}{j\omega 4\pi\varepsilon_0} \tag{3.5}$$

$$\rho = \sqrt{(x - x')^2 + (y - y')^2} \tag{3.6}$$

$$R_1 = \sqrt{\rho^2 + (z - z')^2} \tag{3.7}$$

and ξ is the variable of integration and the related parameters are described in Appendix 3A. The subscripts 1 represents the upper medium which is air and 2 represents the lower medium which is the Earth as seen in Figure 3.12.

For the Hertz potential in the air medium, Π_{1z}, the first term inside the brackets can be interpreted as the particular solution or the direct line-of-sight (LOS) contribution from the dipole antenna source, i.e., a spherical wave or a direct wave originating from the source and reaching the observation point, and the second term can be interpreted as the complementary solution or a reflection term (reflection from the imperfect ground plane). This second term in the potential, Π_{1z}, is responsible for the fields of the *ground wave*, as per IEEE Standard Definitions of Terms for Radio Wave Propagation [18]. Observe in (1) that the second term of this potential is the strongest one near the surface of the earth and decays exponentially as we go away from the interface.

Similarly, the solution for Π_{2z} can be interpreted as a partial transmission of the radiating wave from medium 1 into medium 2.

With these thoughts in mind, the potential Π_{1z} can be split up into two terms, or equivalently in terms of the potentials responsible for the direct and the ground wave as

$$\Pi_{1z} = \Pi_{1z}^{direct} + \Pi_{1z}^{reflected} = P\left(g_0 + g_s\right) \tag{3.8}$$

where

$$\Pi_{1z}^{direct} = P\exp\left(-jk_1 R_1\right)/R_1 = P\,g_0 \tag{3.9}$$

$$\Pi_{1z}^{reflected} = P\int_0^\infty \left(\frac{\varepsilon\sqrt{\xi^2 - k_1^2} - \sqrt{\xi^2 - k_2^2}}{\varepsilon\sqrt{\xi^2 - k_1^2} + \sqrt{\xi^2 - k_2^2}} \right) \frac{J_0(\xi\rho)\exp\left[-\sqrt{\xi^2 - k_1^2}\,(z+z')\right]}{\sqrt{\xi^2 - k_1^2}} \xi\,d\xi = P g_s \tag{3.10}$$

The expression in (3.9) which is a spherical wave originating from the source dipole is easy to deal with. The difficult problem lies in the evaluation of $\Pi_{1z}^{reflected}$. Therefore, we choose to interpret $\Pi_{1z}^{reflected}$ as a superposition of waves represented by the semi-infinite integral which consists of the summation of the reflection of the various plane waves into which the original spherical wave was expanded. This arises from the identity between a spherical wave (left hand side) into a summation of cylindrical waves (right hand side) resulting in

$$\frac{\exp\left(-jk_1 R_2\right)}{R_2} = \int_0^\infty \frac{J_0(\xi\rho)\exp\left[-\sqrt{\xi^2 - k_1^2}\,(z+z')\right]}{\sqrt{\xi^2 - k_1^2}} \xi\,d\xi \tag{3.11}$$

and

$$R_2 = \sqrt{\rho^2 + \left(z+z'\right)^2} \tag{3.12}$$

The term under the integral sign in (3.10) can be recognized as a multiple plane wave decomposition of the spherical wave source multiplied individually with a reflection coefficient term. Upon reflection of the plane waves from the dipole source as expressed in $\Pi_{1z}^{reflected}$, the amplitude of each wave is multiplied by the reflection coefficient $R(\xi)$ due to the presence of the imperfectly conducting Earth. The complex reflection coefficient $R(\xi)$ takes into account the phase change as the wave travels from the source (x', y', z') to the air-earth boundary and then to the point of observation (x, y, z) in air. The reflection coefficient $R(\xi)$ is then defined as per (3.10) as

$$R(\xi) = \frac{\varepsilon\sqrt{\xi^2 - k_1^2} - \sqrt{\xi^2 - k_2^2}}{\varepsilon\sqrt{\xi^2 - k_1^2} + \sqrt{\xi^2 - k_2^2}} \tag{3.13}$$

where the semi-infinite integral over ξ in $\Pi_{1z}^{reflected}$ takes into account all the possible plane waves. As $\varepsilon \to \infty$, i.e., an approximation as a perfect conductor for the Earth, then g_s of (3.10) reduces to (3.11) and represents a simple spherical wave originating at the location of the image point of the source due to the presence of the Earth. The reflection coefficient takes into account the effects of the ground plane in the decomposition of the spherical wave into the cylindrical waves and sums it up as a ray originating from the image of the source dipole but multiplied by a specular reflection coefficient $\Gamma(\theta)$, where

$$\Gamma(\theta) = \frac{\varepsilon\cos\theta - \sqrt{\varepsilon - \sin^2\theta}}{\varepsilon\cos\theta + \sqrt{\varepsilon - \sin^2\theta}} \text{ as illustrated by (3A.47) and (3A.48).}$$

It is now important to point out that there are two forms of $\Pi_{1z}^{reflected}$ that may be used interchangeably as the two expressions are mathematically identical in nature (but have different asymptotic properties as we shall see).

The first form is defined as

$$\Pi_{1z}^{reflected} = P\left[\frac{\exp(-jk_1 R_2)}{R_2} - 2\int_0^\infty \frac{\sqrt{\lambda^2 - k_2^2}}{\sqrt{\lambda^2 - k_1^2}} \frac{J_0(\lambda\rho)\exp\left[-\sqrt{\lambda^2 - k_1^2}(z + z')\right]}{\varepsilon\sqrt{\lambda^2 - k_1^2} + \sqrt{\lambda^2 - k_2^2}} \lambda\, d\lambda\right]$$

$$= P\left[g_1 - g_{sV}\right]$$

(3.14)

where g_1 represents the spherical wave originating from the image of the source, located in the Earth, and g_{sV} represents the correction factor to accurately characterize the effects of the ground.

Equivalently, one can rewrite the same expression as

$$\Pi_{1z}^{reflected} = P\left[-\frac{\exp(-jk_1 R_2)}{R_2} + 2\varepsilon\int_0^\infty \frac{J_0(\xi\rho)\exp\left[-\sqrt{\xi^2 - k_1^2}(z + z')\right]}{\varepsilon\sqrt{\xi^2 - k_1^2} + \sqrt{\xi^2 - k_2^2}} \xi\, d\xi\right]$$

$$= P\left[-g_1 + G_{sV}\right]$$

(3.15)

Now the image from the source in this case, has a negative sign along with a correction factor. This expansion is useful when both the transmitter and the receiver are close to the ground, since the reflection coefficient $\Gamma(\theta)$, is -1 for grazing angle of incidence where $\theta \approx \pi/2$. Then the direct term g_0 cancels the image term g_1 leaving only the correction factor G_{sv}. This form of the Green's function is useful when both the transmitter and the receiver are located close to the ground.

In order to solve for the total field near the interface, a modified saddle point method as explained in Appendix 3B and 3D is applied to take into account the effect of the pole β_P, which is a zero associated with the denominator of G_{sv} or g_{sv} and are given by (3A.41), (3A.42) and (3A.43). This pole β_P may be located near the saddle point as illustrated in Appendix 3A. G_{sV} in (3.15) can now be expressed (as per equation 3D.2) as

$$
G_{sV} = \varepsilon \exp\left(-j\pi/4\right) \int_{\Gamma_1} \left(\frac{2k_1 \sin\beta}{\pi R_2 \sin\theta}\right)^{1/2} \frac{\exp\left[-jk_1 R_2 \cos(\beta-\theta)\right]\cos\beta}{\varepsilon \cos\beta + \sqrt{\varepsilon - \sin^2\beta}} d\beta
$$

$$
\approx \varepsilon \sqrt{\frac{4\pi k_1 j}{R_2}} \frac{\cos\theta}{\cos\theta - \frac{1}{\sqrt{\varepsilon+1}}} \frac{\sqrt{\varepsilon - \sin^2\theta} - \varepsilon\cos\theta}{\varepsilon^2 - 1} \frac{\exp\left[-jk_1 R_2 - W^2\right] erfc(jW)}{\sqrt{1 + \frac{\cos\theta}{\sqrt{\varepsilon+1}} + \frac{\sqrt{\varepsilon}\sin\theta}{\sqrt{\varepsilon+1}}}}
$$

$$
(3.16)
$$

where

$$
W^2 = -jk_1 R_2 2\sin^2\left(\frac{\theta-\beta_P}{2}\right) = -jk_1 R_2 \left[1 + \frac{\cos\theta}{\sqrt{\varepsilon+1}} - \frac{\sqrt{\varepsilon}\sin\theta}{\sqrt{\varepsilon+1}}\right] \qquad (3.17)
$$

Sommerfeld [13] calls the variable W the numerical distance. When W is small, G_{sV} is given by (as per 3D.4)

$$
G_{sV} \approx -\sqrt{2\pi k_1 j} \frac{(z+z')\exp\left[-jk_1 R_2\right]}{R_2^{1.5}}. \qquad (3.18)
$$

Equation (3.18) thus illustrates that when $\theta \approx \pi/2$ the dominant term of the potential $\Pi_{1z} \propto R_2^{-1.5}$ and therefore the leading term for the fields will be varying approximately as $\rho^{-1.5}$, if $(z + z')$ is small compared to ρ in (3.18). It is interesting to observe that Eq. (3.18) is not a function of the ground parameters, as we have seen in section 3.2 from the measured experimental data. So under these conditions the path loss exponent factor should be 3 near the ground, as the fields vary as $\rho^{-1.5}$. Also, *the reflection coefficient method described by (A3.47) and often used in the classical propagation modeling literature to characterize the radiating fields near the interface is not really applicable for antennas located close to the ground, as described in the Appendix 3A.*

Hence, the reflection coefficient method cannot be applied in a cellular communication environment where the transmitter and the receiver are located close to the Earth. This conclusion is in accordance with the case of an imperfectly conducting ground as mentioned by Stratton [19] since the reflection coefficient is approximately +1 for a perfectly conducting ground when the fields are observed far from the ground and it transforms to −1 when the fields are observed near the ground, i.e., for $\theta \approx \pi/2$. This particular variation of the field near an imperfect ground representing the Earth, will be verified by an accurate numerical analysis and using additional experimental data in the next few sections. **The other important conclusion which the theory illustrates is that the propagation path loss as a function of the horizontal distance from the antenna is independent of the material parameters of the Earth and to a large extent does not vary with frequency in the cellular band.**

The previous results are valid for small values of W. However, as W becomes large then (as per 3D.6)

$$G_{sV} \approx 2\sqrt{\varepsilon}\, \exp\left[-jk_1 R_2\right] \frac{(z+z')}{R_2^2}\left[1 - \frac{\varepsilon}{jk_1 R_2}\right] \tag{3.19}$$

It is interesting to observe that in this case the propagation path loss varies as $1/R^2$ as the various experimental data predicts as illustrated earlier. Thus for the total Hertz potential in medium 1, which is valid near the interface, for $|\varepsilon| >> 1$ and $\theta \approx \pi/2$ one obtains $W^2 \approx \dfrac{-jk_1 R_2}{2\varepsilon}$ and therefore

$$\Pi_{1z} \approx \begin{cases} P\left[\dfrac{\exp(-jk_1 R_1)}{R_1} - \dfrac{\exp(-jk_1 R_2)}{R_2} - \sqrt{j2\pi k_1}\,(z+z')\dfrac{\exp(-jk_1 R_2)}{R_2^{1.5}}\right], & W < 1 \\[4mm] P\left[\dfrac{\exp(-jk_1 R_1)}{R_1} - \dfrac{\exp(-jk_1 R_2)}{R_2} + 2\sqrt{\varepsilon}\,(z+z')\dfrac{\exp(-jk_1 R_2)}{R_2^2}\left[1 - \dfrac{\varepsilon}{jk_1 R_2}\right]\right], & W > 1 \end{cases}$$
$$\tag{3.20}$$

We have already mentioned the behavior of the Hertz potential for $W < 1$. Let us now focus our attention on the simplified expression for $W > 1$. We can see that as one moves further away from the source, (i.e., for $W > 1$) the dominant term of the field near the interface decays asymptotically as $1/R^2$ as the first two terms cancel each other. A wave with such a decay can be recognized as the popularly called Norton surface wave [18, 19]. It is important to note that the third term for $W > 1$, i.e., the Norton surface wave term, shows up only in the far field region. Note also, that it provides the so called height-gain for the transmitting and receiving antennas and that, consequently, this height-gain only applies to the intermediate and in the far field regions. Also, in the far field, the ground parameters do influence the result. Finally, it is also seen that there is a higher order term which decays as $1/R^3$, which also shows the height-gain and the influence of the ground parameters.

This analytical development using an approximate steepest descent integral can really explain the nature of variation of the fields for all the experimental data presented in Section 3.2.

However, in all fairness, there have been so many approximations made in deriving these simplified expressions as illustrated in Appendices 3D, 3B and 3C, and also the various approximations made therein to further approximate those approximations that it is very difficult to gauge what is the degree of accuracy of the expressions that we have developed so far. That is why it is extremely important to complement this approximate analysis with a more rigorous purely numerical computational methodology which involves integrating the semiinfinite Sommerfeld integrals in a very accurate way as explained in [14–17].

3.4 Accurate Numerical Evaluation of the Fields Near an Earth–Air Interface

This section provides an accurate computation of the fields in a cellular wireless environment using a macro model (as the nature of the terrain and the presence of buildings and trees are omitted as they represent second order effects as illustrated in section 3.2 of this paper, the primary being the propagation path loss due to the effect of the ground) based on an accurate computer program [15] which uses the Green's function based on the Sommerfeld and the Schelkunoff formulations. We then use these numerically accurate method to predict the propagation path loss of the fields in the experiments carried out by Okumura et al. in [6, 17].

In summary, in Section 3.3, various types of approximations have been made to compute the variation of the radiated electric field with distance to study the nature of its decay as a function of the distance along with the height of transmitters and receivers from the ground. Often, some of the approximations that have been made in the earlier sections are quite crude and the bottom line is after all these various approximations and expanding the solutions in terms of asymptotic series it is difficult to judge with confidence the validity of these conclusions from a scientific point of view, as no error bounds for the analytic solutions are readily available. At times, this type of approximate analysis has raised lots of controversies and debate [20–23]. At this point, to avoid these in-depth discussions or speculations, we will take recourse to a purely numerical methodology, which evaluates the integrals presented in (3.10) in an essentially very-accurate way. There have been several methodologies and user friendly codes that have been published in the literature to numerically compute the fields from a transmitting to a receiving antenna using the exact Sommerfeld formulation and commercial codes are available that can perform these computations on personal computers [15]. To find the total solution, one needs to solve for the correct current distribution on the transmitting antenna as it is radiating in the presence of the ground and then use that current distribution to compute the radiated fields in various regions of interest.

The electric field $\vec{E}_A$ in medium 1 due to a $\vec{z}$ directed current element I_A of length ℓ_A radiating over Earth (as shown in Figure 3.12) is given by

$$\vec{E}_A = \left[\vec{z}k^2 + \vec{\nabla} \frac{\partial}{\partial z} \right] \Pi_{1z} = \frac{-j\omega\mu_0}{4\pi} \int_{\ell A} \vec{z} I_A (g_o + g_s) dz'_A + \frac{\vec{\nabla}}{j\omega 4\pi \varepsilon_0} \frac{\partial}{\partial z} \int_{\ell A} I_A (g_o + g_s) dz'_A$$

(3.21)

where g_0 and g_1 are defined in (3.9) and (3.10), respectively and $\vec{\nabla}$ is the gradient operator. Also $\dfrac{dg_0}{dz} = -\dfrac{dg_0}{dz'_A}$ and $\dfrac{dg_s}{dz} = \dfrac{dg_s}{dz'_A}$, and it is assumed that the current goes to zero at the ends of the open wires. Then the derivative $\dfrac{\partial}{\partial z}$ operation on $(g_o + g_s)$ can be transformed to $\dfrac{\partial}{\partial z'_A}$ now operating on I_A instead, by applying integration by parts. It is important to note that only one of the derivatives can be interchanged with the integral in the second term in (3.21). Hence

$$\vec{E}_A = \frac{-j\omega\mu_0}{4\pi} \int_{\ell_A} \vec{z} I_A (g_o + g_s) dz'_A + \frac{\vec{\nabla}}{j\omega 4\pi \varepsilon_0} \int_{\ell_A} \frac{\partial I_A}{\partial z_A} (g_o - g_s) dz'_A \qquad (3.22)$$

At this point, it is important to remember that because of the nature of the singularity of the Green's function, the gradient operator on the second term cannot be interchanged with the integral sign. This is because, if the gradient operator is interchanged with the integral sign and it operates on the Green's functions directly, then it would result in a divergent integral.

The mutual impedance Z_{BA} between two $\vec{z}$ directed current elements I_A and I_B of lengths ℓ_A and ℓ_B is expressed as

$$Z_{BA} = -\int_{\ell_B} \vec{E}_A \cdot \vec{I}_B \, dz'_B$$

$$= \frac{j\omega \mu_0}{4\pi} \int_{\ell_B} I_B dz'_B \int_{\ell_A} I_A (g_o + g_s) dz'_A - \frac{1}{j\omega 4\pi \varepsilon_0} \int_{\ell_B} I_B \frac{\partial}{\partial z_B} \left[\int_{\ell_A} \frac{\partial I_A}{\partial z'_A} (g_o - g_s) dz'_A \right] dz_B$$

(3.23)

Transferring the derivative operation on I_B in the second term of the above expression and assuming I_B to go to zero at the open ends of the wires, one obtains through integration by parts,

$$Z_{BA} = \frac{j\omega\mu_0}{4\pi}\int_{\ell_B} I_B\,dz'_B\int_{\ell_A} I_A\left(g_o + g_s\right)dz'_A + \frac{1}{j\omega\,4\pi\,\varepsilon_0}\int_{\ell_B}\frac{\partial I_B}{\partial z_B}\left[\int_{\ell_A}\frac{\partial I_A}{\partial z'_A}\left(g_o - g_s\right)dz'_A\right]dz_B$$

(3.24)

In this presentation, observe that the Green's function in (3.24) is never differentiated and therefore one can obtain a stable accurate solution using this methodology. This procedure has been implemented in a general purpose computer code to analyze arbitrary shaped and oriented wire antennas over an imperfect ground plane. The methodology chosen has also been implemented in the commercially available software package titled *Analysis of Wire Antennas and Scatterers* (AWAS, V. 2) [15], that computes accurately the fields from a wire of arbitrary shape and radiating over an imperfectly conducting ground. AWAS is a complete electromagnetic field simulator for wire-like structures, utilizing the accurate Sommerfeld formulation for taking into account the effects of an imperfect ground plane.

3.5 Use of the Numerically Accurate Macro Model for Analysis of Okumura et al.'s Measurement Data

The numerically accurate evaluation of (3.10) containing the Sommerfeld integrals is now used for the analysis of the propagation data measured by Okumura et al. [6] in their classic propagation measurements in the city of Tokyo. Okumura et al. placed a transmitting antenna at different heights above the ground. We will choose a specific height of 140 m from the measurement data set. The transmitted signal was received by another vertically polarized antenna located on top of a van 3 m above the ground. The receiving antenna had a gain of 1.5 dB. The transmitting antenna was a 5 element Yagi having a gain of approximately 11 dB and radiating 150 W of power. The van was then driven in the city of Tokyo from 1 km to 100 km from the transmitting antenna. First, we consider the measurements done at 453 MHz. Since the 5-element Yagi-Uda antenna is composed of wires, in our simulations we used an optimized 5-element Yagi-Uda antenna array which had a gain of 11 dB. First the integral equation using the Green's function was used to solve for the current distribution on the transmitting antenna and then these currents were used to compute the radiating fields [14, 15]. The way we solve for the fields is using both an alternate Green's function (Schelkunoff formulation) [16, 17] and the classical Sommerfeld Green's function [13, 14] in a code that has been already developed for the analysis of radiation over imperfect ground planes based on

Sommerfeld integrals [15]. The three propagation path loss figures at three different operating frequencies are plotted in Figure 3.13.

The parameters for the urban ground were chosen as: relative permittivity of $\varepsilon_r = 4$, and $\sigma = 2 \times 10^{-4}$ mhos/m [5]. We used a simple 1 V as an excitation source at 453 MHz for the transmitting Yagi-Uda antenna. This is because we did not know how Okumura et al. matched their antennas and how it was exactly excited and with how much power. To remove this uncertainty in the radiated fields, we shifted all of our computations for the fields by a constant value in decibels (98 dB), so that the two plots (the numerically computed and the experimental data) exactly matched at a single point at 7 km from the base station antenna as shown in Figure 3.13 (a). We then overlaid the two plots of the theoretical predictions by both the new Schelkunoff formulation [17] and the classical Sommerfeld formulation [15], where 98 dB has been added to all the values computed by the theoretical macro model and then compared them with Okumura et al.'s experimental data on the same plot. The three figures in Figure 3.13 (a) show remarkable similarity between all the three plots. It is interesting to note that the simulation using the Schelkunoff Green's function provided a more stable qualitative plot of the field from 10 km to 100 km from the base station antenna than for the Sommerfeld analysis for reasons explained in [16, 17]. This is due to the Sommerfeld integral tails problem which is totally eliminated in the new formulation [16]. It is also important to point out that for both the theoretical and experimental data, the slope for the path loss exponent between 2 km and 20 km is about 30 dB per decade, which was expected from the theoretical analysis using the saddle point method in [3, 4]. The slope between 20 km and 40 km is 12 dB per octave which equals 40 dB per decade, as predicted by the Norton ground wave for the far field. This illustrates that an accurate electromagnetic macro modeling of the environment is sufficient to predict the path loss as evidenced by the comparison between theory and experiment. Such comparison between theory related to an antenna operating over an Earth and the experimental data measured in a complicated environment is meaningful because the path loss for propagation over imperfect ground is much bigger than the variations in the path loss introduced by buildings, trees and the like.

To compare with the experimental data, a power adjustment was made to the theoretical graphs to compensate for the different input power levels as carried out in Figure 3.13(a) between the theoretical predictions and the experimental data. In Figure 3.13(b), 130 dB is added to the theoretical data and in Figure 3.13(c), 125 dB is added to the theoretical data, so that the theory and the experiment matched at a single point at around a distance of 7 km from the antenna. Note that these numbers include the adjustment of the reference from 1 V/m to 1 μV/m. For the last two frequencies, the plots in Figures 3.13(b) and (c) are not a perfect match with the experimental environment because in our simulations, at these two frequencies, we used simply a half wave dipole as the transmitting antenna whereas Okumura et al. [6] used a highly directive antenna like a parabolic reflector whose dimensions were not reported in their paper.

(a)

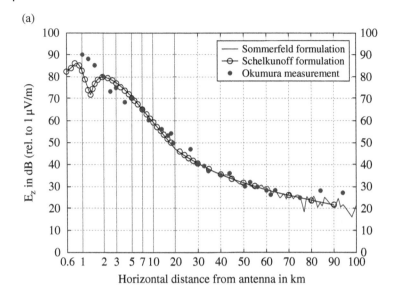

(b)

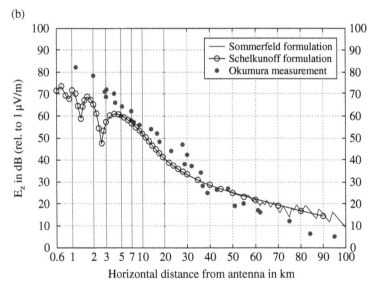

Figure 3.13 Comparison between the experimental (Okumura et al.) and theoretical predictions (Schelkunoff and Sommerfeld formulations) computed through a macro model for predicting propagation path loss in an urban environment at (a) 453, (b) 922, and (c) 1920 MHz.

(c)

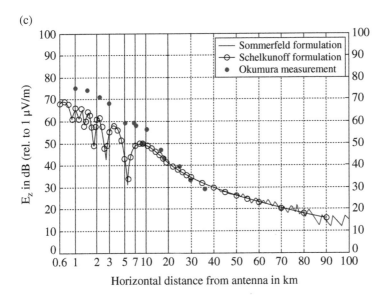

Figure 3.13 (Cont'd)

This illustrates that similar results can be obtained from all the different characterizations of the radiating fields in a cellular environment, using an approximate theoretical analysis, an accurate numerical analysis of the antennas over an imperfect ground plane using a numerical electromagnetics code and from measurements. They all agree well with each other within engineering accuracy. The three separate results obtained using different methodologies confirm that a macro modeling of the environment is sufficient to accurately predict the path loss and that a micro modeling of the environment dealing with buildings, trees and so on, may be an overkill.

Furthermore, the data presented in Figures 3.13 raises an important question and that has to do with the phenomenon of slow fading. How come neither the theoretical data obtained from a numerical code nor the experimental data show the slow fading characteristics, namely oscillatory variation of the field strength with distance. It is important to point out that all these data represent phasor quantities and therefore represent the complex phasor voltages in the frequency domain. The time variation does not show up in the phasor notation in electrical engineering as illustrated in Chapter 1! Also, in our plots of the phasor voltages, for all the cases the radiated fields are plotted in the far field region where the decay is monotonic!

To search for the origin of fading we generated an expanded version of the theoretical plots as shown in Figure 3.14. In this figure, we consider the fields radiated by a half wave dipole antenna transmitting at a frequency of 1 GHz. The height of the transmitting dipole from the ground is varied. We choose to

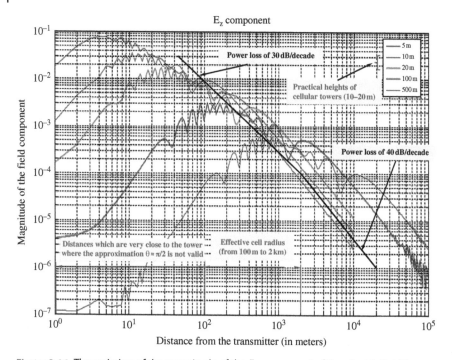

Figure 3.14 The variation of the magnitude of the E_z component of the electric field from a half-wavelength dipole as a function of distance, at an operating frequency of 1 GHz. The height of the observation point was 2 m from an urban ground with a permittivity $\varepsilon_r = 4$ and a conductivity of $\sigma = 2 \times 10^{-4}$.

analyze the radiated fields for different heights of 5 m, 10 m, 20 m, 100 m, and 500 m of the transmitting antenna from the imperfectly conducting Earth. A relative permittivity of $\varepsilon_r = 4.0$ and a conductivity of $\sigma = 2 \times 10^{-4}$ mhos/m (S/m), representing the properties of a typical urban ground as described in [2–5] was chosen for the ground plane. Electric fields were calculated at a height of 2 m from the ground. Figure 3.14 shows the variation of the received field due to different heights of the transmitter above the urban ground. We see that when the field is computed at a distance of approximately $\dfrac{8 H_{TX}\, H_{RX}}{\lambda}$ from the transmitting antenna [5], the magnitude of the field increased with the height of the transmitting antenna above the ground, verifying the dictum of *height gain* in deploying antennas as prevalent in the literature. In addition, the decay of the vertical component of the field was monotonic. However, for distances less than $\dfrac{8 H_{TX}\, H_{RX}}{\lambda}$ from the base station, the field strength actually decreased with the increase of the height. In addition, in the near field regions there appears to be an interference pattern between the direct wave and its various image components (due to the imperfect ground) leading to the phenomenon of slow fading. Beyond this interference pattern, the field decays quite smoothly with distance which generally starts around $\dfrac{4H_{TX}\,H_{RX}}{\lambda}$. The reason, as Sommerfeld explained, the ground waves from the images decay exponentially and so, far away from the antenna their contributions are negligible. Initially the decay is 30 dB per decade increasing to 40 dB per decade in the far field. Therefore, the path loss exponent inside the cell radius of a few kilometers is 3 and in the fringe regions it goes to 4 as expected. As illustrated in [5] the higher the height of the antenna from the ground the farther is the starting point of the far field.

In summary, what we are observing is that when the fields decay as $R^{-1.5}$ then we are in the intermediate region as predicted by the theory. In the fringe regions, where the fields decay as R^{-2} as predicted, then we are in the far-field region. A space division multiple access (SDMA) through beam forming is only possible in the far field region of the antennas.

Let us now focus our attention when the transmitting antenna is located at a height of 5 m from the ground. If we now look at the decay of the fields from a distance of 200 m to 400 m we observe approximately a decay of the fields in amplitude as a function of distance by a factor of 3 which is close to the expected value of $2^{1.5} = 2.83$. For the decay of the amplitudes of the fields from 4 km to 8 km, we observe that the reduction in the amplitude of the fields is about 4.1, which closely follows the $2^2 = 4$ law.

If we now consider, another case, when the transmitting antenna is located 20 m from the ground and we look at the decay of the fields from a distance of 1 km to 2 km from the transmitter, we observe approximately a decay in the amplitude of the fields by a factor of 3.19 which is close to $2^{1.5} = 2.83$. And if we focus our attention from 5 km to 10 km in the same plot, the decay of the fields is about 4 in magnitude approaching the $2^2 = 4$ law.

And finally, if we consider a transmitting dipole 500 m above the imperfectly conducting ground and observe the fields from 20 km to 40 km from the transmitting antenna we observe that the decay in the fields is approximately by a factor of 3.06 which is close to $2^{1.5} = 2.83$.

From these sparse samples it appears that once the slow fading region disappears then the fields manifest a decay of 30 dB per decade and then going to 40 dB per decade. The illustrations of Figure 3.14 show that inside a cell the wave with a decay rate of $1/R^{1.5}$ which increases to $1/R^2$ in the fringe regions.

In summary, there is a height gain in the far field of the antenna but in the near field which is of importance in cellular communication there is actually a height loss. Hence, it is proposed that a better solution will then be to deploy the transmitting antenna closer to the ground. In that case the region of the variation in the field strength would be quite small and the field strength will decay monotonically inside the remainder of the cell minimizing fading. Since there will not be any interference pattern then it is possible to reduce the transmitting power at least by a factor of 10 (say) providing a better safe and cheaper system as in most cases the tower costs more than the antenna system. However, one could deploy more base stations as the power is reduced. An interesting scenario of this can be seen perhaps in some South American cities where Wi-Fi is delivered to individual houses for free by deploying base station antennas on every lamp post.

Such a discussion of deploying the base station antenna close to the ground, is quite relevant as there is a second channel from the mobile to the base station in which there is no height gain as the mobile is near the ground and the mobile is transmitting a fraction of the power of the base station. Moreover, the antenna on the mobile can be oriented in any direction and hence one should question the validity of the state of the art rules of thumb developed for deploying base station antennas as they do not consider the environment of the second mobile channel which is quite important and which does not satisfy any of the rules of thumb associated with the base stations!

Figure 3.15 shows another simulation result for a radiating vertical half wave dipole located 20 m above different types of imperfect grounds. The fields were

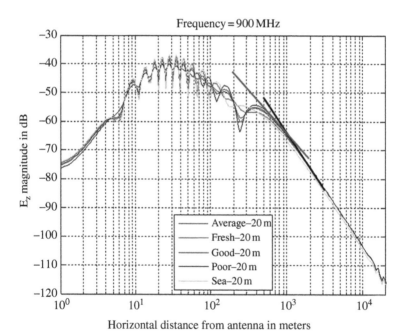

Figure 3.15 Magnitude of the z-component of the electric field in dB radiated from a half-wave vertical dipole antenna over different imperfect grounds, height of the antenna is 20 m, different curve belongs to different imperfect ground, the frequency of operation is 900 MHz.

observed 2 m above the ground. The frequency of operation was 900 MHz. The ground parameters chosen were for the poor ground with a dielectric permittivity and a conductivity of $\varepsilon_r = 4$ and $\sigma = 0.001\,S/m$, for the average ground, $\varepsilon_r = 15$ and $\sigma = 0.005\,S/m$, for the good ground $\varepsilon_r = 25$ and $\sigma = 0.02\,S/m$, for the sea water $\varepsilon_r = 81$ and $\sigma = 5\,mhos/m$, and for the fresh water $\varepsilon_r = 81$ and $\sigma = 0.01S/m$. The ground parameters were taken from [5]. By observing the plots in Figure 3.15 it is seen that within the cell the electrical properties of the ground has very little effects as for a fixed height of the transmitting antenna radiating over an imperfect ground the ground parameters do not change the nature of the distant fields whereas near the antenna the shape of the interference pattern can be slightly different. The slope of 30 dB/decade is marked by a thick purple straight line and the slope of 40 dB/decade is marked by a thick black straight line for all the data. This implies that whether we consider propagation in urban, suburban, industrial, rural or over water areas, the results for the propagation path loss should not differ too much from each other and furthermore, an electromagnetic macro model can accurately make such predictions!

Next we illustrate the nature of the radio wave propagation mechanism related to the physics over a two layer medium.

3.6 Visualization of the Propagation Mechanism

How does the signal propagate physically in a two-layer problem can be visualized by inspecting the path of each of the waves represented by the integrand in equation (3.1), for example, especially, those waves represented by the main part of the semi-infinite integral going from 0 to k_1. Those waves when they hit the imperfectly conducting ground, are multiplied by different reflection coefficients that depend on their respective angles of incidence. On reflection, the rays diverge forming a semi-infinite image. The rays will never converge to form a perfect image of the same size as the source except in case that the second medium is perfectly conducting and perfectly smooth. This fact was predicted in 1935 by Van der Pol [24] which explains the elongated image of the sun/moon on a wavy lake. This can be visualized in the picture shown in Figure 3.16.

An alternate way to look at this problem from a physics point of view is to recognize that the fields from a point source decays as $1/\rho^2$ and it produces sharp shadows as can be seen from lights coming out of a bulb in the projector. On the other hand the tube lights deployed in large rooms are rather long in size and they approximate a line source and so the shadows cast by them are diffused as the fields decay in this case as $1/\rho$. In contrast in a football field or in an operation theater the light sources are planar in nature with the philosophy that a planar source does not cast any shadows. So in the operating rooms the doctors will not see any shadows or in the football fields the players

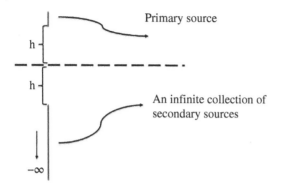

Figure 3.16 Equivalent image sources that generate the desired radiation fields at the interface. The image starts at −*h*.

hopefully will not see the image of the ball in addition. The current LED lights that are now being deployed in almost all different type of environments are trying to mimic a surface source. So now the question in cellular communication then is what type of a source produces a field that decays as $1/\rho^{1.5}$ and what are their properties? That is the question that will be addressed in this section.

Van der Pol [24] points out that the results of Sommerfeld [25] and Weyl [26] were, as a rule, not very transparent on account of the fact that the approximations or developments used were more of a mathematical in nature and a physical interpretation of them have been missing. This lead Van der Pol to recast the same problem without any approximation whatsoever to lead to the form of a solution in terms of a simple space integral which allows a direct physical interpretation. The integration in this new form extends over the part of space occupied by the second medium below the geometrical image. This scenario is seen in Figures 3.16 and 3.17, which is similar to Figure 3.12 in many respects. It is shown that the fields in the first medium where the dipole is located, apart from the direct radiation from an elementary dipole, can be described by a secondary wave originating in the integration space extending from its image from −*h* to −∞. In other words, the image of a source over an imperfect ground consists of a line source starting from its image starting at −*h* and continuing to −∞ as shown in Figure 3.16. The amplitude of the wave from the image is determined by the amplitude of the primary wave. It can be considered to spread from the geometrical image of the point source with the propagation constant and absorption of the second medium. The higher the conductivity of this second medium, the more the primary wave is concentrated near the image of the point source until, for infinite conductivity, it is wholly concentrated at the image itself.

This result, that the fields produced by a dipole over an imperfect ground plane are due to the direct contribution from the source plus the effect of the ground represented by an equivalent image which consists of a line source that extends from from −*h* to −∞ has been used by others researchers. In fact, the

Figure 3.17 Elongated image of the sun over a rippled lake produced by the wind flow.
(http://fineartamerica.com/featured/alexandria-bay-sunset-steve-ohlsen.html)

same physical picture is available in Booker and Clemmow [27]. Sommerfeld also writes a similar expression [13, Eq. (10d) on p.250]. The detailed derivation is available in Van der Pol's work [24] and one obtains for the Hertz potential due to a normalized point source located at a height z_a, while the observation point is located at z_b, the following expression

$$\Pi_{1z} \approx \frac{\exp(-jk_1 R_1)}{R_1} + \frac{\exp(-jk_1 R_2)}{R_2} - \frac{2jk_1^2}{k_2}\exp\left[-a(z_a + z_b)\right]$$
$$\int_{\zeta=z_a+z_b}^{\zeta=z_a+z_b+\varepsilon\times\infty} \frac{\exp(-jk_1\sqrt{\rho^2+\zeta^2})}{\sqrt{\rho^2+\zeta^2}}\exp(a\zeta)d\zeta \tag{3.25}$$

where $a = -\dfrac{jk_1^2}{k_2}$ and $\zeta = z_a + z_b + \varepsilon z$. As stated by Van der Pol, the third term in (3.25) can be interpreted as a wave spreading from the geometrical image of the point source while all points of the second medium below the level of the geometrical image, apparently send secondary waves to the observer. When $\varepsilon \gg 1$, the distance from an arbitrary point in the second medium to the observer, as given in the exponent of the third term, contains z multiplied by ε, i.e., to the observer the vertical part of the distance below the image of the point source is multiplied by ε. The wave originating from the image will

Figure 3.18 Waves on a wet ground due to the partial reflectivity of the surface. The figures give an impression on the physical propagation mechanism to be expected in cellular environments at frequencies where a wet earth represents a complex impedance surface.

therefore be observed as elongated vertically, like the vertically very elongated image of the sun over a wind rippled lake, as shown in Figure 3.17 [17, 24].

Next, consider a rainy night, when the ground is wet representing a partially reflective surface, and then try to observe how the city lights are stretched along the ground by waves propagating over the surface. Compare such a view as shown in Figure 3.18 to the situation as described by Van der Pol in (3.25). Imagine that the existence of the traffic light is the information to be transmitted by the source, and now notice how a wet ground helps to transfer the information to a point where there is no line-of-sight path. In fact, in cellular systems, the ground is always an imperfectly reflecting surface, and the way light propagates in Figure 3.18 is what we should expect as a propagation mechanism in a cellular system. In other words, the transmitting antennas (usually down tilted to the ground) excite a radiating field in the cell, which in the intermediate region behaves as a radiating field from a line source and in the far field region behaves as a Norton surface wave. This wave then represents one of the main dominant means by which the base station antenna communicates with the mobile device. This mechanism of propagation is shown in Figures 3.16 to 3.18. Hence, the stretching of the lights in Figures 3.17 and 3.18 essentially represents the image of the primary source.

Figures 3.16–3.18 thus give us a physical insight into how waves propagate in mobile communications systems and thus illustrates why smart antennas and beamforming are not very successful in cellular communications till now despite all of the research efforts done in those fields. If the real scenario of propagation is something similar to what we see in Figures 3.16–3.18, then we should change our outlook at the implementation of multiple antennas and adaptive arrays in cellular systems. The important point is that the effect of the ground plays the dominant role. This equivalent line source formed from the images generates a field in the air which decays as 30 dB per decade. The decay of the fields as 30 dB per decade with distance is generated in the intermediate region approximately where the slow fading region ends which is roughly at a distance of about $\dfrac{4H_{TX}\, H_{RX}}{\lambda}$ [5] from the base station antenna. Here H_{TX} and H_{RX} represent the height of the transmitting and the receiving antennas over the ground and λ is the wavelength of operation.

In summary it is seen that a field decaying as $1/\rho^{1.5}$ is generated by a semi-infinite line source.

3.7 A Note on the Conventional Propagation Models

Typically, in the current state of the art, the drive test data is fit in a least squares fashion to a statistical model known as the log-normal model. Sometimes, empirical-based models are given based on the path loss exponent values calculated using the log-normal model. The fact of the matter is that current propagation modeling tools cannot capture the fundamental physics, namely inside a cell the path loss is typically 30 dB/decade [2] or equivalently the electric field strength varies as $(1/\rho^{1.5})$ with the distance ρ from the antenna, and outside the cell the propagation path loss is approximately 40 dB/decade or equivalently the electric field strength varies as $(1/\rho^2)$ as a consequence of the addition of the direct wave from the source and the infinite image that has been discussed. The various statistical based models as we will see in this section really do not capture the basic physics.

Conventionally, in wireless communications text books, such as [28], this problem of propagation modeling is tackled first by explaining the two-ray model over a flat perfectly conducting earth [29–31]. In the two-ray model – the incident and the reflected ray – the reflection coefficient is taken to be −1 and θ (the angle of incidence – as seen in Figure 3.12) is always $\pi/2$ (i.e., perfect reflecting earth is considered). However, the two ray model illustrates that the intermediate path loss to be 20 dB per decade instead of 30 dB per decade as we have presented. After explaining the two-ray model, empirical models are usually presented, such as the well-known Okumura-Hata model [31]. Although empirical models have been extensively applied with good results, they suffer from some disadvantages. The main disadvantage is that empirical models provide no physical insight into the mechanism by which propagation occurs. In addition, these models are limited to the specific environments and parameters used in the measurements.

Next, we provide a brief description of the Hata model (one of the examples for the empirical models) and the two ray model. Then we can make a comparison between our proposed macro model and these two models.

Hata Model is an empirical formulation of the graphical path loss data provided by Okumura et al.'s experimental measurements. The formula for the median path loss in urban areas is given by

$$L(dB) = 69.55 + 26.16 \log f_c - 13.82 \log h_{te} - a(h_{re}) + \left(44.9 - 6.55 \log h_{te}\right) \log d$$

$$(3.26)$$

where f_c is the frequency in MHz and varies from 150 to 1500 MHz, h_{te} and h_{re} are the effective height of the base station and the mobile antennas (in meters), respectively, d is the distance from the base station to the mobile antenna in kilo meters, typically a value assumed to be in the far field of the radiating antenna, and $a(h_{re})$ is the correction factor for the effective antenna height of the mobile which is a function of the size of the area of coverage. For small to medium-sized cities, the mobile antenna correction factor is given by

$$a(h_{re}) = (1.1 \log f_c - 0.7) h_{re} - (1.56 \log f_c - 0.8) \, dB \qquad (3.27)$$

For a large city, it is given by

$$a(h_{re}) = 8.29 (\log 1.54 h_{re})^2 - 1.1 \, dB \quad \text{for} \quad f_c \le 300 \text{ MHz} \qquad (3.28)$$

$$a(h_{re}) = 3.2 (\log 11.75 h_{re})^2 - 4.97 \, dB \quad \text{for} \quad f_c \le 300 \text{ MHz} \qquad (3.29)$$

The path loss in a suburban area is obtained by modifying the standard Hata formula as

$$L(dB) = L(urban) - 2\left[\log(f_c/28)\right]^2 - 5.4 \qquad (3.30)$$

The path loss in open rural areas is expressed through

$$L(dB) = L(urban) - 4.78(\log f_c)^2 - 18.33 \log f_c - 35.94 \qquad (3.31)$$

If we are interested in calculating the received power instead of the path loss, then we can use the following formula to calculate the received power in dBm knowing the path loss $L(dB)$ and the transmitted power P_t (dBm) and the gain G_t (dB) as:

$$P_r(d) = P_t + G_t - L \qquad (3.32)$$

This model is quite suitable for large cell mobile systems, but not for personal communications systems, which cover a circular area of approximately a few kilometers in radius [28–31].

The two ray model is based on ray tracing techniques. In the two ray model, we consider only the direct ray and the ray reflected off the surface of the earth by considering the transmit-receive system as point sources. We then consider the reflection coefficient to be −1 (perfect conductor). The formula for the two ray model is given by:

$$P_r(d) = P_t G_t \left[\frac{\lambda}{4\pi} \right]^2 \left| \frac{1}{\sqrt{(h_t - h_r)^2 + d^2}} - \frac{e^{-j2\pi\left(\sqrt{(h_t + h_r)^2 + d^2} - \sqrt{(h_t - h_r)^2 + d^2}\right)}}{\sqrt{(h_t + h_r)^2 + d^2}} \right|^2 \tag{3.33}$$

where P_t is the transmitted power, G_t is the transmitting antenna gain, λ is the wavelength, h_t is the transmitting antenna height in meters, h_r is the receiver antenna height in meters and d is the horizontal distance from the transmitting antenna in meters.

Figure 3.19 plots the results for the electromagnetic macro model generated by AWAS for a vertical half-wave dipole radiating over an urban ground located at a height of 20 m above the ground assuming that the receiving antenna is located at 1.5 m above the ground. The frequency is 900 MHz. Included in Figure 3.19 is also the results for the two ray model and that from the Hata

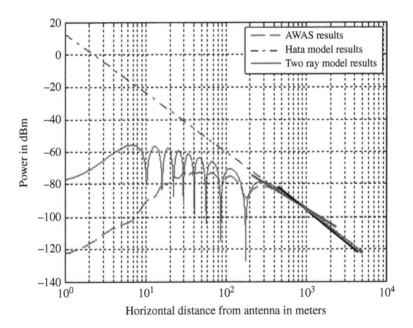

Figure 3.19 The dashed red curve is the received power from a vertical half-wave dipole at 20 m above an urban ground using the electromagnetic analysis code AWAS. The dashed black curve is the results from the Hata model for the same setup as used in the AWAS simulation. The blue curve is the results for the two-ray model for the same setup as used in the AWAS simulation.

model for the same transmitting and receiving antenna locations. One can clearly see that neither the two ray model nor the Hata model predicts a decay of 30 dB/decade and so these models do not reflect the correct physics. Furthermore, the nulls in the two ray model are much sharper than the AWAS [15] results, and this is due to the assumption of a perfect ground used in the two ray model. In fact, if we assume a vertical half-wave dipole over a perfect ground using the code AWAS to generate the results we will get a slope of 20 dB/decade in the far field and not the usual 40 dB/decade as the two ray model suggests!! Again this is because in the two ray model one assumes that the reflection coefficient is −1 while the truth is that the reflection coefficient is +1 for a vertical dipole on top of a perfect ground and it is −1 for a horizontal dipole over perfect ground. In Figure 3.19, the purple thick line represents the slope of 30 dB/decade and the black thick line represents the slope of 40 dB/decade.

So to find more satisfactory models, researchers usually follow one of two paths. Either they choose more sophisticated physical models which include other propagation mechanisms such as diffraction, scattering and ray tracing [7, 29–31], or they delve into statistical modeling [28]. We quote a very interesting conclusion from [7]: "*Although the plane earth model has a path loss exponent close to that observed in actual measurements (i.e., 4), the simple physical situation it describes is rarely applicable in practice. The mobile is always almost operated (at least in macrocells) in situations where it does not have a line-of-sight path to either the base station or to the ground reflection **point,** so the two-ray situation on which the plane earth model relies is hardly ever applicable. To find a more satisfactory physical propagation model, we examine diffraction as a potential mechanism*".

Instead of examining diffraction as a potential mechanism or going to statistical modeling, the work presented in this chapter gives a rigorous physics based mathematical solution using the exact Sommerfeld formulation of the two-ray model but with imperfectly reflecting earth taken into consideration. This approach directly implies that the physical model of propagation in the cellular environment described above is the radiation field associated with a line source. Namely, the power decreases with the distance from the transmitter by 30 dB per decade for most of the practical area within a typical cell, as the antennas are located quite high from the ground. Then the multipath fading and shadowing due to buildings and large obstacles appear as variations around the 30 dB per decade slope line, as seen in the various measured data described in section 3.2. Almost none of the existing physical models take the fields produced by a line source into consideration such as the dielectric canyon model, flat edge model and sophisticated ray tracing models [7, 29–31].

In summary, the ray theory can never predict a decay of the fields as $1/R^{1.5}$. The analysis of Van der Pol [24] presented here provides a physical picture of how the wave, which is not a surface wave, propagates over an imperfect ground from the base station antenna to the mobile device. In short, besides the direct ray from the source, there are the fields from a line source generated by the image of the original source over the imperfectly conducting ground which can be easily visualized in Figures 3.16 and 3.18. In addition, the reflection coefficient method is not applicable when the receiving antenna is close to the ground as illustrated in Appendix 3A.

3.8 Refinement of the Macro Model to Take Transmitting Antenna's Electronic and Mechanical Tilt into Account

To illustrate the effect of the tilt (mechanical or electronic) of the transmitting antenna on the propagation path loss in the cellular networks, simulations were performed using the macro model described and analyzed using AWAS [15]. The objective here is to study whether the mechanical or the electronic tilt of the base station antenna towards the ground minimizes the effects of fading. However, if this phenomenon of fading is due to the interference between the direct wave and the reflected wave from the ground then the tilt of the beam towards the ground will definitely increase the signal levels but will not diminish the effects of the interference between the direct wave and the reflected wave. On the contrary it may increase the dynamic range between the maximum and the minimum of the radiated fields. This is illustaarted next.

We chose the Xpol 45° (cross polarized) model for the transmitting antenna as shown in Figure 3.20. This model actually simulates a practical scenario of a transmitting antenna in cellular networks [4]. We used five half-wave dipole antennas to increase the gain to 8 dBi so as to match the actual gain recorded for the base station antennas under study. In our simulation, we did not include the cross dipoles in the model which are used as separate receiving antennas in the cellular base station antenna panel. Our Xpol 45° (cross polarized) model

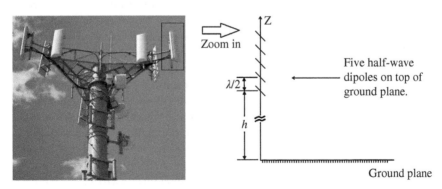

Figure 3.20 Xpol 45° tilted antenna model.

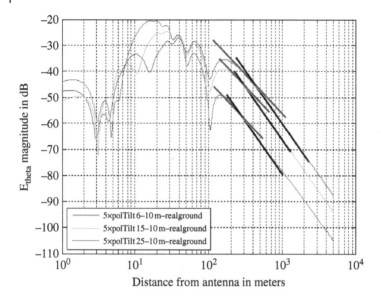

Figure 3.21 Magnitude of the theta component of the electric field in dB radiated from five half-wave dipoles located at a height 10 m above an urban ground.

consists of five half-wave dipoles (the centers of which are separated by half wavelength along the z-axis, the closest dipole to the imperfect ground is h meters away). Each dipole is mechanically tilted 45° from the z-axis. We assumed some progressive phase shifts between the excitations of the five half-wave dipoles to control the tilt of the main beam electronically.

Our first simulation was performed using AWAS. We used the previously described Xpol 45° model where we set h equal to 10. The imperfect ground is assumed to be urban with $\varepsilon_r = 4$ and $\sigma = 0.0002$ mhos/m [2]. The observation points are 1.5 m above the Earth. The frequency of operation is 1 GHz. Figure 3.21 plots the signal strength as a function of distance from the base station in log-scale for various electronic tilts of the antenna main beam towards the ground. This is in addition to the mechanical 45° tilt of each dipole from the z-axis. In Figures 3.21–3.22, the red curve represents a 25° electronic tilt of the main beam towards the ground. For the green curve, there is a 15° electronic tilt of the main beam towards the ground.

The blue curve represents a 6° electronic tilt (the depression angle) of the main beam towards the ground. It is seen that near the antenna the greater the tilt of the main beam towards the ground, the higher the field strength is inside the cell. In addition, the interference pattern between the direct space wave and the fields form the image produced by the imperfect ground is roughly the same for the various antenna tilts. In Figures 3.21–3.22, the purple line represents a slope of −30 dB/decade whereas the black line represents a slope of −40 dB/decade. If the case under study was vertically oriented

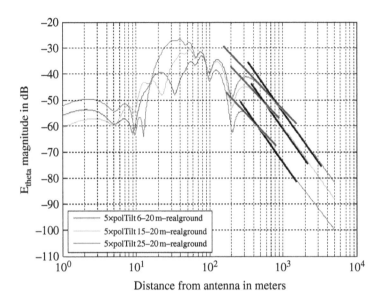

Figure 3.22 Magnitude of the theta component of the electric field in dB radiated from five half-wave dipoles at a height of 20 m above an urban ground.

antennas then we expect that the oscillations in the field strength plots disappear after a distance of $\dfrac{4H_{TR}H_{RX}}{\lambda} = \dfrac{4 \times 11 \times 1.5}{0.3} = 220$ m (1 m added to the height of 10 m to take into account the size of the array) [3]. However, in contrast to the vertically oriented antennas, the slope of approximately -30 dB/decade (-9 dB/octave) starts slightly closer than 220 m when the tilt of the main beam of the antenna is greater toward the ground. Similarly, the monotonic slope of -40 dB/decade (-12 dB/octave) starts slightly closer than $\dfrac{8H_{TR}H_{RX}}{\lambda} = \dfrac{8 \times 11 \times 1.5}{0.3} = 440$ m [3] when the tilt of the antenna is greater towards the ground. Eventually all the curves settle down to -40 dB/decade as expected for a vertically oriented dipole above an imperfect ground plane. Tilting the main beam electronically towards the ground decreases the distance where the break point occurs with respect to a strictly vertically oriented antenna over the ground.

As a second example, consider the Xpol 45° model where we set h equals 20 m with the rest of the parameters being the same as in the previous example. Figure 3.22 shows the results from AWAS. It is seen that near the antenna the greater the tilt of the main beam towards the ground the higher the field strength inside the cell. Furthermore, the interference pattern between the direct space wave and the fields form the images produced by the imperfect ground is roughly the same for various antenna tilts. Also, we can see that the oscillations in the field plots disappear before a distance of $\frac{4H_{TR}H_{RX}}{\lambda} = \frac{4 \times 21 \times 1.5}{0.3} = 420\,\mathrm{m}$ [3] and the monotonic slope of –40 dB/decade (–12 dB/octave) starts before the distance of $\frac{8H_{TR}H_{RX}}{\lambda} = \frac{8 \times 21 \times 1.5}{0.3} = 840\,\mathrm{m}$ [3]. From Figure 3.22, we again conclude that tilting the main beam electronically towards the ground decreases the distance where the break point occurs with respect to a strictly vertically oriented antenna over the ground. So, the previously presented macro model can thus be refined to take into account the antenna tilt factor and evaluate its impact in the cell area. Hence, a slight modification for the values may be required to deal with an antenna mechanically rotated with its main beam electronically tilted towards the ground.

The vertical tilt of the main beam and the physical 45° tilt of the antenna from the *z*-axis thus slightly modify the previously developed macro model for a straight wire to predict the propagation path loss. But this modification is very slight and the previously established rule of thumb is quite appropriate. In addition, the electronic tilt does increase the signal level near the transmitter but the effects of inference between the directed wave and the reflected wave form the ground which characterizes the principles of slow fading is enhanced thereby confirming that the phenomenon of slow fading is due to the effects of the ground and also as Figures 3.21 and 3.22 illustrate that the nulls are also deeper when compared with Figures 3.14 and 3.15 and so by the tilt of the antenna towards the ground the effects of fading is not minimized!

3.9 Refinement of the Data Collection Mechanism and its Interpretation Through the Definition of the Proper Route

The novelty of using a macro model based on the underlying physics lies in its ability to match the simulation and measurement results without any statistical or empirical curve fitting or an adhoc choice of a *reference distance* [2]. Typically for good curve fitting this *reference distance* should be located in the far field of the base station antenna. However, in most models this *reference distance* is chosen to be in the near field resulting in questionable analysis and conclusions [2].

In addition, a new concept called *proper route* is now introduced to enhance the quality of the measured data. It is demonstrated that it is advisable not to combine all the data together in the measurements along different routes and then take an average as this increases the uncertainty in the model. This is illustrated by monitoring the signal levels from the cellular base stations in western India and Sri Lanka and comparing the observed results with the simulated results. The goal here is to illustrate that these numerical simulation tools can accurately predict the propagation path loss in a cellular environment without tweaking some non-physical models based on statistical modeling or heuristic assumptions as the major component of the loss is due to propagation over an imperfect ground and the effects of buildings, trees and so on have second order effects.

In the present study, an extensive experimental campaign had been conducted involving several base stations operating in the 900 MHz band. The nature of the environment for the base stations varied between rural, urban, suburban, dense urban and industrial areas. The transmitting power of all the base stations was around 40 dBm. The transmitting gain was 8 dBi for all the base stations. The receiving antenna was located 1.5 m above the ground. The transmitting antenna is made by Kathrein and the receiving antenna is an omnidirectional external WHIP antenna. The receiver used was a Coyote dual modular receiver from BVS (Berkeley Varitronics Systems) which was used in drive-in tools for field trials. The sensitivity of the receiver was −118 dBm to −30 dBm ±1 dB at 10 KHz IF bandwidth. The position of the mobile is determined from the Global Positioning System (GPS) receiver and this information with the co-ordinates of the base station was utilized to deduce the distance traveled by the mobile from the base station. The average received power values were recorded in dBm. In general, there is a variety of receiver models which can be used for collecting drive test data. The receivers can be set to take samples every S seconds; where S is the sampling interval in seconds. In this case, the receiver is time triggered. In some receiver models, one can set the receiver to take samples every D distance; where D is the sampling interval in centimeters. In this case, the receiver is location triggered. After setting the device to the desired option, the vehicle is driven in the cellular coverage area while measuring and recording the average received power. In addition, GPS data was also recorded; this makes it easy to determine the separation distance from the base station antenna associated with each received power measurement. In our measured drive test, the data was recorded with 512 samples in one second on a laptop computer.

At the first step, our goal is to isolate, as much as possible, the small-scale effects from the large-scale effects both of which are included in the current drive-test data. In most cases, the driving routes are not fixed along a particular direction that goes away from the transmitter as it is restricted by the available streets. This means that the average power at the same separation distance might be repeatedly recorded at different times. Therefore, using the entire data set will increase the variance of the results. Here we introduce the *proper route* concept to limit the small scale effects which are included in our drive test data. We will use the term "route index" which means the order of the sample data as it is taken from the processing computer. *The proper route* is defined as the driving route which is fixed along a radial direction that goes away from the transmitter and at the same time covers most of the available separation distances from the base station antenna.

As a first example consider Figure 3.23 which represents the drive test data taken in the city of Galewela in Sri Lanka where the environment is considered to be rural. Figure 3.23(a) and (b) contains more than 160,000 spatial sample measurements for the received power as a function of distance. It has two plots. One plot is related to the horizontal distance versus the route index as it is taken from the data file which comes out directly from the processing computer of the raw data. The second plot is the data for the measured received power versus the route index. Also in addition 512 samples per second are taken for the measured received power, which implies that multiple samples are recorded at each spatial location. From a theoretical point of view it is difficult to see why there should be any variation of the measured received power at the same location for different time samples as everything is stationary! To limit the effect of the shadowing variations, we introduce the proper route concept. We chose the measured received power data which has index number between 81000 – 93000 as our proper route. This portion of the measured data is taken from a driving route which is fixed along a radial direction that goes away from the transmitter as shown in Figure 3.23. Also, it covers most of the available separation distances from the base station antenna. Then, we plot this portion of the measured data versus separation distance in a log-scale as shown in Figures 3.24 (a) and (b).

Figure 3.24 (a) represents the plotted data at each spatial sample point for this radial route. Figure 3.24 (b) represents the data after averaging it on this

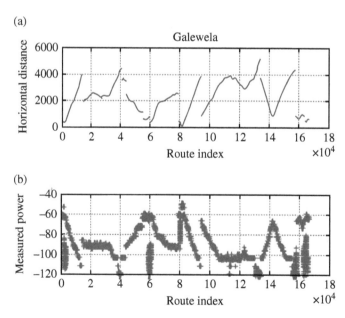

Figure 3.23 Horizontal distance in meters versus (a) route index and (b) measured power versus route index for Galewela, Sri Lanka, where the environment is considered to be rural.

(a)

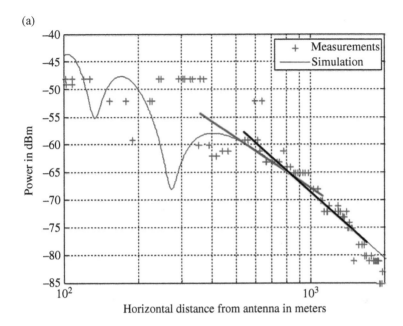

(b)

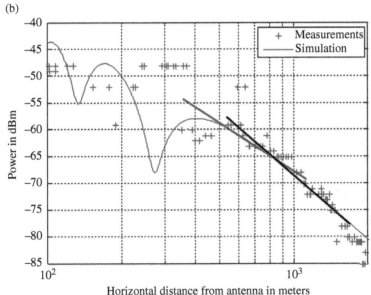

Figure 3.24 Comparison between the drive-test data taken from Galewela and the macro modeling-based simulation results. (a) The measured data along the selected radial route and (b) The measured data along the selected radial route after averaging it at each spatial sample point.

route at each spatial sample. The slope of 40 dB/decade is marked by the black line and the 30 dB/decade by the purple line on these figures. In the simulation, we assumed the base station antenna consisted of five half-wave dipoles (the centers are separated half wavelength along the z-axis, the closest dipole to the imperfect ground was 30 m away). Each dipole was mechanically tilted $45°$ upwards with respect to the ground plane. The imperfect ground was assumed to have $\varepsilon_r = 4$ and $\sigma = 0.0002$ mhos/m. The electronic tilt of the main beam was $15°$ towards the ground. The receiving antenna was located at 1.5 m above the imperfect ground. The frequency of operation was 900 MHz. As seen in Figures 3.24 (a) and (b), the plotted data matches the simulation result obtained using AWAS. Our simulation tool assumes flat imperfect and infinite ground plane. The slope of -30 dB/decade starts slightly before

$$\frac{4H_{TR}H_{RX}}{\lambda} = \frac{4 \times 31 \times 1.5}{1/3} = 558 \text{ m}$$ and the slope of -40 dB/decade starts much

before $\dfrac{8H_{TR}H_{RX}}{\lambda} = \dfrac{8 \times 31 \times 1.5}{1/3} = 1116$ m as expected. For this data set it is seen

that the transition region from -30 dB/decade to -40 dB/decade occurs over a larger distance. However, fitting a straight line to the entire drive test data taken from the base station in Galewela will result in the slope is approximately -51 dB/decade. This is quite large from a physics point of view. Hence selecting a particular radial route and considering the measured data for that route only minimize the small scale variation of the data and provide a result that is expected from a physics point of view.

As a second example, we consider the drive test data in the city of Udubeddawa in Sri Lanka where the environment is considered to be rural. Figure 3.25 contains more than 160,000 spatial sample measurements for the received power as a function of distance. For this data set, to minimize the shadowing variations, we considered the proper route concept so we used the measured power data that has the index number between 1-17000. This portion of the measured data is taken from a drive along a route which is fixed along a radial direction and it goes away from the transmitter as shown in Figure 3.25. Also, it covers most of the available separation distances from the base station antenna. This portion of the measured data versus distance in a log-scale is plotted in Figures 3.26 (a) and (b). Figure 3.26 (a) represents the plotted data along the selected radial route. Figure 3.26 (b) represents the plotted data after averaging the entire data set at each spatial sample. In these figures, the slope of -40 dB/ decade is marked by the black line and the -30 dB/decade by the purple line. In the simulation, we assumed the scenario to be of five half-wave dipoles (the centers are separated half wavelength along the z-axis, the closest dipole to the imperfect ground was 28 m away). Each dipole was mechanically rotated $45°$ upwards with respect to the ground plane. The imperfect ground was assumed to have $\varepsilon_r = 4$ and $\sigma = 0.0002$ mhos/m. The frequency of operation was 900 MHz. The electronic tilt of the main beam was $6°$ towards the ground.

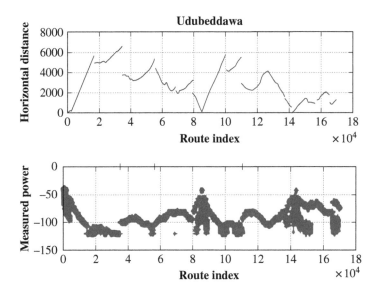

Figure 3.25 Horizontal distance versus route index and measured power versus route index for Udubeddawa where the environment is considered to be rural.

The observation points were at 1.5 m above the ground. As seen from Figures 3.26 (a) and (b), the plotted data approximately matches the simulation result obtained using AWAS. The slope of 30 dB/decade starts before $\frac{4H_{TR}H_{RX}}{\lambda} = \frac{4 \times 29 \times 1.5}{1/3} = 522$ m and the slope of 40 dB/decade starts slightly before $\frac{8H_{TR}H_{RX}}{\lambda} = \frac{8 \times 29 \times 1.5}{1/3} = 1044$ m. However, fitting a straight line to the entire drive test data taken from the base station in Udubeddawa will result in the slope of approximately 40 dB/decade, which is the correct expected result in the far field. For this example averaging of the data at each spatial sample provides a finer structure for the propagation path loss which agrees with the macro model.

It is seen that the shadow variations are consistent along the entire route. This implies that if we look at the data along a particular radial route, the processed data will have less variation, leading to more accurate predictions which are also consistent with the data generated from the electromagnetic macro models [2]. Therefore, to limit the effect of these variations, either one can take the data over a single radial route (the proper route) without any averaging or take the data over a single radial route (the proper route) and then average the data at each spatial location. Such a procedure actually leads to a good match between the measured data and the data generated from the electromagnetic macro model using AWAS.

(a)

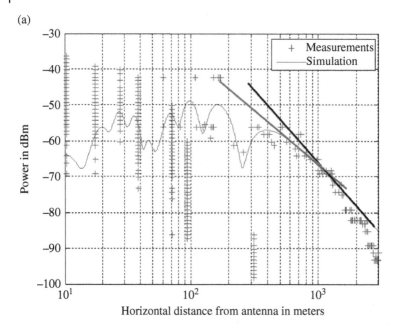

(b)

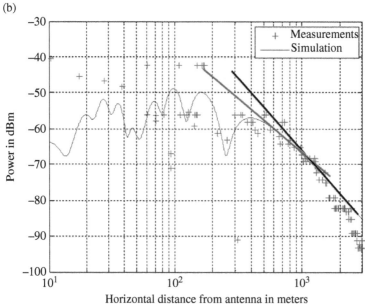

Figure 3.26 Comparison between the drive-test data taken from Udubeddawa and the macro modeling-based simulation results. (a) The measured data along the selected radial route and (b) The measured data along the selected radial route after averaging it at each spatial sample point.

3.10 Lessons Learnt: Possible Elimination of Slow Fading and a Better Way to Deploy Base Station Antennas

It is observed that the major component of the path loss is due to the radio wave propagation over an imperfect ground and that the buildings, trees, and the like provide a second order effect. It is also seen that for an antenna operating in free space, the free space propagation path loss is 20 dB per decade as the fields decay with the radial distance ρ, as $1/\rho$. However, when the antenna is radiating in the presence of an imperfect ground, namely the Earth, the path loss inside the cell is increased to 30 dB per decade as the fields decay in the near fields of the antenna as $1/\rho^{1.5}$ and outside the cell in the far field the path loss is 40 dB per decade as the fields decay as $1/\rho^2$. In addition, one observes that the large scale fading is due to the interference between the direct wave from the antenna and the reflected wave from the ground. The conclusion is that the presence of the Earth increases the propagation path loss. Under the following circumstances, then a good way to deploy the base station antenna will be to bring it close to the ground and at the same time slightly tilt it up, towards the sky so that the effects of the image will be to generate waves which will go deeper into the ground and that very little signal from the image will appear in the upper layer, which is air. If such is the nature of the deployment then the propagation path loss will approach more close to 20 dB per decade of distance instead of the 30 dB per decade. It is also seen from the Figures 3.13 that slow fading exists only in the near field of the radiating antenna and in the far field the slow fading phenomenon disappears.

Figure 3.27 plots the far field radiation pattern for a half wave dipole located at different heights above an urban ground. The operating frequency is 1 GHz. Figure 3.27 contains six different plots which correspond to locations of the base station antenna at different heights of 0.01, 0.25, 0.5, 0.75, 1.0, 2.0 m from the Earth. The end of the radiating dipole is located at that height. Each of these figures contain three different plots, namely for a strictly vertical oriented dipole and then the dipole is tilted 5° and 10° with respect to the vertical z-axis. The tilt is in the *yoz* plane is in the counterclockwise direction.

It is interesting to observe that the front and the back lobes become different in magnitudes with the different angles of tilt and as expected the field is greater along the direction of (φ = +90°), along the obtuse angle made by the dipole with the Earth. The difference between the front and the back lobes located along (φ = +90° and φ = −90°) is larger when the dipole is closer to the ground and this difference diminishes as the dipole is more elevated from the Earth. The other point to note is that the tangential component of the field is the largest along the ground when the dipole is located over a perfect ground plane but it becomes zero when the perfectly cconducting ground is replaced

(a)

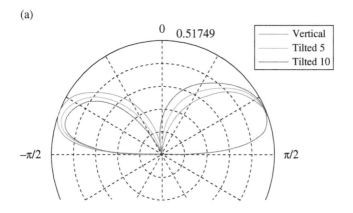

(b)

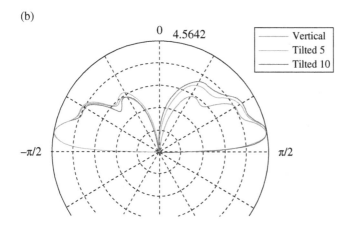

(c)

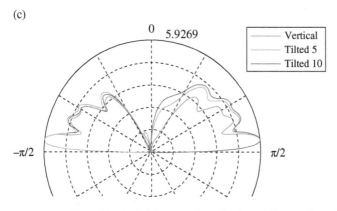

Figure 3.27 Plot of the far-field pattern for a dipole at different tilt angles and located at different heights above earth. The bottom of the dipole is located from the ground at (a) 0.01, (b) 0.25, (c) 0.5, (d) 0.75, (e) 1.0, and (f) 2.0 m.

(d)

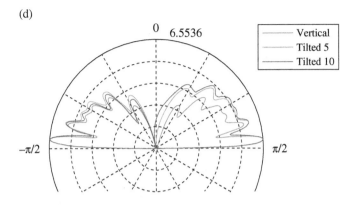

(e)

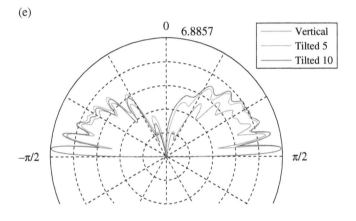

(f)

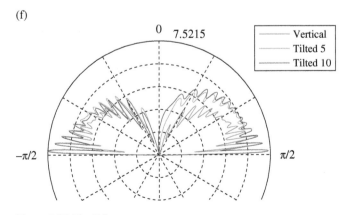

Figure 3.27 (Cont'd)

by an imperfectly conducting ground as the losses during propagation need to be supplied for by the propagating wave.

Two other interesting point to observe is that the field pattern becomes highly lobed when the antenna is located high above the ground and this is due to the inteference pattern produced by the actual antenna and its images below the ground as shown in Figure 3.12. Also the peak of the lobe is shifted upwards when the antenna is closer to the ground. In the far field the gain of the antenna system increases with the height of the antenna above earth and this is known as the *height–gain* in the literature.

However, in a cellular wireless communication system one is interested in the fields near the antenna and not what happens in the far field, as a cell typically does not exceed beyond say 5 km. Therefore the goal is to know what happens in the near fields and how does it vary with the radial distance along with the variation of the height of the antenna above the ground. It is really not of primary interest in this case, as to what the far radiation fields looks like as seen in Figure 3.27.

Table 3.1 lists the values of the electric field for a dipole antenna radiating over a perfect ground plane and over an urban ground. The dipole is half a wavelength long operating at 1 GHz and is of 1 mm radius. The electric field strengths are all normalized so that the total power input to the dipole is always 1 W. Also the dipole is conjugately matched so that it radiates the maximum power density. The received fields are collected by a receiving dipole located at

Table 3.1 Values for the received power obtained at a stationary half-wave receiving dipole and radiated by a half wavelength vertical half-wave dipole located at different radial distances and for different heights over the ground.

Transmit antenna orientation	Nature of ground Radial distance PEC 100 m Received power (µW)	PEC 1000 m Received power (µW)	Urban 100 m Received power (µW)	Urban 1000 m Received power (pW)
$\lambda/2$, 20 m, vertical	0.034	0.0014	0.110	1564.1
$\lambda/2$, 20 m, tilted down 11°	0.036	0.0013	0.124	1506.2
$\lambda/2$, 10 m, vertical	0.080	0.0026	0.140	490.87
$\lambda/2$, 5 m, vertical	0.077	0.003	0.174	131.41
$\lambda/2$, 2 m, vertical	0.139	0.0031	0.144	21.53
$\lambda/2$, 1 m, tilted up 0.58°	0.260	0.0031	0.046	5.47
$\lambda/2$, 1 m, vertical	0.259	0.0031	0.046	5.50
$\lambda/2$, 0.5 m, tilted up 0.58°	0.297	0.0031	0.013	1.42
$\lambda/2$, 0.5 m, vertical	0.297	0.0031	0.013	1.42

different horizontal distances from the transmitting antenna. The receiving antenna for all cases is located at a height of 2 m from the ground and is also conjugately matched to extract the maximum energy from the progating wave. Over a PEC ground it is observed that the recived signals tend to increase as the radiating antenna is brought closer to the ground irrespective of the radial separation distance. For the case of the urban ground, the signal strength tends to increase in the near field as the antenna is brought closer to the ground but then it decreases. The last few rows indicate that when the receiving antenna is located in the far field there is the usual *height gain* as seen in Table 3.1. For the radial distance of 100 m, the receiving antenna is in the near field of a transmitting antenna high above the ground and there is the usual height loss. However, when the transmitting antenna is brought closer to the ground the far field starts also at a closer distance and that is why the field strentgth first increases with the lowering of the antenna closer to the ground but then it decreases after some optimum location. For a distance of 1000 m almost all the measurements are in the far fields and that is why there is the usual height gain. Hence, an optimization need to be carried out to find the best location for the transmitter height for a given cell size for a proper deployment.

If instead of a half wave vertical dipole we use a vertical square loop with half a wavelength on the side then the received power obtained by the same dipole is given in Table 3.2. The power radiated by the loop is slightly higher probably because it has a larger area. However, the point to be made here is that

Table 3.2 Values for the received power obtained at a stationary half-wave receiving dipole and radiated by a loop of half wavelength on each side located at different radial distances and for different heights over the ground.

Nature of ground Radial distance	PEC 100 m	PEC 1000 m	Urban 100 m	Urban 1000 m
Transmit antenna orientation	Received power (µW)	Received power (µW)	Received power (µW)	Received power (pW)
Loop, 20 m, vertical	0.0389	0.0015	0.1049	1700
Loop, 20 m, tilted down 11°	0.0374	0.0014	0.1014	1600
Loop,10 m, vertical	0.0743	0.0028	0.1505	529.71
Loop, 5 m, vertical	0.0908	0.0032	0.1776	144.26
Loop, 2 m, vertical	0.1397	0.0033	0.1626	24.895
Loop, 1 m, tilted up 0.58°	0.2679	0.0033	0.0555	6.7383
Loop, 1 m, vertical	0.2681	0.0033	0.0555	6.7383
Loop, 0.5 m, tilted up 0.58°	0.3297	0.0035	0.0178	2.0126
Loop, 0.5 m, vertical	0.3297	0.0035	0.0178	2.012

providing a slight tilt to the radiating dipole improves the strength of the radiated fields when the radiating element is near the ground.

Next we study the reason as to why the results produced appear this way. This will give to a better understanding on how to deploy base station antennas.

The plot of the horizontal and the vertical componenets of the fields from a half wave dipole conjugately matched and radiating at 1 GHz are considered. The field strengths are plotted as a function of the radial distance and the orientation and height of the transmitting antenna over an imperfect ground. Figures 3.28(a) plots the horizontal component and Figure 3.28(b) plots the vertical component of the radiated fields as a function of the horizontal distance for different height of the base station antenna over the Earth and their orientation with respect to the vertical. It is observed that when the antenna is closer to the ground and the antenna is tilted upwards towards the sky, the effect of fading as expected is greatly reduced as it results in a smoother decay of the fields.

In addition, the signal strength is much stronger in the near vicinity of the antenna when it is located close to the Earth than when the radiating antenna is located on top of a 10 or a 20 m tower even with the antennas tilting towards the ground. As expected, close to the antenna, i.e., inside the cell, the path loss is 30 dB per decade and far away from the antenna it is 40 dB per decade for a perfectly vertically oriented antenna. However, for a tilted antenna whether it be tilted towards the ground and located at a considerable height or tilted towards the sky but located near the ground, under either of these conditions, one can see that it provides a strong horizontal component of the fields. More interestingly and surprisingly, the horizontal component of the fields has a rate of decay of 20 dB per decade of distance when the radiating antenna is tilted, which is the smallest loss that can be achieved under any circumstances, as seen in the plots of Figure 3.28. However, bringing the radiating antenna closer to the ground and also providing a slight tilt with respect to the vertical, one obtains the best scenario as then the near fields in the vicinity of the radiating antenna is also enhanced. What is most surprising is that at some distances from the radiating antenna the horizontal components of the fields become stronger than the vertical components. Hence, this unusual mode of deployment of an antenna by lowering it close to the ground and tilting it upwards towards the sky may actually reduce the fading components as shown in the plots of Figures 3.28.

In summary, it is seen that the major component of the path loss is due to the propagation in free space and it is generally in excess of 90 dB for a propagation distance greater than 1 km. Buildings and trees do have effects on the propagation path loss, but these effects are of second order when compared to the free space propagation loss. The signal loss associated with the horizontal components seems to be lower than the vertical components of the fields. Hence, a surprising result is that better signal strength can be received when the

(a)

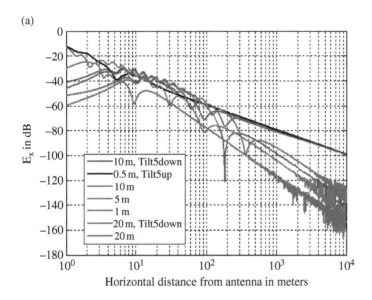

(b)

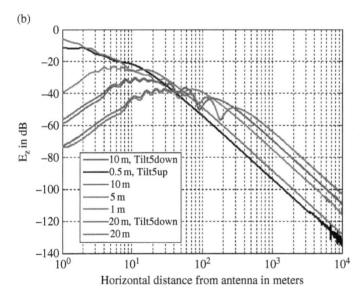

Figure 3.28 (a) Plot of the horizontal component of the electric field as a function of the height of the transmitting antenna above the ground and for a particular plot (the black line) the antenna is tilted upwards towards the sky and away from the ground, resulting in elimination of fading. (b) Plot of the vertical component of the electric field as a function of the height of the transmitting antenna above the ground, and for a particular plot (the black line) the antenna is tilted upwards towards the sky and away from the ground, resulting in elimination of fading.

transmitting antenna is brought closer to the ground and tilted slightly towards the sky which is quite a nonintuitive solution but it manifests itself through the Maxwellian physics!

Next we verify these statements from an experimental point of view and carrying out actual measurements. The measurements were done in an urban environment embedded in trees.

3.10.1 Experimental Measurement Setup

The statements made in the previous section are now verified using experimental results. The experiments were was carried out by Prof. Monai Krairiksh and his students from King Mongkut's Institute of Technology, Ladkrabang, Bangkok, Thailand. In these set of experiments the goal was to study the variation of the received signal strength as a function of the height of the transmitting antenna over the Earth for a fixed location of the receiving antenna and also as a function of frequency over the cellular band. The question of the tilt of the transmitting antenna with respect to the vertical axis was also considered. The measurement setup is shown in Figure 3.29 which illustrates how the transmitting antenna was excited and the mechanism of the receiver located at a fixed distance. The transmitting antenna was a Schwarzbek Mes-Elektronik Log periodic antenna USLP9143B operating from 200 MHz to 7 GHz with a isotropic gain factor of 1.3 dB and an array factor of 11-44 dB/m with a typical VSWR of < 1.5. It had a beamwidth of 45-65° in the E-plane and 90-120° in the H-plane. The receiving antenna is a double ridge guide horn antenna of model SAS-571 manufactured by A. H. Systems Incorporated operating from 700 MHz to 18 GHz with a gain of 1.4-15 dBi with a typical VSWR of 1.6:1. It had a typical beamwidth of 48° in the E-plane and 30° in the H-plane.

At 800 MHz the input power to the base station antenna was- 4 dBm. The gain of the low noise amplifier (LNA) was 32 dB and the insertion loss of the receiver was 2 dB. At 2.4 GHz the input power to the base station antenna was −12 dBm. The gain of the low noise amplifier (LNA) was 28.5 dB and the insertion loss of the receiver was 2 dB. Figure 3.30 plots the received electric field strength for a fixed separation distance between the base station antenna and the receiver at two different frequencies of 800 MHz and 2.4 GHz. It is seen from Figure 3.30 that at both the frequencies the received signal strength increases when the base station antenna is brought closer to the ground. So the received signal strength is increased inside the cell when the transmitting antenna is brought closer to the ground. This apparently hold for most frequencies in the cellular band.

Figure 3.31 plots the electric field strength for two different heights and angular tilt of the base station antenna at 800 MHz and for a fixed separation distance between the transmitter and the receiving antenna. For the first case, the antenna is located close to the ground and the antenna is slightly tilted

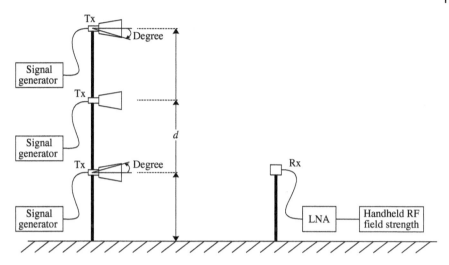

Figure 3.29 Measurement set up for recording the signal strength as a function of the height of the transmitting antenna and also for its various tilts with the vertical axis for a fixed location of the receiver. Three different experimental setups are displayed in a single figure.

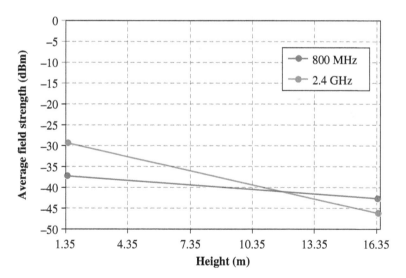

Figure 3.30 Measurement of the electric field strength as a function of the height of the base station antenna for two different frequencies. The receiver is located 1.5 m off the ground.

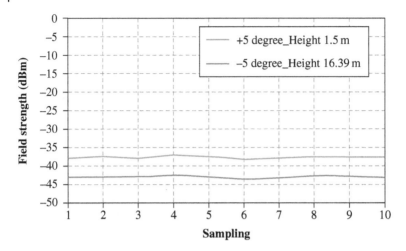

Figure 3.31 Measurement of the electric field strength for two different heights and angular tilt of the base station antenna at 800 MHz.

by +5° towards the sky. For the second case, the antenna is high above the ground and it is tilted by −5° towards the ground so as to apparently increase the signal strength towards the receiver. The conclusion that is easily reached is that when the antenna is closer to the ground and tilted towards the sky – a non-intuitive proposition – the received signal strength is higher than when the antenna is tilted towards the ground and located high above the ground. Ten different samples are taken over different time intervals just to make sure that the measurements are consistent. Thus it illustrates as theorized earlier that lowering the base station antenna closer to the ground and giving it a slight tilt towards the sky is a better solution.

Figure 3.32 plots the electric field strength for two different heights and angular tilt of the base station antenna at 2.4 GHz and for a fixed separation distance between the transmitter and the receiving antenna. For the first case, the antenna is located close to the ground and the antenna is slightly tilted by +5° towards the sky. For the second case, the antenna is high above the ground and is tilted by −5° towards the ground so as to increase the signal strength towards the receiver. This is the same situation as for the 800 MHz lower frequency case illustrated earlier. It is seen that at high frequency even when the antenna is closer to the ground and tilted towards the sky, the received signal strength inside the cell is higher than when the antenna is tilted towards the ground and located high above the ground. Ten different samples are taken over different time intervals just to make sure that the measurements are consistent. Thus it illustrates as theorized earlier that lowering the base station antenna closer to the ground and giving it a slight tilt towards the sky is a better solution independent of the operating frequency in the cellular band.

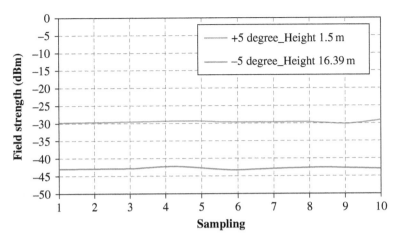

Figure 3.32 Measurement of the electric field strength for two different heights and angular tilt of the base station antenna at 2.4 GHz.

This smaller number of experimental samples does support the theory presented earlier in the section. A possible explanation for the increase in the received signal strength due to the tilt can be perhaps explained by examining Figure 3.27. There it is seen that when the antenna is operating in free space and over ground there is a null in the pattern along the ground due to the ground losses. The maximum value for the pattern, i.e., the main lobe occurs slightly higher in angle than along the horizon. So if the radiating element is slightly tilted towards the sky it may be supplying more energy to the main lobe and hence an increase in the signal strength than when the antenna is strictly vertical. However, when the antenna is located high above the ground the downward tilt will increase somewhat the signal strength as seen in Figures 3.21 and 3.22, but the interference pattern becomes stronger.

3.11 Cellular Wireless Propagation Occurs Through the Zenneck Wave and not Surface Waves

In this section we characterize the nature of the transmitted propagating wave by labelling it with a name to illustrate its behavior. We state that in a cellular wireless communication system the radio wave propagation takes place through the Zenneck wave, which has appeared many times in different contexts in the various scientific literature. This is first illustrated through the first principles and then through numerical simulations.

Consider a transverse magnetic (TM) plane wave that is incident at an angle θ with respect to the normal at the interface on the boundary between two semi-infinite, nonmagnetic, media separated by a planar boundary with the magnetic field parallel to the interface. The two media are air and a material medium having a permittivity ε. Part of the incident wave will be reflected and part of it will be transmitted through the interface. The TM reflection coefficient termed Γ_{TM} will be given by [32, 33] $\Gamma_{TM} = \dfrac{\varepsilon \cos\theta - \sqrt{\varepsilon - \sin^2\theta}}{\varepsilon \cos\theta + \sqrt{\varepsilon - \sin^2\theta}}$. At a grazing angle when $\theta \to 90°$ then $\Gamma_{TM} \to -1$. Also, there exists an incident angle at which the transmission is total and that is called the Brewster's angle. At the Brewster angle [32] defined by $\tan\theta_B = \varepsilon$ the reflection coefficient $\Gamma_{TM} = 0$ and there will be an incident field and a transmitted field but no reflected wave as shown in Figure 3.33a [32, p.270]. This is also the case for the Zenneck and Sommerfeld wave types. The Brewster angle is independent of the frequency of the incident wave and occurs at a *zero of the TM reflection coefficient*.

At a zero of the denominator of the reflection coefficient $\Gamma_{TM} = \infty$, there will be a reflected field and a transmitted field without any incident field as shown in Figure 3.33b [32, p.272]. This results in the generation of the surface wave. Therefore this surface waves cannot be generated using an incident transverse Electromagnetic (TEM) wave but rather through using quasi-particles or evanescent waves. This is because in this case there is no incident wave and the evanescent wave causes a response. This is because an evanescent wave behaves as a quasiparticle and can tunnel through the medium [33] exciting the electrons (this phenomenon of tunneling is called the Hartmann effect). In that way, no incident electromagnetic wave is required to generate the transmitted and the reflected fields as per Figure 3.33b. Also for a surface wave to occur, typically the dielectric constants of the two medium should be of opposite sign

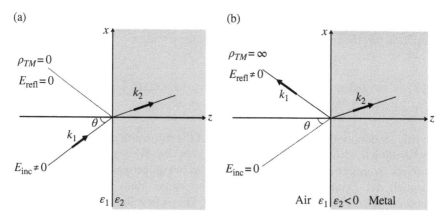

Figure 3.33 Brewster-Zenneck (TM reflection coefficient $\rho_{TM} = 0$) and surface plasmon (TM reflection coefficient $\rho_{TM} = \infty$) cases. (a) Brewster–Zenneck wave, reflection coefficient is zero and (b) surface wave, reflection coefficient is infinity.

and with the dielectric medium having a small loss, compared to the real part. A negative value for the dielectric constant with a small imaginary part occurs in metals generally in the petahertz (and not in terahertz) regions.

If the material medium in Figure 3.33a has a small dissipation then $\varepsilon = \varepsilon_r - \dfrac{j\sigma}{\omega\varepsilon_0}$ where ε is the complex permittivity of the medium. ε_r and ε_0 are the real permittivity of the medium and of vacuum, respectively. σ is the conductivity of the medium. ω is the frequency of the incident wave. For low loss, i.e., $\sigma/\varepsilon_r\omega\varepsilon_0 \ll 1$ the effect of σ on the Brewster angle turns out to be of the second order [34]. If the incident waves are uniform plane waves, that is, their amplitude is constant in any plane perpendicular to the rays, there will be a small reflection [34]. In this case, the Brewster angle is the angle of minimum reflection. However, if the incident plane waves are inhomogeneous then this will be the condition for no reflection. It is known that the rays of inhomogeneous plane waves in a vacuum are straight lines along which the phase change is maximum and the amplitude remains constant. The planes perpendicular to the rays are equiphase planes [34]. Along these planes there exists one direction in which the amplitude does not vary, while in the perpendicular direction the amplitude change is maximum. The exponential rate of attenuation with the increase of distance is small but it is independent of the frequency. This is the case for the Zenneck wave. And indeed if one plots the variation of the reflected field in the air medium then it will not change as a function of frequency. However, for a surface wave as the frequency increases the variation of the fields will become more concentrated near the boundary [34].

We now demonstrate that indeed this is the case for the cellular wireless communication as it takes place through the Zenneck/Sommerfeld wave type and not through a surface wave. Consider the propagation of a wave radiated by a half wave dipole located at a height of 10 m over an urban ground. This is close to a TM wave incident at the air-Earth interface. We now plot the variation of the magnitude of the **reflected field only** as a function of z for a fixed distance of ρ = 100 m from the transmitter. The vertical variation of the reflected field as a function of the height of upto 100 m from the ground is seen in the plot of Figure 3.34. It is seen that even though the frequency changes from 453 MHz, to 906 MHz and then to 1350 MHz, the magnitude of the decay of the reflected fields in the vertical direction to the planar air-earth interface remain practically the same. The fields tend to decay monotonically resembling an evanescent wave. As there is no appreciable variation of the fields even though the frequency changed by a factor of three, indicates that the wave in this case relates to a Zenneck wave and not to a surface wave as correctly suggested by Schelkunoff [34]. This also applies to the waves propagating over a two-layer medium like radio wave propagation in an urban environment. Also, in the plot of Figure 3.34 in the absolute value of the reflected fields, there is an indication of a dip in the strength of the reflected fields at a height of 40 m from

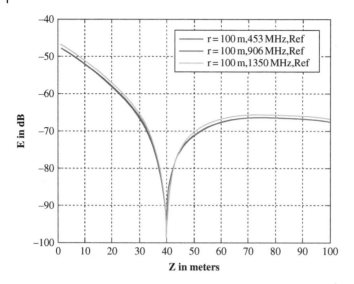

Figure 3.34 Plot of the transverse component of the reflected electric fields at $\rho = 100$ m from the source.

the ground. Considering urban ground which has a $\varepsilon_r = 4$, will yield a Brewster's angle given by $\tan^{-1}(\sqrt{4}) = 63.4°$. From the geometry of the problem, it is seen that the angle subtended by the antenna with respect to the ground will be given by $\tan^{-1}(10/20) = \tan^{-1}(40/80) \approx 26.6°$. Therefore, the angle with respect to the vertical is $90°-26.6° = 63.4°$, indicating that the dip in the field strength is occurring exactly at the Brewster angle of $63.4°$ even though the permittivity of the ground is complex. This indicates as Schelkunoff [34] had predicted that the effect of small conductivity on the Brewster's angle is of the second order and the location of the Brewster's angle does not change with frequency even though it varies over a factor of 3 (from 453 MHz to 1350 MHz).

In summary, the plot of the reflected field strength from the dipole over an imperfect ground reflects the strong influence of the Brewster angle which is due to a zero of the reflection coefficient. The plot also demonstrates the practical invariance of the reflected fields as a function of frequency which is also true at the Brewster's angle as it is also independent of frequency. Finally, the effect of an infinite value of the reflection coefficient resulting in a surface wave (a reflected and a transmitted field without an incident field) is not seen in Figure 3.34. The conclusion that is easily reached is that for propagation over urban ground the surface wave phenomenon is nonexistent. Therefore, this plot confirms that the reflected field is a Zenneck–Sommerfeld type of a radiating wave strongly influenced by the zero of the reflection coefficient. Also the Norton surface wave is not related to any of these two types of waves [34]. As Schelkunoff states: *This wave (Norton Wave) was defined as the difference*

*between the exact field of a dipole above an imperfect ground and the field cal-
culated by the rules of geometrical optics. This Norton surface wave does not
satisfy Maxwell's equations. There is also the term ground wave which is used by
radio engineers to denote the total wave which would have existed on the ground
surface if the Kennelly-Heaviside layer were absent. The wave reflected from the
ionospheric layer is called the sky wave. In the primary service area for all
broadcasting stations operating in the low and medium frequency ranges, the
sky wave is very weak and only the ground wave is important. This ground wave
has some relations to the Norton surface wave but is not identical to it. The
latter vanishes for a perfect ground when the ground wave of the radio engineers
is the strongest. The ground wave has also been confused with the Zenneck
surface wave* [34].

The century old controversy about what should be the nature of the reflection
coefficient: zero or infinity is looked at next. In the original 1909 Sommerfeld
formulation [13] there was no error in the sign, but the presentation by
Sommerfeld was not complete as opposed to his later presentation of 1926 [25].
First, of all Sommerfeld demonstrated in his earlier expression that there was a
pole in the reflection coefficient and he wrote the partial solution – the contribution from the pole and associated with that some physical properties.
However, as has been pointed out in [35] when $\varepsilon_r > 0$, the pole should lie within
the circle defined by the locus whose center is at $\lambda = 0$ and of radius k_1. In addition, when the integration along the branch cut is evaluated using the saddle
point method part of this branch cut integral actually cancels the contribution
from the pole and in the final solution the effect of the pole is not seen! This has
been clearly demonstrated by Collin [20] and others [34, 35] and so the conjecture that there was an error in sign in Sommerfeld's formulation as initially
intimated by Norton is a myth! Also it has been shown by Collin *and others that
when the branch cut contribution was taken into account the effect of the surface
wave pole is cancelled by the branch cut contribution and so there are no surface
waves that can be generated by a dipole source radiating over a two layer media
in the final total solution. In addition, it was shown later by experiments that
such kind of wave does not exist and cannot be excited in a two-layer problem
(for example, air and ground having a single interface) as illustrated by the
experimental results performed on Seneca Lake by Burrows in* [9].

This has been also demonstrated by Schelkunoff in 1959 [34]. Finally, the
presence of $J_0(\lambda\rho)$ in the expression (3.1) for the evaluation of the potential
makes the numerical computation very unstable. For large values of ρ the
integrals in (3.1) are quite difficult/tedious and/or cumbersome to evaluate
numerically as we are dealing with a highly oscillatory function, which in
the literature is labeled as the *Sommerfeld tail problem*. This numerical drawback can be avoided by using the Schelkunoff formulation [16, 17].

Historically, in 1909 Sommerfeld computed the integral along the positive
real axis by first applying the Cauchy principal integral method to close the

contour by a large semicircle at infinity given by the semi- circular contour with indentations in the third and the fourth quadrants. As seen in his book [36], this closed contour of integration is equivalent to two integrals around the branch cuts associated with the branch points at $+k_1$ and $+k_2$ and a contour integration around the pole $\lambda = \lambda_P$ where $\lambda_P = \dfrac{k_1 k_2}{\sqrt{k_1^2 + k_2^2}}$. The other branch points $-k_1$ and $-k_2$ and the pole located at $-\dfrac{k_1 k_2}{\sqrt{k_1^2 + k_2^2}}$ are of no concern as they are located in the upper half plane where the contour is not closed as seen in the discussions of Appendix 3A in Figure 3.32. Sommerfeld then evaluated the residue at the pole and showed that it has the form of a *surface wave* (This wave is defined in Appendix F). Then Sommerfeld showed that the residue at the pole is

$$\Pi_{1z}^{pole} = -2\pi \, jP \left[\frac{k_2^2 H_0^{(2)}\left(\lambda_P \rho\right) \exp\left(-z\sqrt{\lambda_P^2 - k_1^2}\right)}{\dfrac{k_2^2}{\sqrt{\lambda_P^2 - k_1^2}} + \dfrac{k_1^2}{\sqrt{\lambda_P^2 - k_2^2}}} \right] \qquad (3.34)$$

where Π_{1z}^{pole} is part of the solution from the pole contribution. Next, a large argument approximation is made for the Hankel function by following the path of integration of Figure 3.32 and 3A.1 shown in Appendix 3A. For large values of ρ, the asymptotic representation for the Hankel function is used

$$H_0^{(2)}\left(\lambda \rho\right) \xrightarrow[|\lambda \rho| \to \infty]{\text{Limit}} \sqrt{\frac{2}{\pi \lambda \rho}} \exp\left[-j\lambda \rho + j\pi/4\right] \qquad (3.35)$$

resulting in

$$\Pi_{1z}^{pole} = P \left[\sqrt{\frac{2\pi}{j\lambda_P \rho}} \frac{k_2^2 \exp\left(-j\lambda_P \rho - z\sqrt{\lambda_P^2 - k_1^2}\right)}{\dfrac{k_2^2}{\sqrt{\lambda_P^2 - k_1^2}} + \dfrac{k_1^2}{\sqrt{\lambda_P^2 - k_2^2}}} \right] \qquad (3.36)$$

As Sommerfeld in his book [36] then points out: "*this formula bears all the marks of surface waves*". [a true surface wave is a slow wave and the fields become concentrated to the interface as the frequency increases]. *It was the main point of the author's work of 1909 to show that the surface wave fields are automatically contained in the wave complex. This fact has of course, not changed. What has changed is the weight which we attached to it. At that time it seemed conceivable to explain the overcoming of the earth's curvature by radio signals with the help of the character of the surface waves; however we know now that it is due to the ionosphere. In any case, the recurrent discussion in the literature on the reality of the Zenneck waves seems immaterial to us.*

These statements by Sommerfeld indicate that the pole contribution – which results in the surface wave is non-existent in a two layer problem, for most practical cases. As pointed out by Kahan and Eckart [35]: *"Sommerfeld did not notice while computing his asymptotic development of the branch cut integral that this contains beside the space wave, the surface wave with a negative sign and so cancels the residue of the pole".* Furthermore, Kahan and Eckart [35] point out *that the pole should not come into the picture as the singularity does not meet the radiation condition and it appears only through an inadvertency in Sommerfeld's calculations.* This fact has also been illustrated by Baños in his book [22, pp. 55–61] that the pole is not located on the right half plane of the branch cut. This becomes much clearer when the saddle point method is applied as the saddle point path never crosses the pole and so the surface wave contribution as envisaged by Sommerfeld never arises. This was the difference in the solution of Sommerfeld and Weyl [26] as was also pointed out by Baños in his book [22, pp. 55–61].

In summary, Sommerfeld had no error in sign in his paper irrespective of what others have said in the past. It turns out that when the branch cut contours are evaluated in a proper way the pole contribution is cancelled by part of the branch cut contour integration which was demonstrated by Kahan and Eckart [35] and Collin [20]. In addition Ishimaru [21] illustrated that if the branch cuts are chosen to be vertical straight lines instead of the hyperbolic one then the pole moves from the proper Riemann sheet to an improper one. In conclusion a surface wave does not exist for a two medium problem having a single boundary. So for a two media problem there is a Brewster zero and hence a Zenneck type wave propagates over the surface where the vertical distribution of the evanescent field is independent of the frequency. This is the case in a cellular wireless communication system.

3.12 Conclusion

The objective of this chapter has been to illustrate that an electromagnetic macro model can accurately predict the propagation path loss in a cellular wireless environment as the buildings, trees and so on are the secondary source of the propagation path loss, the primary being the mechanism of electromagnetic wave propagation over an imperfect ground. Documentary evidence using both theory and experimental data from multiple sources have been provided to illustrate these subtle points.

In addition, it has been shown that for all the experimental data (this excludes many measurements which deal measurements in the extreme near field very close to the radiating element) collected from various different operating environments illustrate that the presented macro model in this chapter is quite accurate and provide a basic physical understanding of the propagation mechanism. One of the salient features of this mode of propagation is *nature of the ground, such as urban, rural, suburban, or water, which has practically no effect on a cellular environment.*

Also, an optical analog is provided to visualize the propagation mechanism and to illustrate that the image of an antenna radiating over an imperfect ground can be characterized by an image which is essentially a semi-infinite line source which extends form the same height at which the antenna is placed over ground but now it is located below the ground and extending to negative infinity. This line source is responsible for predicting the proper propagation path loss. Such a mechanism cannot be explained either through a statistical model, a multiple ray model or through diffraction as none of these modes of analysis deal with the imperfect ground which is the main mechanism of propagation over ground.

Finally, it is illustrated through numerical simulations, that deploying an antenna closer to the ground and tilting it up, towards the sky will reduce fading as the wave generated from the image penetrates further into the earth rather than coming towards the upper layer and causing interference. In addition, the propagation over imperfect ground loss may also get reduced for non-vertical polarization components of the fields which is generally the case when a mobile responds to a base station.

Appendix 3A Sommerfeld Formulation for a Vertical Electric Dipole Radiating Over an Imperfect Ground Plane

Consider an elementary electric dipole located at (x', y', z') of moment Idz oriented along the z-direction. The dipole is situated over an imperfect ground plane characterized by a complex relative dielectric constant ε as seen in Figure 3.12. The complex relative dielectric constant is given by $\varepsilon = \varepsilon_r - \dfrac{j\sigma}{\omega\varepsilon_0}$ where ε_r represent the relative permittivity of the medium, ε_0 is the permittivity of vacuum, σ is the conductivity of the medium, ω stands for the angular frequency, and j is the imaginary unit, i.e., $j = \sqrt{-1}$. It is possible to formulate a solution to the problem of radiation from the dipole operating in the presence of the imperfect ground in terms of a single Hertzian vector Π_z of the electric type. A time variation of $\exp(j\omega t)$ is assumed throughout the analysis, where, t is the time variable. The Hertzian vector $\hat{u}_z\Pi_z$ in this case satisfies the wave equation

$$\left(\nabla^2 + k_1^2\right)\Pi_{1z} = \frac{-Idz}{j\omega\varepsilon_0}\delta\left(x - x'\right)\delta\left(y - y'\right)\delta\left(z - z'\right) \tag{3A.1}$$

$$\left(\nabla^2 + k_2^2\right)\Pi_{2z} = 0 \tag{3A.2}$$

where

$$k_1^2 = \omega^2 \mu_0 \varepsilon_0 \tag{3A.3}$$

$$k_2^2 = \omega^2 \mu_0 \varepsilon_0 \varepsilon \tag{3A.4}$$

and δ represents the delta function in space. The primed and unprimed coordinates are for the source and field points respectively. The subscript 1 denote the upper half space which is air and the subscript 2 denote the lower half space which is the imperfectly conducting earth characterized by a complex relative dielectric constant ε. The electric and the magnetic field vectors are derived from the Hertzian vector using

$$\vec{E}_i = \vec{\nabla}\left(\vec{\nabla} \cdot \overrightarrow{\Pi_i}\right) + k_i^2 \overrightarrow{\Pi_i} \tag{3A.5}$$

and

$$\bar{H}_i = j\omega\varepsilon_0\varepsilon_i\left(\vec{\nabla} \times \overrightarrow{\Pi_i}\right) \tag{3A.6}$$

respectively, with $i = 1, 2$. In medium 1, $\varepsilon_1 = 1$ and for medium 2, $\varepsilon_2 = \varepsilon$. So that the propagation constants in medium 1 and 2, are called k_1 and k_2, and related by $\dfrac{k_2}{k_1} = \sqrt{\varepsilon}$. At the interface between the air and the Earth characterized by $z = 0$, the tangential electric and magnetic field components must be continuous, conditions which in terms of the Hertzian vector components can be written as

$$\frac{\partial \Pi_{1z}}{\partial y} = \varepsilon \frac{\partial \Pi_{2z}}{\partial y} \tag{3A.7a}$$

$$\frac{\partial \Pi_{1z}}{\partial x} = \varepsilon \frac{\partial \Pi_{2z}}{\partial x} \tag{3A.7b}$$

$$\frac{\partial}{\partial y}\left(\frac{\partial \Pi_{1z}}{\partial z}\right) = \frac{\partial}{\partial y}\left(\frac{\partial \Pi_{2z}}{\partial z}\right) \tag{3A.7c}$$

$$\frac{\partial}{\partial x}\left(\frac{\partial \Pi_{1z}}{\partial z}\right) = \frac{\partial}{\partial x}\left(\frac{\partial \Pi_{2z}}{\partial z}\right) \tag{3A.7d}$$

Since all the boundary conditions must hold at $z = 0$ for all x and y, and therefore the x and y dependence of the fields on either side of the interface must be the same. Therefore

$$\Pi_{1z} = \varepsilon \Pi_{2z} \tag{3A.8a}$$

$$\frac{\partial \Pi_{1z}}{\partial z} = \frac{\partial \Pi_{2z}}{\partial z} \tag{3A.8b}$$

The complete solutions for the Hertz vectors satisfying the wave equations (3A.1) and (3A.2) and the boundary conditions (3A.8) have been derived by many researchers over the last century. A partial list [13–17, 19–22] that will be important to our discussions is provided starting with Sommerfeld [13]. The solutions are

$$\Pi_{1z} = P\left[\frac{\exp(-jk_1 R_1)}{R_1} + \int_0^\infty \frac{J_0(\xi\rho)}{\sqrt{\xi^2 - k_1^2}} \frac{\varepsilon\sqrt{\xi^2 - k_1^2} - \sqrt{\xi^2 - k_2^2}}{\varepsilon\sqrt{\xi^2 - k_1^2} + \sqrt{\xi^2 - k_2^2}} \right.$$
$$\left. \exp\left(-\sqrt{\xi^2 - k_1^2}\,(z + z')\right)\xi\,d\xi \right] \tag{3A.9}$$

and

$$\Pi_{2z} = 2P\int_0^\infty \frac{J_0(\xi\rho)\exp\left(\sqrt{\xi^2 - k_2^2}\,z - \sqrt{\xi^2 - k_1^2}\,z'\right)}{\varepsilon\sqrt{\xi^2 - k_1^2} + \sqrt{\xi^2 - k_2^2}}\xi\,d\xi \tag{3A.10}$$

for $\mathrm{Re}\left[\sqrt{\xi^2 - k_{1,2}^2}\right] > 0$. $J_0(x)$ represents the zero-th order Bessel function of the first kind of argument x. Here

$$P = \frac{I\,dz}{j\omega 4\pi\varepsilon_0} \tag{3A.11}$$

$$\rho = \sqrt{(x - x')^2 + (y - y')^2} \tag{3A.12}$$

$$R_1 = \sqrt{\rho^2 + (z - z')^2} \tag{3A.13}$$

and ξ is the variable of integration and $k_1 = \dfrac{2\pi}{\lambda}$, with λ is the wavelength at the transmitted frequency. Similarly, the solution for Π_{2z} can be interpreted as a partial transmission of the wave from medium 1 into medium 2. With these thoughts in mind Π_{1z}, or equivalently the potential responsible for the ground wave, can be split up into two terms

$$\Pi_{1z} = \Pi_{1z}^{direct} + \Pi_{1z}^{reflected} = P(g_0 + g_s) \tag{3A.14}$$

where

$$\Pi_{1z}^{direct} = P\exp(-jk_1 R_1)/R_1 = P g_0 \tag{3A.15}$$

Figure 3A.1 The contour of integration along the real axis from 0 to ∞ in the complex λ-plane.

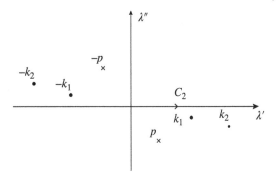

$$\Pi_{1z}^{reflected} = P\int_0^\infty \frac{\varepsilon\sqrt{\xi^2 - k_1^2} - \sqrt{\xi^2 - k_2^2}}{\varepsilon\sqrt{\xi^2 - k_1^2} + \sqrt{\xi^2 - k_2^2}} \frac{J_0(\xi\rho)\exp\left[-\sqrt{\xi^2 - k_1^2}(z+z')\right]}{\sqrt{\xi^2 - k_1^2}}\xi\,d\xi = Pg_s$$

(3A.16)

The path of integration for the semi-infinite integral is labeled C_2 and is depicted in Figure 3A.1 along with the singularities of the multivalued functions, with four branch points at $\pm k_1$ and $\pm k_2$ and two poles located at $\pm p$ arising from the ratio of the first fraction of (3A.16).

A physical explanation to the two components of the Hertz potential Π_{1z} can now be given. The first one Π_{1z}^{direct} can be explained as a spherical wave originating from the source dipole. This term is easy to deal with. The difficult problem lies in the evaluation of $\Pi_{1z}^{reflected}$. Therefore, $\Pi_{1z}^{reflected}$ can be interpreted as a superposition of plane waves resulting from the reflection of the various plane waves into which the original spherical wave has been expanded and multiplied by a different value for the reflection coefficient for each ray. This arises from the identity

$$\frac{\exp(-jk_1R_2)}{R_2} = \int_0^\infty \frac{J_0(\xi\rho)\exp\left[-\sqrt{\xi^2 - k_1^2}(z+z')\right]}{\sqrt{\xi^2 - k_1^2}}\xi\,d\xi$$

(3A.17)

for $\text{Re}\left[\sqrt{\xi^2 - k_1^2}\right] > 0$ and

$$R_2 = \sqrt{\rho^2 + (z+z')^2}$$

(3A.18)

The reflection coefficient $R(\lambda)$ is then defined as

$$R(\xi) = \frac{\varepsilon\sqrt{\xi^2 - k_1^2} - \sqrt{\xi^2 - k_2^2}}{\varepsilon\sqrt{\xi^2 - k_1^2} + \sqrt{\xi^2 - k_2^2}}$$

(3A.19)

where the semi-infinite integral over ξ in $\Pi_{1z}^{reflected}$ takes into account all the possible plane waves. As $\varepsilon \to \infty$, i.e., a perfect conductor for the earth, then g_s of (3A.16) reduces to (3A.17) and represents a simple spherical wave originating at the image point. This physical picture will later be applied in the derivation of the reflection coefficient method. The reflection coefficient takes into account the effects of the ground plane in all the wave decomposition of the spherical wave and sums it up as a ray originating from the image of the source dipole but multiplied by a specular reflection coefficient $R(\theta)$, where θ is interpreted as the angle of the incident wave to the ground.

It is now important to point out that there are two forms of $\Pi_{1z}^{reflected}$ that may be used interchangeably as the two expressions are mathematically identical in nature (but have different asymptotic properties as we shall see) and are defined as

$$
\Pi_{1z}^{reflected} = P \left[\frac{\exp(-jk_1 R_2)}{R_2} - 2\int_0^\infty \frac{\sqrt{\xi^2 - k_2^2}}{\sqrt{\xi^2 - k_1^2}} \frac{J_0(\xi\rho)\exp\left[-\sqrt{\xi^2 - k_1^2}\,(z+z')\right]}{\varepsilon\sqrt{\xi^2 - k_1^2} + \sqrt{\xi^2 - k_2^2}} \xi \, d\xi \right]
$$

$$
= P\left[g_1 - g_{sV}\right]
$$

(3A.20)

where g_1 represents the spherical wave originating from the image of the source, and g_{sV} represents the correction factor to accurately characterize the effects of the ground. Equivalently, one can rewrite the same expression as

$$
\Pi_{1z}^{reflected} = P \left[-\frac{\exp(-jk_1 R_2)}{R_2} + 2\varepsilon\int_0^\infty \frac{J_0(\xi\rho)\exp\left[-\sqrt{\xi^2 - k_1^2}\,(z+z')\right]}{\varepsilon\sqrt{\xi^2 - k_1^2} + \sqrt{\xi^2 - k_2^2}} \xi \, d\xi \right]
$$

$$
= P\left[-g_1 + G_{sV}\right]
$$

(3A.21)

Now the image from the source has a negative sign along with the correction factor. This expansion is useful when both the transmitter and the receiver are located close to the ground and the reflection coefficient is -1 for grazing angle of incidence where $\theta \approx \pi/2$. Then the direct term g_0 cancels the image term g_1 leaving only the correction factor G_{sV}. Now for grazing incidence the fields are obtained using G_{sV}.

In order to evaluate the semi-infinite integrals, first the Bessel function of the first kind and zeroth order is transformed to a Hankel function of the first and second kinds and zeroth order through the use of the following identity by using

$$
J_0(x) = \frac{1}{2}\left[H_0^{(1)}(x) + H_0^{(2)}(x)\right]
$$

(3A.22a)

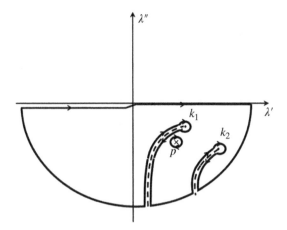

Figure 3A.2 Actual location of the pole in the lower complex λ-plane and the contour C_1 enclosing them.

and, also utilizing

$$H_0^{(1)}\left(xe^{j\pi}\right) = -H_0^{(2)}(x) \tag{3A.22b}$$

where $H_0^{(1)}$ and $H_0^{(2)}$ are the Hankel functions of zeroth order and of first and second kinds, respectively. The integrals in (3A.20) and (3A.21) can be transformed to $-\infty$ to ∞. In the evaluation of (3A.20) and (3A.21), the crux of the problem lies in the characterization of the various branch points and singularities associated with (3A.19) as illustrated in Figure 3A.2 along with the locations of the branch cuts and the poles. The first point to observe is that the second term of the Hertz potential denoted by a complex integral and particularly $R(\xi)$ has four branch points located at $\pm k_1$ and $\pm k_2$. Associated with these branch points are four branch cuts and this give rise to four Riemann sheets. On the four separate Riemann sheets, the following conditions are satisfied:

Sheet 1: $\mathrm{Re}\left(\sqrt{\xi^2 - k_1^2}\right) > 0$ and $\mathrm{Re}\left(\sqrt{\xi^2 - k_2^2}\right) > 0$

Sheet 2: $\mathrm{Re}\left(\sqrt{\xi^2 - k_1^2}\right) < 0$ and $\mathrm{Re}\left(\sqrt{\xi^2 - k_2^2}\right) > 0$

Sheet 3: $\mathrm{Re}\left(\sqrt{\xi^2 - k_1^2}\right) > 0$ and $\mathrm{Re}\left(\sqrt{\xi^2 - k_2^2}\right) < 0$

Sheet 4: $\mathrm{Re}\left(\sqrt{\xi^2 - k_1^2}\right) < 0$ and $\mathrm{Re}\left(\sqrt{\xi^2 - k_2^2}\right) < 0$

Sheet 1 is the proper Riemann sheet. Now, the function described in (3A.19) has two zeros corresponding to the zeros of the numerator, usually called zeros associated with the Brewster's angle, and two poles corresponding to the zeros of the denominator, and usually called the surface wave poles. So, the zero of the reflection coefficient in (3A.19) illustrates the Brewster's phenomenon (i.e., the wave goes into the second medium for a particular angle of incidence without reflecting any energy) and an infinite value for the reflection coefficient illustrates the presence of a surface wave (i.e., a wave propagating close to the interface).

The zeros and poles occur exactly at the same location $\xi_S = \pm \dfrac{k_1 k_2}{\sqrt{k_1^2 + k_2^2}}$.

However on some Riemann sheets they appear as poles and on other Riemann sheets as zeros. So, the four poles and zeros are distributed on the four Riemann sheets. The most confusing stuff is that on the proper Riemann sheet whether it will be a pole or zero depends on the value of the dielectric constant ε. For example examining the denominator of $R(\xi)$ one can observe that for real values of ε on the proper Riemann sheet there is a zero at ξ_s. Whereas for a complex value of the dielectric constant ε, that zero may become a pole on the proper Riemann sheet. Therefore unless one specifies the values for the dielectric constant and chooses the proper Riemann sheet, it is difficult to know whether the pole will occur or not in the expression of (3A.19).

The convergence of the integrals in (3A.20) and (3A.21) is assured even in the presence of the Hankel function, when Im $(\xi) \leq 0$ as $\rho \to \infty$. Convergence is also assured for

$$\mathrm{Im}\left(k_1^2 - \xi^2\right)^{\frac{1}{2}} < 0 \tag{3A.23}$$

and for

$$\left(k_1^2 - \xi^2\right)^{\frac{1}{2}} \xrightarrow[\xi \to 0]{} k_1' + jk_1'', \ k_1' > 0, \ k_1'' \leq 0. \tag{3A.24}$$

If

$$\left(k_1^2 - \xi^2\right)^{\frac{1}{2}} = k_z' + jk_z'' \tag{3A.25}$$

then

$$k_z'' = \frac{k_1' k_1'' - \xi'\xi''}{k_z'} \tag{3A.26}$$

and

$$k_z' = \sqrt{\left[\sqrt{\left(\frac{k_1'^2 - k_1''^2 - \xi'^2 + \xi''^2}{2}\right)^2 + \left(k_1'k_1'' - \xi'\xi''\right)^2} + \frac{k_1'^2 - k_1''^2 - \xi'^2 + \xi''^2}{2}\right]} > 0 \tag{3A.27}$$

where

$$\xi = \xi' + j\xi'' = \mathrm{Re}(\xi) + j\mathrm{Im}\ (\xi). \tag{3A.28}$$

The positive sign is chosen for k_z' since $k_z' > 0$ (from (3A.24), (3A.25) and (3A.27)). On the path of integration $k_z'' < 0$ and if $k_1'' \neq 0$, convergence would be assured even if $\xi'' = 0$. If medium 1 is lossless $k_1'' = 0$, then from (3A.26) $\xi' > 0$ when $\xi'' > 0$ and $\xi' < 0$ when $\xi'' < 0$. The first condition $\xi'' > 0$ conflicts with the convergence requirement for the Hankel function $H_0^{(2)}(\xi\rho)$. This problem can be avoided if medium 1 is assumed to be lossy (i.e., $k_1'' \neq 0$) and the "lossless case" then is assumed to be the limiting form of the expression as $k_1'' \to 0$. Since $H_0^{(2)}(\xi\rho)$ can be integrated through the origin ($\xi = 0$) even if $\xi'' = 0$, the path C_2 of Figure 3A.1 can now be modified to the path C_1 of Figure 3A.2, following the real axis from $-\infty$ to $+\infty$. Now (3A.20) can be rewritten as

$$g_{sV} = -j \int_{C_1} \frac{\sqrt{k_2^2 - \xi^2}}{\sqrt{k_1^2 - \xi^2}} \frac{H_0^{(2)}(\xi\rho) \exp\left[-j\sqrt{k_1^2 - \xi^2}(z + z')\right]}{\varepsilon\sqrt{k_1^2 - \xi^2} + \sqrt{k_2^2 - \xi^2}} \xi\, d\xi \tag{3A.29}$$

where the contour of integration C_1 is shown in the Figure 3A.2 along with the location of the branch points at k_1 and k_2 and their associated branch cuts, together with the pole of (3A.29). The presence of the free term ξ will nullify the singularity of the Hankel function at $\xi = 0$. The integral from $-\infty$ to 0 goes slightly below the negative real axis as the Hankel function has a branch cut along that line. g_{sV} is now a spectrum of plane waves travelling away from the ground plane with the vertical component of the propagation constant as $\sqrt{k_1^2 - \xi^2}$. The integral in (A.29) also contain double valued functions $\sqrt{k_{1,2}^2 - \xi^2}$. The proper sheet of the double valued functions are those on which the radiation condition

$$\left(\frac{\partial g_{sV}}{\partial z} - j\sqrt{k^2 - \xi^2}\, g_{sV}\right) e^{-j\sqrt{k_1^2 - \xi^2}\, z}\Bigg|_{z=-\infty}^{\infty} = 0 \tag{3A.30}$$

is satisfied and can occur only if $\mathrm{Im}\left[\sqrt{k_1^2 - \xi^2}\right] > 0$, i.e., $g_{sV} \to 0$ as $\rho \to \infty$ for a fixed z, and $g_{sV} \to 0$ as $z \to \infty$ for a fixed ρ. For large values of ρ, the asymptotic representation for the Hankel function is used

$$H_0^{(2)}(\xi\rho) \xrightarrow[|\xi\rho| \to \infty]{\text{Limit}} \sqrt{\frac{2}{\pi\xi\rho}} \exp\left[-j\xi\rho + j\pi/4\right] \tag{3A.31}$$

Then following Ott's formulation [37] which takes into account the presence of the pole near the saddle point path of integration, the integral in (3A.29) can be further simplified by making the following substitutions:

$$\xi = k_1 \sin\beta \tag{3A.32}$$

$$\rho = R_2 \sin\theta \tag{3A.33}$$

$$z + z' = R_2 \cos\theta \tag{3A.34}$$

The interpretation of the angle θ is shown in Figure 3.12. Hence, the application of (3A.31)–(3A.34) to (3A.29) yields

$$g_{sV} \approx \int_{\Gamma_1} \left[\frac{2k_1\sin\beta}{\pi R_2\sin\theta}\right]^{1/2} \frac{\sqrt{\varepsilon-\sin^2\beta}}{\varepsilon\cos\beta + \sqrt{\varepsilon-\sin^2\beta}}$$

$$\times \exp\left[j\left\{-\pi/4 - k_1 R_2 \cos(\beta-\theta)\right\}\right]d\beta \tag{3A.35}$$

where Γ_1 is a path in the complex β plane as shown in Figure 3A.3. An approximate methodology utilizing the saddle point integration methodology is used to evaluate these infinite integrals.

There is one obvious weakness in the arguments presented to derive (3A.35), namely, that there are points on the path for which the argument of the Hankel function $H_0^{(2)}(\xi\rho)$ used in (3A.29) is not large and may even be zero, so that the asymptotic expansion for large arguments cannot be used. However, as argued by Brekhovskikh [38], the arguments will be rigorous if the large argument approximation is used only after the path of integration has been changed to the path of steepest descent Γ_0 of Figure 3A.3. The result will then be the same.

Assuming medium 1 to be lossless, then k_1 in the transformation of (3A.32) $\xi = k_1\sin\beta$, implies for a complex β ($\beta = \beta' + j\beta'$), the equivalent complex value of ξ is:

$$\xi' + j\xi'' = k_1\left(\sin\beta'\cosh\beta'' + j\cos\beta'\sinh\beta''\right) \tag{3A.36}$$

Or, equivalently,

$$\xi' = k_1\sin\beta'\cosh\beta'' \tag{3A.37}$$

$$\xi'' = k_1\cos\beta'\sinh\beta''. \tag{3A.38}$$

Hence the mapping in (3A.32), $\xi = k_1\sin\beta$, transforms the quadrants of the ξ plane in parallel strips of width $\pi/2$ radians, and the path of integration from $\xi' = -\infty$ to ∞ is transformed to the path Γ_1, where $\xi'' = \mathrm{Im}\ (k_1\sin\beta) = 0$, as shown in Figure 3A.3. The requirement $\mathrm{Im}\left(\sqrt{k_1^2 - \xi^2}\right) < 0$ amounts to $\mathrm{Im}\ (k_1\cos\beta) < 0$ on the path of integration, or for k_1 real,

$$\sin\beta'\sinh\beta'' > 0 \tag{3A.39}$$

as $\cos(\beta' + j\beta'') = \cos\beta'\cosh\beta'' - j\sin\beta'\sinh\beta''$. The script '$\mathcal{U}$' in Figure 3A.3, denotes the strips of the β-plane on which the above inequality (3A.39) is satisfied (upper Riemann sheet). The other strips are denoted by '$\mathcal{L}$' (lower Riemann sheet). The path Γ_1 then totally lies on $\mathcal{U}$.

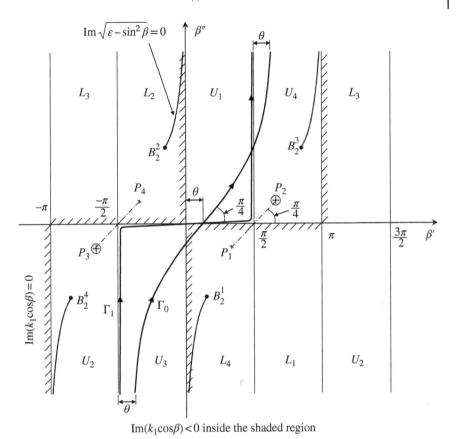

Figure 3A.3 The complex β plane showing possible branch points, branch cuts, poles, and the path of steepest descent for an imperfect ground plane with the material parameter $\varepsilon = \varepsilon'(1-j)^2$ and $|\varepsilon| >> 1$.

The location of the branch points at $\xi = \pm k_1$ in the ξ-plane are now transformed into $\sin \beta_{B_1} = \pm 1$ in the β-plane and are situated at $\pm \pi/2$, $\pm 3\pi/2$ and so on. The branch cuts along Im $\left(\sqrt{k_1^2 - \xi^2}\right) = 0$ are now transformed to Im $(k_1 \cos \beta) = 0$, and begin at the branch points labelled B located at $\beta = \pm \pi/2$. Since Im $(k_1 \cos \beta) = $ Im $[k_1 \sin(\pi/2 \pm \beta)] = 0$, these branch cuts will run parallel to the path Γ_1, [Im $(k_1 \sin \beta) = \lambda'' = 0$] but shifted by $\pm \pi/2$ from the origin along the real axis. So the transformation $\xi = k_1 \sin \beta$ has transformed the upper and lower sheets associated with the branch points $\pm k_1$ into one sheet where certain strips on the sheet belong to the previous upper (U) and lower (L) Riemann sheet on the ξ-plane.

The remaining branch points $\xi = \pm k_2$ are transformed to $\sin \beta_{B_2} = \pm\sqrt{\varepsilon}$ which has solutions

$$\beta_{B_2} = j\ln\left[\pm j\sqrt{\varepsilon} \pm \sqrt{1-\varepsilon}\right].\tag{3A.40}$$

The branch cuts $\text{Im}\left[\sqrt{k_2^2 - \lambda^2}\right] = 0$ are transformed into $\text{Im}\left[\sqrt{\varepsilon - \sin^2 \beta}\right] = 0$. In the β-plane there are then two Riemann sheets connected along the branch cuts $\text{Im}\left[\left(\varepsilon - \sin^2 \beta\right)^{1/2}\right] = 0$ of the branch points β_{B_2}. Finally the poles in the ξ-plane are now given by $\varepsilon\cos\beta_P + \sqrt{\varepsilon - \sin^2 \beta_P} = 0$. Hence,

$$\sin \beta_P = \pm\sqrt{\frac{\varepsilon}{1+\varepsilon}} \quad \text{with} \quad \cos \beta_P = -\sqrt{\frac{1}{1+\varepsilon}}\tag{3A.41}$$

since $\text{Im}\,(k_1 \cos\beta) < 0$. The possible locations for the poles can be approximated by

$$\pm\sin \beta_P = \sqrt{\frac{\varepsilon}{1+\varepsilon}} \approx 1 - \frac{1}{2\varepsilon} \approx \cos\left(\frac{\pi}{2} \mp \beta_P\right)\tag{3A.42}$$

which results in

$$\beta_P \approx \pm\left(\frac{\pi}{2} \mp \frac{1}{\sqrt{\varepsilon}}\right)\tag{3A.43}$$

For the parameters of a highly conducting ground $\varepsilon_r = \varepsilon' (1-j)^2$ the locations of the poles and the branch cuts are pictorially depicted in Figure 3A.3 (not to scale). Out of the possible locations of the branch points and poles, B_2^3, B_2^4, P_2 and P_3 are situated on the upper Riemann sheet of the branch points β_{B_2} on which $\text{Im}\,(k_1 \cos\beta) < 0$. It is also important to note that none of the poles (P_2, P_3) are situated between the original path of the integration Γ_1 and the path of the steepest descent Γ_0. However, when the path of steepest descent Γ_0 lies in close proximity of the pole, special precautions must be taken in the evaluation of the integral of (3A.35) as carried out by Ott [37]. The pole P_1 is of no concern since it lies on the second Riemann sheet of the branch point β_{B_2} on which $\text{Im}\left(\sqrt{\varepsilon - \sin^2 \beta}\right) > 0$. The presence of the branch point B_2^3 should ordinarily be taken into account when ε is close to unity and when deforming the path Γ_1 to the path of steepest descent Γ_0. Often, the contribution along the borders of the branch cut would be a fast decreasing exponential that can be neglected in comparison to the contribution from the saddle point integration.

In short, (3A.35) becomes

$$g_{sv} \approx \int_{\Gamma_1} e^{-j\pi/4} \sqrt{\frac{2k_1 \sin\beta}{\pi R_2 \sin\theta}} \frac{\sqrt{\varepsilon - \sin^2 \beta}}{\varepsilon\cos\beta + \sqrt{\varepsilon - \sin^2 \beta}} \exp\left[-jk_1 R_2 \cos(\beta - \theta)\right] d\beta\tag{3A.44}$$

where Γ_1 is a path in the complex β plane as shown in Figure 3A.3. The path of steepest descent never crosses any of the poles. The contributions along the borders of the second branch cut associated with the branch point k_2, particularly for low values of the dielectric constant ε are not necessary as they would be fast decreasing exponentials [19–22] that can be neglected in comparison to the contribution from the saddle point integration.

Hence, by application of the method of steepest descent [(3B.2) and (3B.7) as explained in Appendix 3B] to (3A.44), one obtains for $\theta < \pi/2$, (i.e., when the pole is not near the saddle point)

$$
g_{sV} \approx \frac{2\exp[-jk_1 R_2]}{R_2} \frac{\sqrt{\varepsilon - \sin^2\theta}}{\varepsilon\cos\theta + \sqrt{\varepsilon - \sin^2\theta}}
$$
$$
\times \left[1 - \frac{1}{2jk_1 R_2} \left\{ \frac{\varepsilon(\varepsilon-1)\left[\begin{array}{l} 2\varepsilon(\varepsilon-1)+\varepsilon\cos^2\theta(3-\cos^2\theta)+ \\ \cos\theta\sqrt{\varepsilon-\sin^2\theta}(2\varepsilon+\sin^2\theta) \end{array} \right]}{\left(\varepsilon-\sin^2\theta\right)^2\left[\varepsilon\cos\theta+\sqrt{\varepsilon-\sin^2\theta}\right]^2} - \frac{1}{4\sin^2\theta} \right\} \right]
$$

$$(3A.45)$$

Hence $\Pi_{1z}^{reflected}$ of (3A.20) can now be written as

$$
\Pi_{1z}^{reflected} \approx P\frac{\exp\left(-jk_1 R_2\right)}{R_2}\left[\left(\frac{\varepsilon\cos\theta - \sqrt{\varepsilon - \sin^2\theta}}{\varepsilon\cos\theta + \sqrt{\varepsilon - \sin^2\theta}} \right) \right.
$$
$$
+ \frac{1}{jk_1 R_2}\left\{ \frac{\varepsilon(\varepsilon-1)\left[\begin{array}{l} 2\varepsilon(\varepsilon-1)+\varepsilon\cos^2\theta\left(3-\cos^2\theta\right)+ \\ \cos\theta\left(2\varepsilon+\sin^2\theta\right)\sqrt{\varepsilon-\sin^2\theta} \end{array} \right]}{\left(\varepsilon-\sin^2\theta\right)^{3/2}\left[\varepsilon\cos\theta+\sqrt{\varepsilon-\sin^2\theta}\right]^3} \right.
$$
$$
\left. \left. - \frac{\sqrt{\varepsilon-\sin^2\theta}}{2\sin^2\theta\left[\varepsilon\cos\theta+\sqrt{\varepsilon-\sin^2\theta}\right]} \right\} \right]
$$

$$(3A.46)$$

The first term of (3A.46) represents a spherical wave originating from the image and can be rewritten as

$$
\Pi_{1z}^{reflected} \approx P\Gamma_{TM}\frac{\exp\left(-jk_1 R_2\right)}{R_2}
$$

$$(3A.47)$$

where Γ_{TM} can be recognized *as the TM reflection coefficient associated with the spherical wave* [14, 19–22, 33], and is given by

$$\Gamma_{TM} = \frac{\varepsilon\cos\theta - \sqrt{\varepsilon - \sin^2\theta}}{\varepsilon\cos\theta + \sqrt{\varepsilon - \sin^2\theta}} \tag{3A.48}$$

The name reflection coefficient method is derived from (3A.47) since $\Pi_{1z}^{reflected}$ is now obtained as the reflection coefficient times the potential from the image of the source. **The method represents a good approximation, as long as the fields are computed far away from the ground plane and away also from the source dipole to ensure** $\theta < \pi/2$. This implies that the use of the reflection coefficient in the computations of the reflected fields are **NOT VALID NEAR THE GROUND**, where $\theta \approx \pi/2$ [4, 14, 34].

The total Hertz potential in medium 1, when the conductivity of the relative permittivity of the lower medium is large, i.e., $|\varepsilon| > 1$, is given by

$$\Pi_{1z} \approx P\left[\frac{\exp(-jk_1R_1)}{R_1} + \frac{\exp(-jk_1R_2)}{R_2}\right.$$
$$\left.\left\{\frac{\sqrt{\varepsilon}\cos\theta - 1}{\sqrt{\varepsilon}\cos\theta + 1} + \frac{2\varepsilon}{jk_1R_2}\left(\frac{1}{\sqrt{\varepsilon}\cos\theta + 1}\right)^3 + \ldots\right\}\right] \tag{3A.49}$$

Note that when $|\varepsilon| \to \infty$, Π_{1z} of (3A.49) goes properly into the form of a source plus an image term due to a vertical electric dipole located above a perfectly conducting ground plane.

However when $\theta \approx \pi/2$, and $\sqrt{\varepsilon}\cos\theta \ll 1$ it becomes

$$\Pi_{1z} \approx P\left[\frac{\exp(-jk_1R_1)}{R_1} - \frac{\exp(-jk_1R_2)}{R_2} + \frac{2\varepsilon}{jk_1R_2^2}\exp(-jk_1R_2) + \ldots\right] \tag{3A.50}$$

It is now important to recognize from (3A.50) that the sum of the first two terms may be smaller than the third term. As a matter of fact when both the transmitter and the receiver are near the ground, i.e.,

$$R_1 = R_2 \approx \rho \tag{3A.51}$$
$$z \approx 0 \approx z' \tag{3A.52}$$

then observe that the fields will solely be determined by the third and higher order terms of (3A.50). Also, there is no surface wave term in the expression and the dominant term behaves as $1/R^2$. The reason for this poor convergence in the vicinity of $\theta \approx \pi/2$ is that the effect of the pole close to $\pi/2$ becomes important. The bottom line is since it is the higher order terms that are responsible for the calculation of the fields along the interface in (3A.50), we need to carry out a different asymptotic expansion, starting using (3A.21) and not (3A.20). This new procedure is described in Appendix 3D. There is another point to be mentioned here, namely, the methodology of Appendix 3B is no longer applicable when a pole is located near a saddle point. This necessitates the development presented in Appendix 3C rather than using the methodology of Appendix 3B. The mathematical details are included in the appendices for completeness of the presentation.

Appendix 3B   Asymptotic Evaluation of the Integrals by the Method of Steepest Descent

The method of steepest descent (or the saddle point method) deals with the approximate evaluation of integrals of the form [39]

$$I(\rho) = \int_C F(\xi) \exp\left[-\rho f(\xi)\right] d\xi \tag{3B.1}$$

for large values of ρ, where the contour C in the complex ξ plane is such that that integrand goes to zero at the ends of the contour. The functions $f(\xi)$ and $F(\xi)$ are arbitrary analytic functions of the complex variable ξ.

The basic philosophy of the method of steepest descent is as follows: A path is selected in the complex ξ plane in such a way that the entire value of the integral is determined from a comparatively short portion of the path. Within certain limits, the contour of integration C may be altered to such a path without affecting the value of the integral. Then the integral is replaced by another, simpler function, which closely approximates the integrand over the essential portions of the path. The behavior of the new integrand outside the important portion of the path is of no concern. For real and positive values of ρ and for a general contour C the quantity $\rho f(\xi)$ is positive on some parts of the path and there are other regions where it is negative. The latter regions are more important since the integrand is larger, and in those regions, where the negative of $\text{Re}[\rho f(\xi)]$ is the largest, it is important to reduce oscillations. A contour is chosen along which the imaginary part of $[\rho f(\xi)]$ is constant in the region where the negative of its real part is largest. The path in the region where $\text{Re}[\rho f(\xi)]$ is greatest may be chosen so that $\text{Im}[\rho f(\xi)]$ varies if this turns out to be necessary to complete the contour. In this way, the oscillations of the integral cause the least trouble. Since the path of integration must pass along the line of most rapid increase and decrease of $\text{Re}[\rho f(\xi)]$ it must coincide with the line $\text{Im}[\rho f(\xi)] =$ constant, which may be a line of constant phase. The point of the path at which $\text{Re}[f(\xi)]$ is an extremum is called the saddle point and the derivative of $\text{Re}[f(\xi)]$ must be zero at this point. Since $\text{Im}[f(\xi)]$ is a constant on this path, then its derivative must also be zero, and therefore

$$\frac{df}{d\xi} = 0 \tag{3B.2}$$

at the saddle point. Thus the most advantageous path of integration must go through the saddle point along the line of the most rapid decrease of the function $\mathrm{Re}[f(\xi)]$, which coincides with the line $\mathrm{Im}[f(\xi)] =$ constant. This path then is called the path of steepest descent. If the saddle point occurs at $\xi = \xi_0$, then it follows that the path of integration will be determined from

$$f(\xi) = f(\xi_0) + s^2 \tag{3B.3}$$

where s is real and $-\infty \le s \le \infty$. The saddle point corresponds to the point $s = 0$.

Now going back to the integral (3B.1) and using (3B.3)

$$I_{SD} = \exp\left[-\rho f(\xi_0)\right] \int_{-\infty}^{\infty} F(\xi) \exp\left[-\rho s^2\right] d\xi \tag{3B.4}$$

If

$$\Phi(s) = F(\xi) \frac{d\xi}{ds} \tag{3B.5}$$

then

$$I_{SD} = \exp\left[-\rho f(\xi_0)\right] \int_{-\infty}^{\infty} \Phi(s) \exp\left[-\rho s^2\right] ds \tag{3B.6}$$

Now if ρ is large, then the integrand in (3B.6) will fall off rapidly with an increasing value of s, the distance from the saddle point. Thus only small values of s will contribute significantly and, therefore, we can expand $\Phi(s)$ in a Taylor series about the saddle points at 0. Therefore we can write,

$$\Phi(s) = \Phi(0) + s\Phi'(0) + \frac{s^2}{2}\Phi''(0) + \dots \tag{3B.7}$$

Substituting (3B.7) into (3B.6) one obtains

$$I_{SD} = \exp\left[-\rho f(\xi_0)\right] \int_{-\infty}^{\infty} \exp\left[-\rho s^2\right] \left\{ \Phi(0) + \frac{s^2}{2}\Phi''(0) + \dots \right\} ds \tag{3B.8}$$

as the odd powers of s will integrate to zero. Since

$$\int_{-\infty}^{\infty} \exp\left[-\rho s^2\right] s^{2n} ds = \frac{\sqrt{\pi}(2n)!}{n! 2^{2n}} \frac{1}{\rho^{n+0.5}} \tag{3B.9}$$

and substituting (3B.9) into (3B.8) yields

$$I_{SD} = \sqrt{\frac{\pi}{\rho}} \exp\left[-\rho f(\xi_0)\right]\left[\Phi(0) + \frac{1}{4\rho}\Phi''(0) + ...\right] \tag{3B.10}$$

Now we need to relate $\Phi(s)$ to $F(\xi)$ as described in (3B.5). To make the connection, we first expand $f(\xi)$ in a Taylor series around the saddle point $f(\xi_0)$, and if $\xi - \xi_0 = x$, then one obtains

$$f(\xi) = f(\xi_0) + \frac{x^2}{2!}f''(\xi_0) + \frac{x^3}{3!}f'''(\xi_0) + \frac{x^4}{4!}f^{IV}(\xi_0) + ... = f(\xi_0) + s^2 \tag{3B.11}$$

The goal here is to relate x, to a power series of s. To this end we get

$$x = a_0 s\left(1 + a_1 s + a_2 s^2 + a_3 s^3 + ...\right) \tag{3B.12}$$

And therefore

$$x^2 = a_0^2 s^2\left[1 + 2a_1 s + \left(2a_2 + a_1^2\right)s^2 + ...\right] \tag{3B.13}$$

$$x^3 = a_0^3 s^3\left[1 + 3a_1 s + 3\left(a_1^2 + a_2\right)s^2 + ...\right] \tag{3B.14}$$

$$x^4 = a_0^4 s^4\left[1 + 4a_1 s + 2\left(2a_2 + 3a_1^2\right)s^2 + ...\right] \tag{3B.15}$$

Rewriting (3B.11) we get

$$s^2 = Ax^2 + Bx^3 + Cx^4 + ... \tag{3B.16}$$

where

$$A = \frac{f''(\xi_0)}{2}; \quad B = \frac{f'''(\xi_0)}{6}; \quad C = \frac{f^{IV}(\xi_0)}{24} \tag{3B.17}$$

Now substituting (3B.12)–(3B.15) into (3B.16), we get

$$s^2 = A\left[a_0^2 s^2 + 2a_1 a_0^2 s^3 + \left(a_1^2 + 2a_2\right)a_0^2 s^4\right] \\ + B\left[a_0^3 s^3 + 3a_1 a_0^3 s^4 + ...\right] + C\left[a_0^4 s^4 + ...\right] \tag{3B.18}$$

from which

$$a_0 = \frac{1}{\sqrt{A}}; \quad a_1 = -\frac{B}{2A^{3/2}}; \quad a_2 = -\frac{C}{2A^2} + \frac{5}{8}\frac{B^2}{A^3} \tag{3B.19}$$

Next we expand the function $F(\xi)$ in a Taylor series around the saddle point ξ_0, and with $\xi - \xi_0 = x$, we get

$$F(\xi) = F(\xi_0)\left[1 + Px + Qx^2 + \ldots\right] \tag{3B.20a}$$

and

$$P = \frac{F'(\xi_0)}{F(\xi_0)} \quad \text{and} \quad Q = \frac{F''(\xi_0)}{2F(\xi_0)} \tag{3B.20b}$$

Using (3B.11), (3B.13)–(3B.15), (3B.19), and (3B.20), we get

$$
\begin{aligned}
\Phi(s) &= F(\xi)\frac{d\xi}{ds} = F(\xi)\frac{dx}{ds} \\
&= F(\xi)\left[a_0\left(1 + a_1 s + a_1 s^2 + \ldots\right) + a_0 s\left(a_1 + 2a_2 s + \ldots\right) + \right] \\
&= F(\xi)\left[a_0 + s(2a_0 a_1) + s^2\left(a_0 a_2^2 + 2a_2 a_0\right) + \ldots\right] \\
&= F(\xi_0)\left[1 + Px + Qx^2 + \ldots\right]\left[a_0 + 2a_0 a_1 s + s^2\left(a_0 a_2^2 + 2a_2 a_0\right) + \ldots\right] \\
&= F(\xi_0)\left[1 + a_0 Ps + a_0 a_1 Ps^2 + a_0 a_2 Ps^3 + a_0^2 Qs^2 + 2a_0^2 a_1 Qs^3 + \ldots\right] \\
&\quad \times\left[a_0 + 2a_0 a_1 s + 3a_0 a_2 s^2 + \ldots\right] \\
&= F(\xi_0)\left[a_0 + s\left(a_0^2 P + 2a_0 a_1\right) + s^2\left(a_0^2 a_1 P + 2a_0^2 a_1 P + 3a_0 a_2 + a_0^3 Q\right) + \ldots\right] \\
&= \frac{F(\xi_0)}{\sqrt{A}}\left[1 + s\left(\frac{P}{\sqrt{A}} - \frac{B}{A^{3/2}}\right) + s^2\left(\frac{Q}{A} - 3\left\{\frac{C}{2A^2} - \frac{5B^2}{8A^3}\right\} - \frac{3}{\sqrt{A}}\frac{BP}{2A^{3/2}}\right) + \ldots\right] \\
&= \frac{F(\xi_0)}{\sqrt{A}}\left[1 + s\left(\frac{P}{\sqrt{A}} - \frac{B}{A^{3/2}}\right) + s^2\left(\frac{Q}{A} + \frac{15B^2}{8A^3} - \frac{3BP}{2A^2} - \frac{3C}{2A^2}\right) + \ldots\right]
\end{aligned}
$$

$$\tag{3B.21}$$

Therefore

$$\Phi(0) = \sqrt{\frac{2}{f''(\xi_0)}}\, F(\xi_0) \tag{3B.22}$$

$$\Phi''(0) = 2\Phi(0)\left[\frac{F''}{Ff''} + \frac{5\left(f'''\right)^2}{12\left(f''\right)^3} - \frac{f'''}{\left(f''\right)^2}\frac{F'}{F} - \frac{f^{iv}}{4\left(f''\right)^2}\right] \tag{3B.23}$$

Hence

$$
I_{sd} = \sqrt{\frac{2\pi}{\rho f''(\xi_0)}} F(\xi_0) \exp\left[-\rho f(\xi_0)\right]
$$

$$
\times \left\{ 1 + \frac{1}{2\rho} \left[\frac{F''}{Ff''} + \frac{5\left(f'''\right)^2}{12\left(f''\right)^3} - \frac{f'''}{\left(f''\right)^2} \frac{F'}{F} - \frac{f^{IV}}{4\left(f''\right)^2} \right] + \ldots \right\}
$$

(3B.24)

The interesting point regarding the result of (3B.24) is that it is a divergent series for a fixed ρ as the number of terms in this expansion increases for a fixed value of ρ. Such a series is called an asymptotic series as introduced by Poincaré.

A divergent series is one

$$
g(\rho) = A_0 + \frac{A_1}{\rho} + \frac{A_2}{\rho^2} + \ldots + \frac{A_n}{\rho^n}
$$

(3B.25)

in which the sum of the first $(n+1)$ terms is $S_n(z)$, given by

$$
S_n(\rho) = \sum_{L=0}^{n} A_i \rho^{-i}
$$

(3B.26)

which is said to be an asymptotic expansion of a function $g(\rho)$ for a given range of argument ρ if the expression

$$
R_n(\rho) = \rho^n \left\{ g(\rho) - S_n(\rho) \right\}
$$

(3B.27)

satisfies the condition

$$
\lim_{|\rho| \to \infty} R_n(\rho) = 0 \quad \text{for } n \text{ fixed.}
$$

(3B.28)

even though

$$
\lim_{n \to \infty} \left| R_n(\rho) \right| = \infty, \quad \rho - \text{ fixed.}
$$

(3B.29)

When this is the case, one can make

$$
\left| R_n(\rho) \right| = \left| \rho^n \left\{ g(\rho) - S_n(\rho) \right\} \right| < \varepsilon
$$

(3B.30)

where ε is arbitrarily small, by making $|\rho|$ sufficiently large. Some of the properties of this definition are:

a) Asymptotic expansions can be multiplied unconditionally.
b) Asymptotic expansions can be integrated unconditionally.

c) An asymptotic expansion of a function is unique.
d) One asymptotic expansion may represent several functions.
e) Asymptotic expansions can be divided providing the divisor contains at least one non-zero coefficient.

The point about the series of (3B.25) is that for sufficiently large values of $|\rho|$ the terms of the series decrease at least initially, and that if the series is truncated before the smallest term, the error is of the order of magnitude of the first discarded term. So, if $\Phi(s)$ is any function in (3B.6) for which $\int_{-\infty}^{\infty} \Phi(s) \exp\left[-\rho s^2\right] ds$ converges for sufficiently large values of the parameter ρ, then the asymptotic expansion of (3B.6) in descending powers of $\sqrt{\rho}$ can be given by replacing $\Phi(s)$ by a Taylor series in ascending powers of s in (3B.7) and then integrating term by term. In that case (3B.24) is the asymptotic expansion of (3B.6). In terms of the problem for our case, the integrals that we will be dealing with are of the following form

$$I(kR) = \int F(\beta) \exp\left[-jkR\cos(\beta - \theta)\right] d\beta \tag{3B.31}$$

so that

$$\rho = jkR \tag{3B.32}$$

$$f(\beta) = \cos(\beta - \theta) \tag{3B.33}$$

The saddle point then occurs at $\beta = 0$ and we get

$$f(\theta) = 1, \ f^{I}(\theta) = 0, \ f^{II}(\theta) = -1, \ f^{III}(\theta) = 0, \ f^{IV}(\theta) = 1.$$

Substituting these values in (3B.24) we get

$$I_{SD}(kR) = \sqrt{\frac{2\pi j}{kR}} F(\theta) e^{-jkR} \left\{ 1 + \frac{j}{2kR}\left(\frac{F^{II}(\theta)}{F(\theta)} + \frac{1}{4} \right) \right\} + \dots \tag{3B.34}$$

Appendix 3C Asymptotic Evaluation of the Integrals When there Exists a Pole Near the Saddle Point

The asymptotic expansion given by (3B.34) is not valid if there is a pole near the saddle point θ. However, the method of steepest descent can be modified in such a way that the presence of poles is taken into account from the very beginning in evaluating these integrals. Of special interest in the analysis will be an integral of the form

$$I(kR) = \int_{\Gamma_1} F_1(\beta) \exp\left[-jkR\cos(\beta - \theta)\right] d\beta \tag{3C.1}$$

where Γ_1 is a path of integration in the complex β plane as discussed in the previous sections. $F_1(\beta)$ now has a pole β_P near the saddle point θ. For large values of kR, the pole can be factored out from $F_1(\beta)$ by writing $F_1(\beta) = \dfrac{F(\beta)}{\sin\left(\dfrac{\beta - \beta_P}{2}\right)}$. It is then argued that since $F(\beta)$ has no singularities in the vicinity of the saddle point, it may be removed from under the integral sign with β equated to θ, as presented by Clemmow [40]. Thus the integral of (3C.1) can be written as

$$I_{SD}(kR) = F(\theta) \int_{\Gamma_1} \frac{\exp\left[-jkR\cos(\beta - \theta)\right]}{\sin\left(\dfrac{\beta - \beta_P}{2}\right)} d\beta = F(\theta) \int_{\Gamma_0} \frac{\exp\left[-jkR\cos\alpha\right]}{\sin\left(\dfrac{\alpha + \theta - \beta_P}{2}\right)} d\alpha \tag{3C.2}$$

where $\alpha = \beta - \theta$. By reversing the sign of α as

$$I_{SD}(kR) = F(\theta) \int_{\Gamma_0} \frac{\exp\left[-jkR\cos\alpha\right]}{\sin\left(\dfrac{\theta - \alpha - \beta_P}{2}\right)} d\alpha \tag{3C.3}$$

and then adding (3C.2) to (3C.3) and then dividing by two will convert (3C.2) to

$$I_{SD}(kR) = 2\sin\frac{\gamma}{2} F(\theta) \int_{\Gamma_0} \frac{\exp\left[-jkR\cos\alpha\right]}{\cos\alpha - \cos\gamma} \cos\frac{\alpha}{2} d\alpha \tag{3C.4}$$

where $\gamma = \theta - \beta_P$. Now by changing the variable of integration from α to τ suchcorrections

$$\tau = \sqrt{2} \exp\left[-j\pi/4\right] \sin\frac{\alpha}{2} \tag{3C.5}$$

the path Γ_0 is now transformed to an integral from $-\infty$ to $+\infty$. Hence

$$I_{SD}(kR) = 2b \exp\left[-jkR + j3\pi/4\right] F(\theta) \int_{-\infty}^{\infty} \frac{e^{-kR\tau^2}}{\tau^2 + jb^2} d\tau \tag{3C.6}$$

where

$$b = \sqrt{2} \sin \frac{\gamma}{2} \qquad (3C.7)$$

since

$$\int_{-\infty}^{\infty} \frac{e^{-kR\tau^2}}{\tau^2 + jb^2} d\tau = \frac{\pi}{b} \exp\left[j\left(b^2 kR - \frac{\pi}{4} \right) \right] \times erfc\left(\sqrt{jkRb^2} \right) \qquad (3C.8)$$

and

$$W^2 = -jkRb^2 = -j2kR\sin^2\left(\frac{\theta - \beta_P}{2} \right) \qquad (3C.9)$$

then (3C.6) becomes

$$I_{SD}(kR) = 2\pi jF(\theta)\exp[-jkR - W^2]erfc(jW). \qquad (3C.10)$$

where *erfc* represents the complementary error function. This completes the derivation.

It is important to stress the point that only Clemmow [40] and Hill and Wait [41], Wait [23] follow the developments given by (3C.2). Others, like Tyras [39], Collin [21] and Karawas [42] follow a different procedure for handling the pole and their formulation thus yields a different expression. We follow a specific procedure because it provides a similar expression as to what we are looking for!

Appendix 3D   Evaluation of Fields Near the Interface

In order to solve for the total fields near the interface [14], a modified saddle point method as explained in Appendix 3C is applied to take into account the effect of the pole β_P near the saddle point. In the expression of both g_{sV} and G_{sV} in (3A.20) and (3A.21) there is a pole β_P which is seen from [4, 14]

$$\frac{1}{\left[\varepsilon \cos\beta + \sqrt{\varepsilon - \sin^2\beta} \right]} = \frac{1}{\varepsilon^2 - 1} \times \frac{\sqrt{\varepsilon - \sin^2\beta} - \varepsilon\cos\beta}{\sin(\beta + \beta_P)\sin(\beta - \beta_P)} \qquad (3D.1)$$

where $\varepsilon\cos\beta_P + \sqrt{\varepsilon - \sin^2\beta_P} = 0$ with $\sin\beta_P = \pm\sqrt{\dfrac{\varepsilon}{\varepsilon + 1}}$ and $\cos\beta_P = -\sqrt{\dfrac{1}{\varepsilon + 1}}$.

Applying (3D.1) and (3B.11) to (3B.21) we obtain

$$G_{sV} = \varepsilon \exp\left(-j\frac{\pi}{4}\right) \int_{\Gamma_1} \left(\frac{2k_1 \sin\beta}{\pi R_2 \sin\theta}\right)^{1/2} \frac{\exp\left[-jk_1 R_2 \cos(\beta-\theta)\right]\cos\beta}{\varepsilon \cos\beta + \sqrt{\varepsilon - \sin^2\beta}} d\beta$$

$$\approx \varepsilon \sqrt{\frac{4\pi k_1 j}{R_2}} \frac{\cos\theta}{\cos\theta - \frac{1}{\sqrt{\varepsilon+1}}} \frac{\sqrt{\varepsilon - \sin^2\theta} - \varepsilon\cos\theta}{\varepsilon^2 - 1} \frac{\exp\left[-jk_1 R_2 - W^2\right] erfc(jW)}{\sqrt{1 + \frac{\cos\theta}{\sqrt{\varepsilon+1}} + \frac{\sqrt{\varepsilon}\sin\theta}{\sqrt{\varepsilon+1}}}}$$

$$(3D.2)$$

where

$$W^2 = -jk_1 R_2 2\sin^2\left(\frac{\theta - \beta_P}{2}\right) = -jk_1 R_2 \left[1 + \frac{\cos\theta}{\sqrt{\varepsilon+1}} - \frac{\sqrt{\varepsilon}\sin\theta}{\sqrt{\varepsilon+1}}\right] \qquad (3D.3)$$

and W was called *the numerical distance* by Sommerfeld [13]. If $|\varepsilon| > 1$ and $\theta \approx \pi/2$, and when W is very small, then we have $\exp[-W^2] erfc(jW) \approx 1$. Under this assumption, applying (3A.34) to (3D.2), one gets a simplified expression for

$$G_{sV} \approx -\sqrt{\frac{2\pi k_1 j}{R_2}} \exp\left[-jk_1 R_2\right]\frac{(z+z')}{R_2} \frac{\varepsilon}{\sqrt{\varepsilon^2-1}} \approx -\sqrt{2\pi k_1 j} \frac{(z+z')\exp\left[-jk_1 R_2\right]}{R_2^{1.5}}$$

$$(3D.4)$$

Equation (3D.4) thus illustrates that when $\theta \approx \pi/2$ the dominant term of the potential $\Pi_{1z} \propto R_2^{-1.5}$ and therefore the leading term for the fields will be also varying as $\rho^{-1.5}$, if $(z + z')$ is small compared to ρ in (3A.18). It is interesting to observe that Eq. (3D.4) is not a function of the ground parameters ε.

However as W becomes large then

$$\exp\left[-W^2\right] erfc(jW) \approx \frac{-j}{W\sqrt{\pi}}\left[1 + \frac{1}{2W^2}\right] \quad \text{for } |W| \to \infty \text{ and } \left|\arg W\right| < \frac{3\pi}{4}$$

$$(3D.5)$$

and for $|\varepsilon| > 1$, $W^2 \approx \frac{-jk_1 R_2}{2\varepsilon}$. Under this condition,

$$G_{sV} \approx 2\sqrt{\varepsilon}\exp\left[-jk_1 R_2\right]\frac{(z+z')}{R_2^2}\left[1 - \frac{\varepsilon}{jk_1 R_2}\right] \qquad (3D.6)$$

Thus the total Hertz potential in medium 1 which is valid near the interface for $|\varepsilon| > 1$ and $\theta \approx \pi/2$ becomes:

$$\Pi_{1z} \approx \begin{cases} P\left[\dfrac{\exp(-jk_1R_1)}{R_1} - \dfrac{\exp(-jk_1R_2)}{R_2} \\ -\sqrt{j2\pi k_1}\,(z+z')\dfrac{\exp(-jk_1R_2)}{R_2^{1.5}}\right], & W<1 \\[4pt] P\left[\dfrac{\exp(-jk_1R_1)}{R_1} - \dfrac{\exp(-jk_1R_2)}{R_2} \\ +2\sqrt{\varepsilon}\,(z+z')\dfrac{\exp(-jk_1R_2)}{R_2^{2}}\left[1-\dfrac{\varepsilon}{jk_1R_2}\right]\right], & W>1 \end{cases} \tag{3D.7}$$

The above simplified expressions illustrate that a Norton surface wave [4, 14] decays asymptotically as $1/R^2$ and this applies only in the far field region, where $W > 1$, as the first two terms cancel in the second expression. Also, it is interesting to note that the third term for $W > 1$ provides the so called height-gain for the transmitting and receiving antennas. However, this height gain applies to both intermediate and far field regions. In the intermediate region, the fields decay as approximately $\rho^{-1.5}$. Also, observe that for $W < 1$, the above expression is independent of the ground parameters. This is confirmed in Figure 3.15, using a more accurate numerical analysis. It is important to note that this Sommerfeld representation for the fields is not valid when z and z' are close to 0.

In summary, Sommerfeld characterizes W as the numerical distance. And when the large argument approximation is invoked for W then the fields decay as $1/R^2$. Interestingly, this is one of the confusing parts in all the discussions as Sommerfeld stated: *"for small values of the numerical distance the spatial-wave type predominates in the expression for the reception intensity; in this case the ground peculiarities have no marked influence and we can make computations using an infinite ground conductivity without introducing great errors. For larger values of W the rivalry between the space and the surface waves is apparent"*. However, our interests in a cellular wireless communication system are for small to intermediate values of W, for which very little information is available. Our observations for the path loss exponent in a cellular wireless communication is 3 for moderate distances where the fields vary as $1/R^{1.5}$ from the base station antenna and in the fringe regions (i.e., further away from the antenna base station) it is 4 which has been verified by a more accurate numerical analysis and experimental results! In addition, in this region, the ground parameters have little effect as seen by Eq. (3D.4).

At this point, it is important to point out that the novelty of our solution, which is not available in the popular literature, lies for the intermediate regions where we have used two different procedures that deviated from the classical formulations. First, we used the second form of the Green's function as shown in (3A.21) to observe the fields near the interface. And secondly, we used a different saddle point method of integration in handling the pole near the saddle point in evaluating the integrals of (3A.21) which was outlined by Clemmow [40] and also used by Hill and Wait [41].

If one applies the modified saddle point method for evaluating the integral, as explained in the Appendix 3C, to the Green's function given by (3A.20) then one obtains

$$
g_{sV} \approx \exp\left(\frac{-j\pi}{4}\right)\sqrt{\frac{2k_1}{\pi R_2}} \frac{\left(\sqrt{\varepsilon - \sin^2\theta}\right)\left(\sqrt{\varepsilon - \sin^2\theta} - \varepsilon\cos\theta\right)}{\left(\varepsilon^2 - 1\right)\left(\cos\theta - 1./\sqrt{\varepsilon + 1}\right)\left(2\sin\frac{(\theta + \beta_P)}{2}\right)}
$$

$$
\int_{\Gamma_1} \frac{\exp\left[-jk_1 R_2 \cos(\beta - \theta)\right]}{\sin\frac{(\beta - \beta_P)}{2}} d\beta
$$

$$
\approx \sqrt{\frac{4\pi k_1 j}{R_2}} \frac{\sqrt{\varepsilon - \sin^2\theta} - \varepsilon\cos\theta}{\cos\theta - 1./\sqrt{\varepsilon + 1}} \frac{\sqrt{\varepsilon - \sin^2\theta}}{\varepsilon^2 - 1} \frac{\exp\left[-jk_1 R_2 - W^2\right]\mathrm{erfc}(jW)}{\sqrt{1 + \frac{\cos\theta}{\sqrt{\varepsilon + 1}} + \frac{\sqrt{\varepsilon}\sin\theta}{\sqrt{\varepsilon + 1}}}}
$$

$$
\tag{3D.8}
$$

Now for small values of W, and when the fields are desired close to the interface then we also require $\theta \approx \pi/2$. In this case we obtain for $|\varepsilon| > 1$

$$
g_{sV} \approx -\sqrt{\frac{2\pi k_1 j}{R_2}} \frac{\exp\left[-jk_1 R_2\right]}{\sqrt{\varepsilon + 1}} \tag{3D.9}
$$

By incorporating (3D.9) into (3A.20), it is seen that the space wave term dominates and the additional contribution of the surface wave term given by (3D.9) and as predicted by Sommerfeld, is small in magnitude.

However when W is large then

$$
g_{sV} \approx \frac{2\exp\left[-jk_1 R_2\right]}{R_2}\left[1 - \frac{\varepsilon}{jk_1 R_2}\right] \tag{3D.10}
$$

Substituting this expression in (3A.20) it is seen than the dominant terms for the space waves cancel each other and the Hertz potential is given by the higher order terms and therefore

$$
\Pi_{1z} \approx \frac{2P\varepsilon \exp\left[-jk_1 R_2\right]}{jk_1 R_2^2} \tag{3D.11}
$$

A similar asymptotic form was obtained previously, as seen in (3A.50). That is why we expanded the Hertz potential in a different form given by (3A.21) which cancelled the space waves and provided the dominant ground wave term when both the transmitter and the receiver are close to the ground. The rational for using this special treatment for the potential has been explained by Stratton [19] as the reflection coefficient is +1 for a perfect ground when the fields are evaluated far from the interface but then it transforms to −1 when the fields are evaluated near the interface. This second form was also used originally by Sommerfeld.

In short, there are two unique features of this presentation as it differs from the other researchers' work. First, it is the use of (3A.21) in the modified saddle point method to calculate the fields in the regions both near and far from the base station antenna. And second, use of a different mathematical form when applying the modified saddle point method when there is a pole near the saddle point as explained in Appendix 3C.

Appendix 3E   Properties of a Zenneck Wave

The main contribution of Zenneck was the development of a specific type of solution of Maxwell's equations in a three dimensional space. This solution is an inhomogeneous type of plane wave and generally occurs at a zero of the TM reflection coefficient – the Brewster angle – whose field components can be derived as follows: In the three dimensional rectangular coordinates, consider that the plane z = 0 is the boundary between medium 1, free space, and medium 2 which is of arbitrary parameters $(\varepsilon, \mu, \sigma)$. Zenneck showed that there exists a solution for Maxwell's equation in this two-layer problem. This solution represents a wave that has progressive phase propagation in the x-direction, while at the same time decays exponentially in the positive and negative z-directions. This wave has to be a TM wave with respect to the x-z plane. The field components for such a wave in medium 1 are given by:

$$E_{1x} = E_1 \exp\left(-jk_x x - k_{1z} z\right), \qquad z > 0 \qquad\qquad (3E.1)$$

$$H_{1y} = H_1 \exp\left(-jk_x x - k_{1z} z\right), \qquad z > 0 \qquad\qquad (3E.2)$$

where a harmonic time variation of frequency ω is assumed. The corresponding forms in medium 2 are given by:

$$E_{2x} = E_2 \exp\left(-jk_x x + k_{2z} z\right), \qquad z < 0 \qquad\qquad (3E.3)$$

$$H_{2y} = H_2 \exp\left(-jk_x x + k_{2z} z\right), \qquad z < 0 \qquad\qquad (3E.4)$$

where E_1, E_2, H_1 and H_2 are constant values representing the amplitude of the field components. k_x is the magnitude of the propagation vector component in the direction tangential to the boundary. k_{1z} and k_{2z} represent the propagation constants in the positive and negative z-directions respectively. According to Zenneck, the real parts of k_{1z} and k_{2z} should be positive. Thus the wave given in (3E.1)–(3E.4) decays exponentially away from the boundary which lies at $z = 0$ plane, but is oscillatory in addition. The dispersion relation relating the propagation vector components is determined by the value of the frequency ω and the parameters of the medium (ε, μ, σ). The field of a Zenneck wave decays exponentially in amplitude and suffers a progressive advance in phase with increasing distance above the surface. Also, the wave is attenuated in the direction of propagation whilst subject to a progressive lag in phase along the interface.

Appendix 3F   Properties of a Surface Wave

Schelkunoff [34] wrote in 1959: "For obvious reasons the same word conveys different meanings to different individuals. Hence, some "noise" in communication between us is unavoidable. As long as the noise level is relatively low, we manage to understand each other reasonably well. When the noise level becomes high, serious misunderstandings are inevitable, and needless as well as wasteful controversies may arise. Such a situation has arisen in microwave theory in connection with the so called Surface waves". Schelkunoff recognized eleven types of different electromagnetic phenomena, all of which are called "surface waves". He concluded: "The loose use of the term surface wave is unfortunate and causes a great deal of unnecessary confusion. If it is continued, the best that one could hope for is that the term will become entirely devoid of meaning. This writer (Schelkunoff) hopes however, that the classical definition of the term (Lord Rayleigh's) will be restored. Sommerfeld and Zenneck adhered to it, although they have made an unfortunate slip in their analysis which subsequently confused the issue".

In the original Sommerfeld formulation [13] there was no error in the sign, but the presentation by Sommerfeld was not complete. The effect of the pole does not contribute to the total solution makes sense as in some cases the poles may not exist on the proper Riemann sheet, as we have seen particularly for real values of the dielectric constant. Also, when the hyperbolic branch cuts in the Sommerfeld's solution is replaced by vertical branch cuts the poles migrate into a different Riemann sheet and are not relevant [21, 22]! Therefore, since the choice of the branch cuts are arbitrary and so then are the inclusion of the poles in the computations and hence the surface waves are also not clearly present as they depend on the location of the poles on the proper Riemann sheet which not only depends on the value of ε but also how the branch cuts are arbitrarily chosen. Hence one cannot associate physics with objects that are not clearly defined. It is therefore important that we look at the total solution and not the component solutions. The situation was exactly illustrated by Goubau. The very elegant remark made by Goubau in his IRE Transactions paper [43] on *Waves on interfaces* in 1959 is quite revealing in clarifying the situation: "*The existence of the Zenneck wave has been disputed so much in the literature that it may be excusable if a few more remarks are added to this subject. The dispute originated in the attempt to extract from the mathematical solution of a problem more answers than there were questions when the problem was formulated mathematically.*

If one formulates the problem of a dipole radiating above a plane interface, one can only expect a solution which describes the total field of the dipole, no matter what mathematical method one uses. There is no obligation from the physical point of view to use complex variables for solving the problem. The fact that the problem is more accessible to a rigorous treatment, when solved in the complex plane, is a purely mathematical matter and one cannot expect this method to yield more information than a treatment with only real quantities. If Sommerfeld had solved the problem by use of real quantities only, it is unlikely that the question as to a surface wave and a space wave would ever have arisen (except perhaps now) since the actual field of a dipole at the interface differs substantially from that of the surface wave.

The fact that the answer to a given problem can be written in terms of a complex integral whose integrand comprises a pole, does not a priori mean that this pole has physical meaning. Even if this is the case, there is still no obligation that the integration has to be performed in such a manner that the pole is included.

In order to separate the field into two components, it is necessary to define both these components. Defining only one, namely the surface wave, is not enough. However, I believe that the orthogonality relations quoted in this paper should fill this gap and present a satisfactory basis for the separation". In addition, a surface wave is a slow wave and hugs closer to the surface with an increase in frequency. This is not true for the Zenneck wave which is a fast radiating wave and its variation in the transverse direction to the propagation is essentially independent of frequency [32, 34]. Surface Plasmon Polaritons (SPP) are excited not by transverse electromagnetic (TEM) waves but by quasi particles as the reflected and the transmitted wave exist without an incident electromagnetic wave [33]. These SPP are excited by an evanescent wave which behaves as a quasiparticle [44] in stirring up the electron oscillations in a metal.

Surface plasmons are excited in metal foils of special thickness by quasi particles and not by a transverse electromagnetic wave since surface waves are TM waves with a longitudinal field component. Surface plasmons should not be related to Zenneck waves and the exciting conditions for the SPP should be strictly defined. Unfortunately, a clear statement that SPP wave is indeed a TM surface wave with a longitudinal field component and not a TM Zenneck wave is missing in most recent works on terahertz surface plasmons. SPPs are true surface waves with a longitudinal field component and are generated when the two medium has permittivity of opposing signs. However, this is not sufficient for the existence of the pole as many metals has a negative permittivity at terahertz frequencies but the conductivity is still too large. For metals at petahertz frequencies, when the conductive losses become smaller than the permittivity then only the surface wave pole of the TM reflection coefficient will manifest itself. Surface waves are thus in no way related to a Zenneck wave which is generated due to a Brewster zero – the zero of the TM reflection coefficient.

References

1 D. Gabor, "Communication Theory and Physics," *IRE Transactions on Information Theory*, Vol. 1, No. 1, pp. 48–59, Feb. 1953.

2 M. N. Abdallah, W. Dyab, T. K. Sarkar, M. V. S. N. Prasad, C. S. Misra, A. Garcia Lampérez, M. Salazar-Palma, and S. W. Ting, "Further Validation of an Electromagnetic Macro Model for Analysis of Propagation Path Loss in Cellular Networks Using Measured Driving-Test Data," *IEEE Antennas and Propagation Magazine*, Vol. 56, No. 4, pp. 108–129, Aug. 2014.

3 T. K. Sarkar, W. Dyab, M. N. Abdallah, M. Salazar-Palma, M. V. S. N. Prasad, S. Barbin, and S. W. Ting, "Physics of Propagation in a Cellular Wireless Communication Environment," *Radio Science Bulletin*, No. 343, pp. 5–21, Dec. 2012. http://www.ursi.org/files/RSBissues/RSB_343_2012_12.pdf. Accessed on November 21, 2017.

4 T. K. Sarkar, W. Dyab, M. N. Abdallah, M. Salazar-Palma, M. V. S. N. Prasad, S. W. Ting, and S. Barbin, "Electromagnetic Macro Modeling of Propagation in Mobile Wireless Communication: Theory and Experiment," *IEEE Antennas and Propagation Magazine*, Vol. 54, No. 6, pp. 17–43, Dec. 2012.

5 A. De, T. K. Sarkar, and M. Salazar-Palma, "Characterization of the Far Field Environment of Antennas Located over a Ground Plane and Implications for Cellular Communication Systems," *IEEE Antennas and Propagation Magazine*, Vol. 52, No. 6, pp. 19–40, Dec. 2010.

6 Y. Okumura, E. Ohmori, T. Kawano, and K. Fukuda, "Field Strength and Its Variability in VHF and UHF Land Mobile Service," *Review of the Electrical Communication Laboratory*, Vol. 16, No. 9–10, pp. 825–873, 1968.

7 K. Fujimoto, *Mobile Antenna Systems Handbook*, Third Edition, Artech House, Norwood, MA, 2008.

8 H. Zaghloul, G. Morrison, D. Tholl, M. G. Fry, and M. Fattouche, "Measurement of the Frequency Response of the Indoor Channel," *Electronics Letters*, Vol. 27, No. 12, pp. 1021–1022, 1991.

9 C. R. Burrows, "The Surface Wave in Radio Propagation over Plane Earth," *Proceedings of the IRE*, Vol. 25, pp. 219–229, Feb. 1937.

10 M. Lazarus, "The Great Spectrum Famine," *IEEE Spectrum*, Vol. 47, No. 10, pp. 26–31, Oct. 2010.

11 R. W. McMillan, Terahertz Imaging, Millimeter-Wave Radar. www.nato-us.org/sensors2005/papers/mcmillan.pdf. Accessed on November 21, 2017.

12 Y. Zhang, T. K. Sarkar, X. Zhao, D. Garcia-Donoro, W. Zhao, M. Salazar-Palma, and S. Ting, *Higher Order Basis Based Integral Equation Solver (HOBBIES)*, John Wiley & Sons, Inc., Hoboken, NJ, 2012.

13 A. N. Sommerfeld, "Propagation of Waves in Wireless Telegraphy," *Annals of Physics*, Vol. 28, pp. 665–736, Mar. 1909.

14 T. K. Sarkar, "Analysis of Arbitrarily Oriented Thin Wire Antennas over a Plane Imperfect Ground," *AEÜ*, Band 31, Heft 11, pp. 449–457, 1977.

15 A. R. Djordjevic, M. B. Bazdar, T. K. Sarkar, and R. F. Harrington, *AWAS Version 2.0: Analysis of Wire Antennas and Scatterers, Software and User's Manual*, Artech House, Norwood, MA, 2002.

16 W. M. Dyab, T. K. Sarkar, and M. Salazar-Palma, "A Physics-Based Green's Function for Analysis of Vertical Electric Dipole Radiation over an Imperfect Ground Plane", *IEEE Transactions on Antennas and Propagation*, Vol. 61, No. 8, pp. 4148–4157, Aug. 2013.

17 T. K. Sarkar, W. M. Dyab, M. N. Abdallah, M. Salazar-Palma, M. V. S. N. Prasad, and S. W. Ting, "Application of the Schelkunoff Formulation to the Sommerfeld Problem of a Vertical Electric Dipole Radiating Over an Imperfect Ground," *IEEE Transactions on Antennas and Propagation*, Vol. 62, No. 8, pp. 4162–4170, Aug. 2014.

18 "IEEE Standard Definitions of Terms for Radio Wave Propagation," IEEE Std. 211-1997, 1998. See also: http://ieeexplore.ieee.org/stamp/stamp.jsp?tp=&arnumber=705931&userType=inst&tag=1. Accessed on November 21, 2017.

19 J. A. Stratton, *Electromagnetic Theory*, McGraw-Hill Book Company, New York, 1941.

20 R. E. Collin, "Hertzian Dipole Radiating over a Lossy Earth or Sea: Some Early and Late 20th-Century Controversies," *IEEE Antennas and Propagation Magazine*, Vol. 46, No. 2, pp. 64–79, Apr. 2004.

21 A. Ishimaru, *Electromagnetic Wave Propagation, Radiation, and Scattering*, Prentice Hall, Englewood Cliffs, NJ, 1991, Chapter 15 and Appendix to Chapter 15.

22 A. Baños, Jr., *Dipole Radiation in the Presence of a Conducting Half-Space*, Pergamum Press, Oxford, England, 1966.

23 J. R. Wait, "The Ancient and Modern History of EM Ground-Wave Propagation," *IEEE Antennas and Propagation Magazine*, Vol. 40, No. 5, pp. 7–24, Apr. 1998.

24 B. Van der Pol, "Theory of the Reflection of Light from a Point Source by a Finitely Conducting Flat Mirror with Application to Radiotelegraphy," *Physics*, Vol. 2, pp. 843–853, Aug. 1935.

25 A. N. Sommerfeld, "Propagation of Waves in Wireless Telegraphy," *Annals of Physics*, Vol. 81, pp. 1135–1153, Dec. 1926.

26 H. Weyl, "Propagation of Electromagnetic Waves over a Plane Conductor," *Annals of Physics*, Vol. 60, pp. 481–500, Nov. 1919.

27 H. G. Booker and P. C. Clemmow, "A Relation between the Sommerfeld Theory of Radio Propagation over a Flat Earth and the Theory of Diffraction at a Straight Edge," *Proceedings of IEE*, Vol. 97, Part III, No. 45, pp. 18–27, Jan. 1950.

28 A. Goldsmith, *Wireless Communications*, Cambridge University Press, Cambridge, UK, 2005.

29 Z. Ji, B. H. Li, H.X. Wang; H. Y. Chen, and T. K. Sarkar, "Efficient Ray-Tracing Methods for Propagation Prediction for Indoor Wireless," *IEEE Antennas and Propagation Magazine*, Vol. 43, No. 2, pp. 41–49, 2001.

30 Z. Ji, T. K. Sarkar, and B. H. Li, "Methods for Optimizing the Location of Base Stations for Indoor Wireless Communications," *IEEE Transactions on Antennas and Propagation*, Vol. 50, No. 10, pp. 1481–1483, 2002.

31 T. K. Sarkar, Z. Ji, K. Kim, A. Medouri, and M. Salazar-Palma, "A Survey of Various Propagation Models for Wireless Communication," *IEEE Antennas and Propagation Magazine*, Vol. 45, No. 3, pp. 51–82, 2003.

32 S. J. Orfanidis, *Electromagnetic Waves and Antennas*, Rutgers University, Piscataway, NJ, 2012. http://www.ece.rutgers.edu/~orfanidi/ewa/. Accessed on November 21, 2017.

33 T. K. Sarkar, M. N. Abdallah, M. Salazar-Palma, and W. M. Dyab, "Surface Plasmons/Polaritons, Surface Waves and Zenneck Waves: Clarification of the Terms and a Description of the Concepts and Their Evolution," *IEEE Antennas and Propagation Magazine*, Vol. 59, No. 3, pp. 77–99, June 2017.

34 S. Schelkunoff, "Anatomy of a Surface Wave," *IRE Transactions on Antennas and Propagation*, AP-7, pp. 133–139, 1959.

35 T. Kahan and G. Eckart, "On the Existence of a Surface Wave in Dipole Radiation over a Plane Earth," *Proceedings of the IRE*, Vol. 38, No. 7, pp. 807–812, 1950.

36 A. Sommerfeld, *Partial Differential Equations in Physics*, Academic Press, New York, 1949.

37 H. Ott, "Die Sattelpunktsmethode in der Umgebung eines Pols mit Anwendungen auf die Wellenoptic und Akustik," *Annals of the Physics*, Vol. 43, pp. 393–404, 1943.

38 L. Brekhovskikh, *Waves in Layered Media*, Academic Press, New York, 1973.

39 G. Tyras, *Radiation and Propagation of Electromagnetic Waves*, Academic Press, New York, 1969.

40 C. Clemmow, *The Plane Wave Spectrum Representation of Electromagnetic Fields*, Pergamon Press, New York, 1966, pp. 46–58.

41 D. A. Hill and J. R. Wait, "Excitation of the Zenneck Surface Wave by Vertical Apertures," *Radio Science*, Vol. 13, pp. 969–977, 1978.

42 G. K. Karawas, Theoretical and Numerical Investigation of Dipole Radiation over a Flat Earth, Ph.D. dissertation, Case Western Reserve University, Cleveland, OH, 1985.

43 G. Goubau, "Waves on Interfaces," *IRE Transactions on Antennas and Propagation*, Vol. 7, No. 5, pp. 140–146, 1959.

44 T. E. Hartman, "Tunneling by a Wave Packet," *Journal of Applied Physics*, Vol. 33, No. 12, pp. 3427–3433, 1962.

4

Methodologies for Ultrawideband Distortionless Transmission/Reception of Power and Information

Summary

Broadband antennas are very useful in many applications as they operate over a wide range of frequencies. The objective of this chapter is to study the transient responses of various well-known antennas over broad frequency ranges. It is illustrated that the well-known frequency domain principles quite prevalent in many antenna text books are not applicable for time domain applications. Specifically, the impulse response of an antenna on transmit is the time derivative of the impulse response of the same antenna on receive and hence it is not possible to transmit and receive a broadband pulse without distortion using the same antenna. As such, the phase responses of these antennas as a function of frequency are of great interest. In the ensuing analysis, each antenna is excited by a monocycle pulse. Many antennas show resonant properties, and numerous reflections exist at the antenna outputs. The first part of this chapter deals with ways of converting various resonating antennas to traveling-wave antennas by using resistive loading. Appropriate loading increases the bandwidth of operation of the antennas. But the drawback is the additional loss in the load applied to the antenna structure, leading to loss of efficiency to around fifty percent. However, some of the antennas are inherently broadband, up to a 100:1 bandwidth. Hence, they can be suitable in replacing multiband antenna systems. The second part of the chapter illustrates the radiation and reception properties of various conventional ultrawideband (UWB) antennas in the time domain. An antenna transient response can be used to determine the suitability of the antenna in wideband applications with a low cut-off frequency. Experimental results are provided to verify the various time domain properties of some selected antennas carried out by Dr. J. R. Andrews of Picosecond Pulse Laboratory. It is shown how to transmit and

The Physics and Mathematics of Electromagnetic Wave Propagation in Cellular Wireless Communication, First Edition. Tapan K. Sarkar, Magdalena Salazar Palma, and Mohammad Najib Abdallah.
© 2018 John Wiley & Sons, Inc. Published 2018 by John Wiley & Sons, Inc.

receive a 40 GHz bandwidth waveform without any distortion when propagating through space. How to generate a time limited ultrawideband pulse fitting the FCC mask is presented and a transmit/receive system is described which can transmit and receive the pulse without little distortion. Finally, simultaneous transmission of power and information is also illustrated and it is shown how performances can be optimized.

4.1 Introduction

To understand the performance of antennas over a broadband it is necessary to consider the analysis of the problem in the time domain as transient signals by themselves constitute waveforms covering a band of frequencies. Since we excite a system by a transient signal which for realizibility purposes must be of finite duration, it guarantees that that this waveform will not be strictly bandlimited even though the energy after some high frequencies may be quite small. This arises from the principles of Fourier techniques which states that a function may not be simultaneously of finite support in both time and frequency domains. If a function is limited in time then its spectrum cannot have a finite frequency support and vice versa. In summary, as we are going to consider the responses of antennas over a finite time duration the exciting pulse will have frequencies in theory from DC to daylight as the response in the time domain has a finite support and therefore in the frequency domain will have an infinite support. If the highest frequency becomes unbounded then the wavelength λ associated with that waveform approaches zero. The far field which is supposed to start at $2D^2/\lambda$ now has a value of infinity! Hence in the time domain analysis, an antenna has no far field regions and it is operating fully in the near field, unless the waveforms are strictly bandlimited. In addition, as we shall see a classical antenna pattern cannot be described in the time domain as an antenna pattern is only defined in the far field. In addition, the reciprocity theorem which deals with a transfer impedance and not a transfer function will change from a simple product of the voltage and the current in the frequency domain to a convolution in the time domain. Furthermore, many other interesting phenomenon shows up which we illustrate next.

Consider the simple circuit consisting of a capacitor and resistor connected in series as shown in Figure 4.1. The circuit has four terminals and the pairs of terminals for applying an input and observing the output are represented by terminals T1 and T2. Let us apply a square

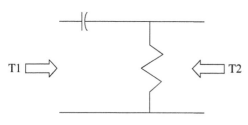

Figure 4.1 A simple C-R circuit.

pulse to the terminals of T1 and connect an oscilloscope at the terminals T2. It is assumed that the input impedance of the oscilloscope is very high. So if a square pulse is applied at T1 the usual differentiation of the pulse made by this circuit will be observed at the terminals T2 as shown in the upper part of Figure 4.2. Now if we switch the location of the source – the square pulse to terminal T2 and connect the oscilloscope to the terminals T1 – to observe the effect of the propagation of the pulse across this circuit then we see that it will behave as a DC block circuit and the wave shape will be given by the lower set of waveforms displayed in Figure 4.2. This behavior of this simple circuit should not be confused with the principle of reciprocity as first of all this theorem on reciprocity in the time domain is a convolution and secondly it deals with the transfer impedance and not a transfer function! Furthermore, the reason we considered this circuit is because this is an approximate equivalent circuit of an antenna. So for broadband applications the same antenna cannot be used for both as a transmitter and also as a receiver. This is because the impulse response of an antenna on transmit is the time derivative of the impulse response of the same antenna acting as a receiver as illustrated in the Figures of 4.2. Hence for broadband distortion less application different antennas should be used for transmit and receive.

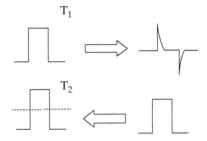

Figure 4.2 Response of the circuit.

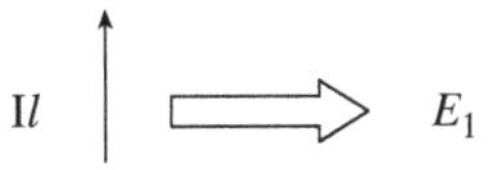

$$E_1 = -\frac{j\omega\mu}{4\pi} Il \frac{e^{-jkR}}{R}$$

Figure 4.3 Response on transmit.

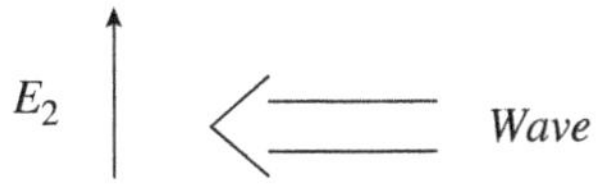

$$\int E_2 Idl = Idl \frac{e^{-jkR}}{4\pi R}$$

Figure 4.4 Response on receive.

This is further illustrated in Figures 4.3 and 4.4. For Figure 4.3 a dipole is radiating an electric field and one is looking at the expression of the radiated fields given in terms of the dipole moment. For Figure 4.4, the dipole is acting as a receiver where an electromagnetic field is incident on the dipole and the expression for the induced open circuit voltage on the dipole is given. It is seen that there is a difference of the $j\omega$ term in the transmit transfer function which is missing in the receiving path.

In the frequency domain when we are considering antennas operating at a single frequency this $j\omega$ term is just a complex constant and so the same response is observed functionally in the frequency domain when the same antenna is used for transmit and on receive. However, in the transform domain –i.e., operating in the time domain this additional $j\omega$ term behaves as a derivative and it deserves a special attention!

The far field radiated by a straight current element along the broadside direction $J(z')$ is given by

$$\mathbf{E}_{far} = -\frac{j\omega\mu_0}{4\pi}\frac{e^{-jkr}}{r}\int_{z'}J(z')e^{-jkz'}dz'.\tag{4.1}$$

Now if we consider a long straight wire irradiated by a plane wave of constant amplitude, then the open circuit voltage generated at the gap between the two terminals located on the wire will be

$$V_{rec} = \frac{1}{I(0)}\int I(z')e^{-jkz'}dz',\tag{4.2}$$

where $I(z')$ is the current distribution on the wire when it is in the receive mode. The term $j\omega$ in $\mathbf{E}_{far}$ indicates that it contains a derivative operation in the temporal domain in contrast to V_{rec} as it is multiplied by a linear function of ω. This represents a derivative operation in the time domain. Therefore, the impulse response of an antenna in the transmit mode is the time derivative of the impulse response of the same antenna in the receive mode. This happens to be true for any antenna.

4.2 Transient Responses from Differently Sized Dipoles

In this section we observe the transient responses from different sized dipole structures due to a pulse of finite duration. Because the impulse responses of antennas are very complex in nature, it is not possible to obtain an analytic solution as was carried for a small wire in Chapter 1, but it is necessary to perform a numerical simulation. Such numerical analysis is quite possible nowadays through the use of computer programs defined in [1, 2], which can carry out analysis of electromagnetic radiation from any arbitrary-shaped composite metallic and dielectric structures. The general-purpose code described in [1] can perform the analysis in the time domain, whereas for frequency domain analysis the computer code described in [2] can handle any arbitrary-shaped composite geometries containing frequency dependent losses. Here we use these two codes to present typical results for radiated fields from different arbitrary shaped antennas. A frequency domain code contained in [2] is useful when dealing with electromagnetic structures containing frequency dependent losses like skin effect, as the methodology becomes quite complex in the time domain.

In Chapter 1, we assumed the current distribution on the structure and then developed the expression for the radiated fields from those assumed current distributions. However, in a real situation, one needs to solve for the actual current distribution on the antennas, and so what we carried out in Chapter 1 is only an approximate analysis. It is difficult to obtain the physics of the problem from a purely numerical solution, and the simplified expressions of Chapter 1 may provide more intuitive feeling as an approximation to the actual nature of the radiation fields from the antenna structures.

As an example, consider a Gaussian pulse of the form given in Figure 4.5, which is used to excite an antenna. The spectrum of this pulse is shown in Figure 4.6 and it contains a wide band of frequencies. This pulse is applied at the feed point located at the center of a strip shaped dipole oriented along the z-direction, as shown in Figure 4.7. The strip is 1 m long and 0.04 m wide. It is subdivided into small triangular patches. The center point corresponds to edge 41, where the Gaussian shaped waveform of Figure 4.5 is applied. We now observe the radiated fields along various elevation directions θ from this structure. Figure 4.8 illustrates the different waveshapes that are radiated along different elevation directions and the fields oscillate for a long time even after the excitation has died down. As expected, the fields are stronger along the

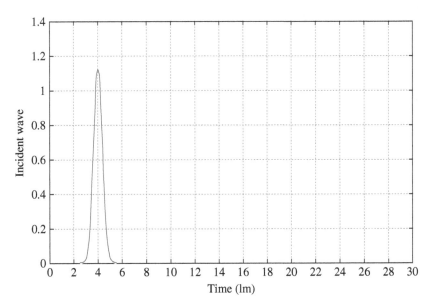

Figure 4.5 Gaussian-shaped pulse used either as a voltage source or as an incident wave. The unit of time is in light-meter (lm).

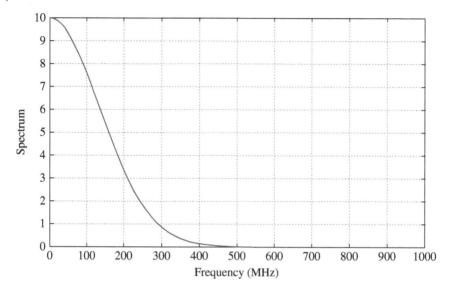

Figure 4.6 Spectrum of the Gaussian pulse which is used as an input excitation. The vertical axis represents the amplitude of the spectrum.

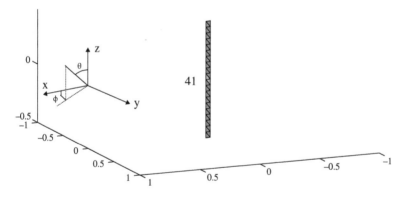

Figure 4.7 Single dipole of length 1m and width 0.04 m.

broadside direction. It is important to note that the usual *donut* shaped pattern in the form of a *figure of 8* of the far field radiation pattern of a dipole which is prevalent in the frequency domain does not exist in the time domain. In the time domain the antenna pattern is not defined. The shape of the radiated transient field is different along different directions. The antenna pattern in the time domain displays no distance-independent nulls along angular directions which are characteristics of all antenna far-field radiation patterns in the frequency domain. Also, the radiated signal varies as a function of time as

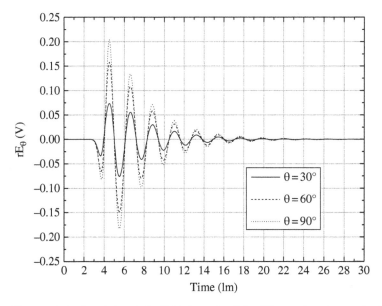

Figure 4.8 Radiated electric far-field along $\phi = 0°$ for different elevation angles for the incident Gaussian pulse.

expected. Throughout the chapter the unit of time is chosen as a *light meter* (lm), a measure of the time required by light to travel 1 m. So 1 lm = (speed of light)$^{-1}$ and therefore 1 lm = 3.3333×10^{-9} sec.

Next, we consider this dipole not as a transmitting antenna but as a receiving antenna. In this case we have an incident electric field coming from different elevation angles and we want to observe the form of the current that is induced at the center point of the dipole due to different types of incident fields. The nature of the incident pulse is exactly the same as in Figure 4.5. The plot of the current induced at the feed point due to different angles of incidence of the Gaussian pulse is shown in Figure 4.9. In summary, from the plots of Figures 4.8 and 4.9, it is seen that the impulse response of a dipole on transmit is the time derivative of the response on receive.

For the next example we consider two strip dipoles of length 1 m and separated from each other by the same distance, as shown in Figure 4.10. The widths of the dipoles are 0.04 m. We apply the same Gaussian pulse at edge 41, as shown in Figure 4.5 to the first dipole. The other dipole then acts as a parasitic element. We observe the various polarizations of the radiated fields and study its temporal shapes along different elevation and azimuth angles. In Figure 4.11 the transient waveshapes for the E_θ component are observed for a fixed azimuth angle of 45° and for different elevation angles. Figure 4.12 provides the other component of the electric field E_ϕ for the same

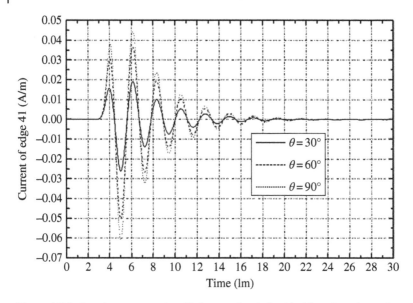

Figure 4.9 Induced current at edge 41 due to a θ-polarized incident Gaussian pulse arriving from $\phi = 0°$ and for different incident elevation angles.

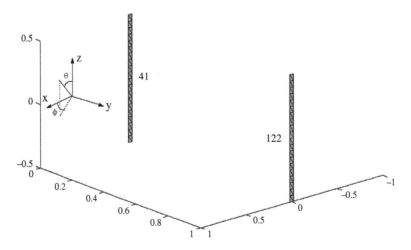

Figure 4.10 Two dipoles separated by 1 m along the y-direction.

set of angles. Figures 4.13 and 4.14 provide the electric radiated fields for the azimuth angle of 90° and for different elevation angles for the E_θ and E_ϕ components, respectively. Finally, Figures 4.15 and 4.16 provide the radiated electric fields for E_θ and E_ϕ components of the electric field for the azimuth angle of 135°. The interesting point here is that even for such simple structures

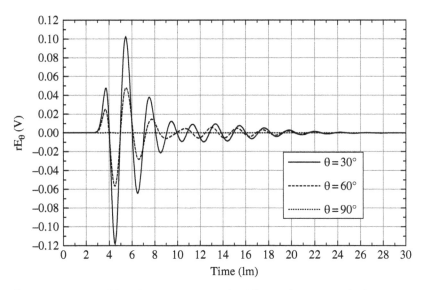

Figure 4.11 Radiated field for E_θ along $\phi = 45°$ for different elevation angles from the two dipoles when one of them is center fed with a Gaussian pulse of Figure 4.5.

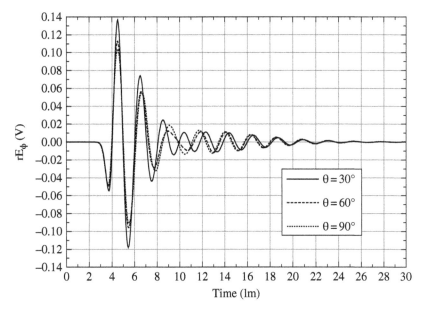

Figure 4.12 Radiated field for E_ϕ along $\phi = 45°$ for different elevation angles from the two dipoles when one of them is center fed with a Gaussian pulse of Figure 4.5.

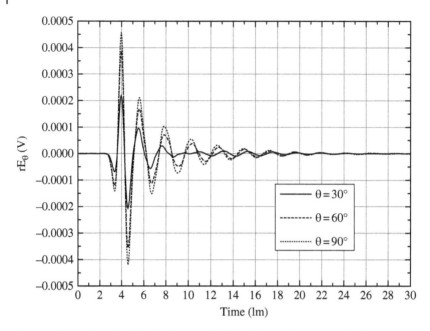

Figure 4.13 Radiated field for E_θ along $\phi = 90°$ for different elevation angles from the two dipoles when one of them is center fed with a Gaussian pulse of Figure 4.5.

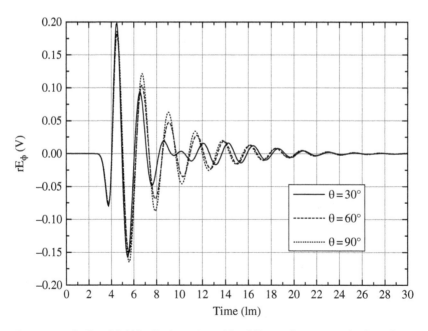

Figure 4.14 Radiated field for E_ϕ along $\phi = 90°$ for different elevation angles from the two dipoles when one of them is center fed with a Gaussian pulse of Figure 4.5.

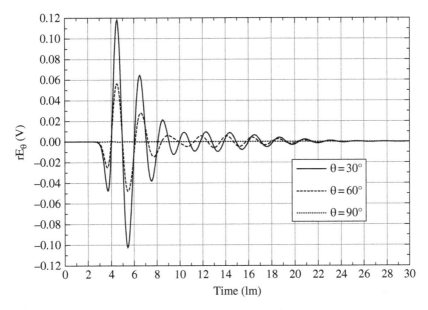

Figure 4.15 Radiated field for E_θ along $\phi = 135°$ for different elevation angles from the two dipoles when one of them is center fed with a Gaussian pulse of Figure 4.5.

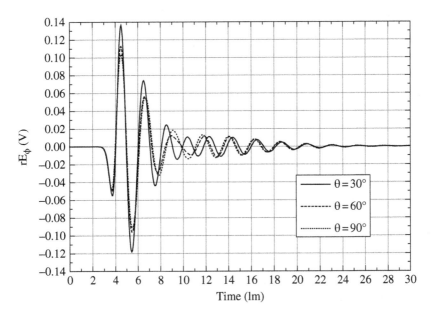

Figure 4.16 Radiated field for E_ϕ along $\phi = 135°$ for different elevation angles from the two dipoles when one of them is center fed with a Gaussian pulse of Figure 4.5.

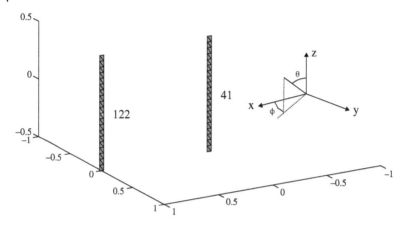

Figure 4.17 Two dipoles separated by 1 m along the *x*-direction.

without a numerical electromagnetic code as described in [1, 2], it will be almost impossible to predict the nature of the radiated electromagnetic fields accurately.

For the receiving case, we observe the induced current at edge 41 on one of these strip dipoles when they are illuminated by the Gaussian pulse of Figure 4.5 from different angles of elevation. In this case the broadside of the strips, as shown in Figure 4.17, are oriented along the direction of polarization of the incident field; otherwise, the induced current on the strip will be zero. The induced current along edge 41 for different elevation angles is presented in Figure 4.18.

Next we illustrate the transient response of a dipole as a function of its length when it is excited by an ultrawideband pulse.

4.3 A Travelling Wave Antenna

Most antennas display resonant like properties. In the time domain, the output waveform has components from numerous reflections on the antenna structure when it is excited by a pulse of finite duration. The waves traveling outward from the feed point get reflected from the discontinuities of the structure so that a standing wave is formed. To obtain a structure that can operate over a broadband a travelling wave current distribution is necessary. To obtain a traveling wave current distribution on the structure, one needs to focus the energy in a single pulse by preventing the reflections. This can be achieved by using matched loads at the ends of the structure. But, if the antenna is used over a broad band of frequencies, this approach is not useful because matched

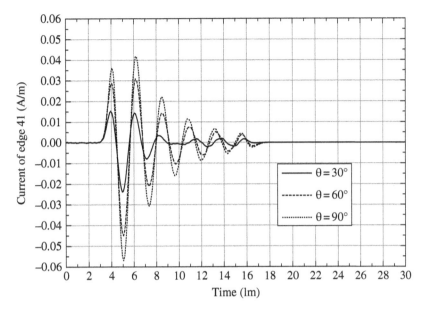

Figure 4.18 Induced current at edge 41 for one of the dipoles due to a θ-polarized incident Gaussian pulse arriving from $\phi = 0°$ for different elevation angles.

loads are effective only over a very narrow range of frequencies. The most effective way to reduce the reflections is to attenuate the outward traveling current towards the ends of the structure. This can be achieved by loading the antennas with a tapered resistive profile so that the resistance increases from the feed point to the ends of the structure. Various loading profiles have been analyzed in [3], where it is shown that a resistive loading is most effective in producing pulsed radiation. **Therefore a travelling wave antenna will truly be an ultrawideband antenna.**

The creation of an appropriate resistive profile was first proposed by Wu and King in 1965 [4]. Consider an antenna with a length $2h$ and radius r. One can define a dimensionless parameter ψ for which the current on the antenna has a maximum, as

$$\psi = 2\left(\sinh^{-1}\frac{h}{r} - C\left(2kr,2kh\right) - jS(2kr,2kh)\right) + \frac{j}{kh}\left(1 - e^{-j2kh}\right) \tag{4.3}$$

where $C(a, x)$ and $S(a, x)$ are the generalized cosine and sine integrals defined by

$$C\left(a,x\right) = \int_0^x \frac{1-\cos W}{W}\,du \qquad S\left(a,x\right) = \int_0^x \frac{\sin W}{W}\,du \tag{4.4}$$

with

$$W = \left(u^2 + a^2\right)^{1/2} \tag{4.5}$$

Wu and King represent the input impedance of a linear dipole antenna through

$$Z_0 = R_0 - j\frac{1}{\omega C_0} = \frac{\psi \zeta_0}{2\pi} - \frac{j}{\omega \varepsilon_0 h} \approx 60\psi \tag{4.6}$$

with

$$\zeta_0 = \sqrt{\frac{\mu_0}{\varepsilon_0}} = 120\pi \tag{4.7}$$

Generally R_0 and C_0 are complex functions, and hence are not the resistance and capacitance except in some special cases. Also when $k_0h \gg 1$; $R_0 \gg 1/(\omega C_0)$. However, in keeping with the notation of Wu and King, define the continuously varying loading profile along dimension κ by

$$z^i(\kappa) = r^i(\kappa) - j\frac{1}{\omega c^i(\kappa)} \approx \frac{\zeta_0 \psi}{2\pi} \frac{1}{h - |\kappa|} = \frac{60\psi}{h - |\kappa|} \tag{4.8}$$

The profile in (4.8) contains poles at $\kappa = \pm h$. However in practical applications, the antenna is divided into a finite number of subsections, where the impedance is calculated at the midpoint of each section and so the poles may not be encountered in practice. Since κ actually never approaches h in (4.8), the presence of the poles does not limit practical applications. Moreover, the frequency dependence appears only in the form of a logarithm for small values of kh, so the antenna shows very broad frequency characteristics when compared to an antenna loaded with a lumped resistance as proposed earlier by Altshuler [5]. Because the reactive part of the impedance is very small compared to the resistive part, one does not need to implement the capacitive profile presented in (4.8) which can be neglected. Consequently, we use only the resistive loading profile

$$r^i(\kappa) = \frac{60 \,\mathrm{Re}\{\psi\}}{h - |\kappa|} \tag{4.9}$$

given by the real part of (4.8). Re $\{\bullet\}$ represent the real part of the function.

To reduce the end reflections, we employ the resistive loading specified by (4.9). However, utilization of (4.9) to broaden the spectral content of the impulse response of the antenna is achieved at the expense of radiating efficiency and the gain of the loaded antenna. The radiation **efficiency** of an antenna is calculated by taking the ratio of the power delivered to the antenna by the power radiated from the antenna. Hence, the radiation efficiency is linked to the power lost in the resistive loading on the antenna along with the power radiated from the antenna. As the reader will observe in the ensuing analysis, the radiation efficiency reduces by almost 50–60% in the case of a loaded antenna when compared to the unloaded case, where the radiation efficiency is much higher, often close to 90–100%.

4.4 UWB Input Pulse Exciting a Dipole of Different Lengths

For radiating antennas in the microwave range, the input must have a very short-duration pulse transition, typically in the picosecond regime. Commonly used baseband pulses are the impulse and the monocycle, the input of choice for this work. The monocycle is basically a doublet formed by differentiating an impulse or by doubly differentiating the unit step. As the monocycle has both positive and negative subpulses, it is useful for driving both halves of symmetrical antennas. Additionally, the anti-symmetrical nature of the monocycle does not permit the generation of dc currents, so the output of an antenna excited by a monopulse has no dc component.

The simulations considered in this section are carried out using the monocycle pulse as an input. The monocycle is obtained by taking the time derivative of a short-duration Gaussian pulse described by:

$$
E^{inc}(t) = \hat{\mathbf{u}}_i \, \frac{E_0}{\sigma\sqrt{\pi}} \frac{d}{dt} \left\{ \exp\left(-\frac{(t - t_0 - \mathbf{r} \cdot \mathbf{k})^2}{\sigma^2} \right) \right\}
\tag{4.10}
$$

where $\hat{\mathbf{u}}_i$ is the unit vector that defines the polarization of the incoming plane wave, E_0 is the amplitude of the incoming wave (chosen to be 377 V/m), σ controls the width of the pulse, t_0 is the delay that is used to ensure the pulse rises smoothly from 0 at the initial time to its value at time t, $\mathbf{r}$ is the position of an arbitrary point in space, and $\mathbf{k}$ is the unit wave vector defining the direction of arrival of the incident pulse. The frequency spectrum of (4.10) is given by

$$
E(j\omega) = \hat{\mathbf{u}}_i \, E_0 \, j\omega \, \exp\left(-\frac{\sigma^2 \omega^2}{4} - j\omega(t_0 + \mathbf{r} \cdot \mathbf{k}) \right)
\tag{4.11}
$$

where f is the frequency of the signal and $\omega = 2\pi f$.

(a)

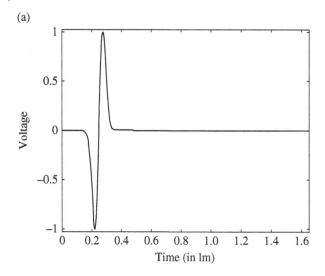

(b)

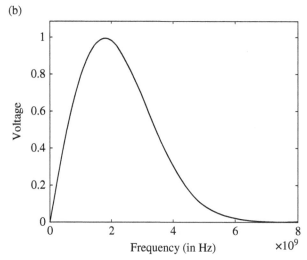

Figure 4.19 (a) A typical monocycle input. (b) The frequency spectrum of the input.

Figures 4.19a and 4.19b, show the shape of a typical monocycle pulse in the time domain and its associated spectrum, respectively. The duration of the pulse is chosen corresponding to the frequency range of operation.

The transmitting and receiving properties of different antennas are simulated in the frequency and time domains. The antennas have been modeled with a program that utilizes the Electric Field Integral Equation (EFIE) to evaluate the currents on the structures [2]. The time-domain data is obtained by taking the inverse Fourier transform of the frequency domain data via the Fast-Fourier-Transform (FFT) technique [6]. The FFT process imposes certain limits on the time-domain data. In particular, the time domain response is periodic and the total time span of the response is controlled by the lowest frequency of operation. The proper choice of the frequency bands and the frequency spacing makes this transformation procedure very helpful for simulating an experimental setup, as will be shown throughout the chapter. The efficiency of the structure has also been included in many cases. Alternatively one can also obtain the results directly in the time domain [1, 7].

4.5 Time Domain Responses of Some Special Antennas

Several classic antennas are simulated in this chapter. In each case, the resulting radiated field is calculated from the antenna that is excited by a monocycle voltage pulse (Figures 4.19a and 4.19b) at the feed point, which is short circuited after the passing of the initial pulse. For reception, the same antenna is used as a scatterer that is illuminated by a monocycle pulsed wave, and the induced current at the feed point is observed. The simulated results are described in the following subsections. Throughout the chapter the unit of time is chosen as a *light meter* (lm), a measure of the time required by light to travel 1 m. So 1 lm = (speed of light)$^{-1}$ and therefore 1 lm = 3.3333×10^{-9} sec.

We study the performance of various common antennas like the dipole, the bicone, the TEM horn in the time domain. Although all of these antennas exhibit time (phase) dispersions, they can be made suitable for UWB applications by various design modifications and sometimes by proper loading. Each antenna is separately simulated as a radiating element and as a receiving element. The resulting electric fields are normalized to an absolute scale in order to illustrate their wave shapes.

4.5.1 Dipole Antennas

Consider a center-fed dipole antenna of length $2h$ and radius r and define the transit time of an input signal from the feed point to each end as $\tau = h/c$ (the half length of the antenna divided by the velocity of light). The effective illumination of the antenna is determined by the parameter σ/τ, where σ is the width of the input pulse. To observe the effects of these parameters, two thin dipoles are considered: $h = 1$ m, $r = 1$ cm, and $\sigma/\tau < 1$ for Antenna 1; and $h = 0.05$ m, $r = 0.05$ cm, and $\sigma/\tau > 1$ for Antenna 2. The simulations are done for a σ of 0.15 lm over the frequency range from 10 MHz to 8 GHz. The time support of the response determined by the lowest frequency of operation is 30 lm.

(a)

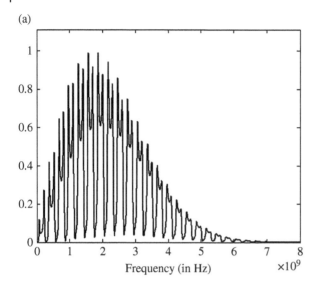

(b)

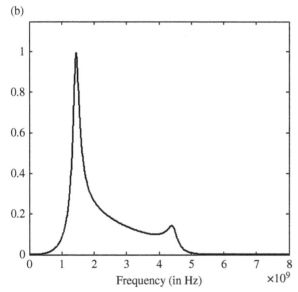

Figure 4.20 (a) Frequency spectrum with $\sigma/\tau < 1$. (b) Frequency spectrum with $\sigma/\tau > 1$.

This artifact comes from the application of the FFT as it computes a Fourier series and not a real transform.

The spectral magnitudes of the radiations along the broadside direction for both transmitting antennas are shown in Figures 4.20a and 4.20b. Antenna 1 ($\sigma/\tau < 1$) has significantly more spectral peaks, which results in sharper peaks

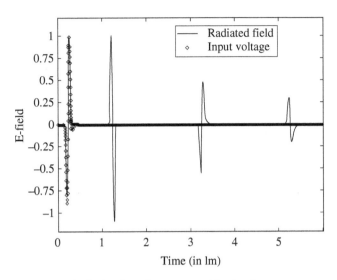

Figure 4.21 Radiation from the antenna with $\sigma/\tau < 1$.

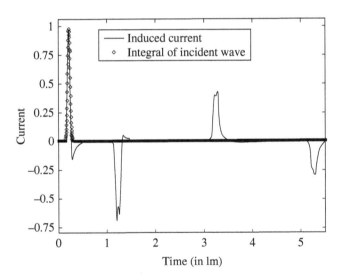

Figure 4.22 Reception of a monocycle pulse by the antenna with $\sigma/\tau < 1$.

and less oscillation of the time-domain radiated and received fields (Figures 4.21 and 4.22) for the unloaded case and in Figures 4.23 and 4.24 for a loaded dipole antenna. In Figures 4.25 and 4.26 the radiated and received fields for Antenna 2 ($\sigma/\tau > 1$) are presented.

For the case of an infinitely long antenna, the transmit transfer function is almost flat with respect to frequency [8]. So in the case of Antenna 1, an

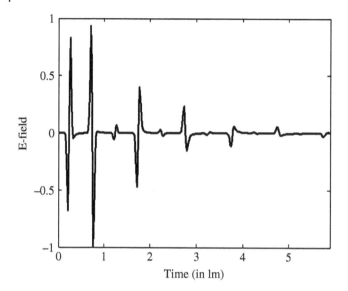

Figure 4.23 Radiation from the antenna 1 with a 50 ohms load at the feed.

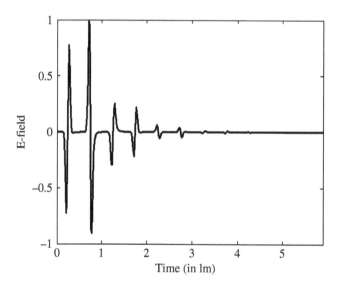

Figure 4.24 Radiation from the antenna 1 with a 500 ohms load at the feed.

approximation to an infinitely long antenna, the radiated field in Figure 4.21 is nearly a replica of the driving-point voltage, as expected. Kanda has shown that the transient response of an antenna in the transmit mode is proportional to the time derivative of the impulse response of the same antenna when it is

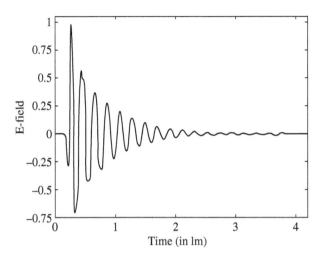

Figure 4.25 Radiation from the antenna with $\sigma/\tau > 1$.

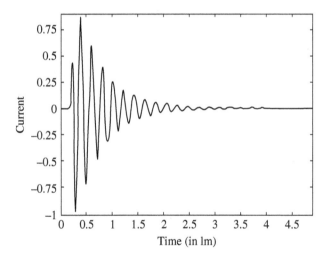

Figure 4.26 Reception of a monocycle pulse by the antenna with $\sigma/\tau > 1$.

operating in the receive mode [9]. Thus, it follows that the induced current will be an integral of the incident field as shown in Figure 4.22. On the other hand, for the shorter dipole, Antenna 2 ($\sigma/\tau > 1$), the radiated far field is proportional to the second temporal derivative of the input voltage on the structure [10], whereas the received open-circuit voltage will be an approximate time derivative of the incident field [7, 11]. The above observations are valid for the case when an antenna is carrying a purely traveling wave of current. However, as a consequence of reflections from the finite size of the dipole, the radiated output from the antenna has a number of reflected pulses in addition to the initial pulse. In the radiated far field, the initial observed pulse is a replica of the input voltage that is radiated directly from the feed point; whereas subsequent pulses are radiations and reflections of the initial pulse from the end and feed points of the dipole [7, 10, 11]. Specifically, the second and third pulses are radiations from the end points that occur at h/c seconds (the time it takes the signal to travel from the feed point to each end) after the initial pulse. Moreover, these two signals arrive at the far away observation point at different times because the corresponding distances to the observation point are distinct, unless the observation point is along the broadside direction.

In the case of Antenna 1, the pulse emanates from the feed 1 lm (6.67 pulse durations) before it radiates from the ends, since $\tau = 6.67\sigma$. Consequently, the radiations from the endpoints do not overlap in time with the initial pulse from the feed. An observer located far away from the antenna in any direction sees a replica of the source, followed by an aggregate radiation from the endpoints, followed by a reduced aggregate radiation from the endpoints 3 lm (20 pulse durations) after the initial emanation from the feed. Depending on the observation direction, the replicas from endpoints could overlap, especially near the broadside. As expected, the radiated field exhibits separately identifiable pulses with the anticipated temporal separations (Figure 4.25). If one considers only the first pulse, then the properties of the infinitely long dipole antenna are verified; that is, the first radiated pulse is a replica of the input. When Antenna 1 is receiving the monocycle, the induced current at the feed shows a number of reflected waves (Figure 4.22). It is important to note that the first pulse of the received current is identical to the integration of the incident wave.

In the above analysis, the feed of the antenna is short circuited after the initial radiation which effectively eliminates the feed as a subsequent location of emanating radiation. Thus a second pulse from the feed, as presented in [7, 10, 11], does not appear in this analysis. For the case when the feed is not short circuited after the initial radiation, instead a load is applied at the feed, reflections from the feed point are observed in addition to the reflections from the ends. The radiation from Antenna 1 when a load of 50 ohms is connected at its feed point is shown in Figure 4.23. The radiated field shows a replica of the source followed by an aggregate radiation from the endpoints occurring 1 lm, an aggregate radiation from the feed point at 2 lm, a reduced aggregate radiation from the endpoints at 3 lm after initial emanation from the feed, and so on. When the 50 ohms load at the feed point is replaced by a load of 500 ohms, Figure 4.24 shows the radiated field. As the loading at the feed point is increased, the reflection from the feed point also increases in magnitude as is evident from Figures 4.23 and 4.24.

The fourth and fifth pulse like radiations in the plot of the time-domain fields are second radiations from the endpoints that occur at $3h/c$ seconds after the initial pulse as a result of reflections from the dipole's endpoints. A second pulse from the feed, presented in [7, 10, 11] does not appear in this analysis, because short circuiting the feed after the initial radiation effectively eliminates the feed as a subsequent source of radiation.

For Antenna 2 (σ /τ > 1), the pulse emanates from the feed one third of a pulse duration (0.05 lm) and one pulse duration (0.15 lm) before the first and second endpoint radiations, respectively, since $\tau = \sigma$ /3.

To an observer located far away from the antenna, the temporal supports of radiation from the feed and the first radiations from the ends have a 67% overlap, and the supports of the first and second radiation from the end points overlap 33% of the time. Consequently, the first pulse cannot be distinguished from the reflected pulses as they are too closely spaced in time, which results in the longer continuous interference signal of Figure 4.25. The time-domain received signal behaves similarly (Figure 4.26), when the monocycle field is incident on Antenna 2.

To reduce the reflections, a tapered loading is applied along the length of the antenna as discussed in Section 4.3. According to (4.3), one can calculate the value of the parameter ψ at the frequency for which the dipole is a half-wavelength long. The surface resistance is calculated using (4.9), which depends on the length of the antenna and its radius, and is plotted for both cases in Figures 4.27 and 4.28 as a function of location along the antenna. To apply the loading, divide the antenna into a number of smaller sections, calculate the value of $r^i(\kappa)$ at the midpoint of each section, and apply the calculated resistance as a constant surface resistance on that section. This surface resistance

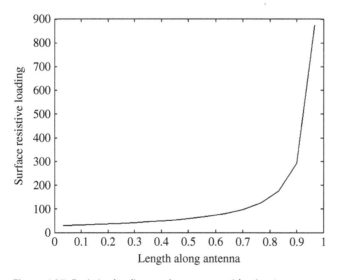

Figure 4.27 Resistive loading on the antenna with $\sigma/\tau < 1$.

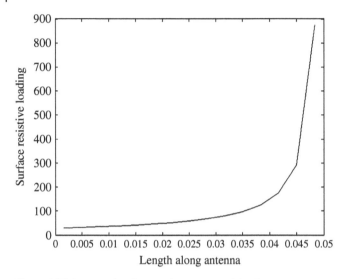

Figure 4.28 Resistive loading on the antenna with $\sigma/\tau > 1$.

can be implemented by coating the antenna with a resistive material. If the number of sections is sufficiently large, the step-function variation of $r^i(\kappa)$ will closely resemble the continuously varying resistive profile.

The results after applying resistive loading to Antennas 1 and 2 for transmit and receive modes are depicted in Figures 4.29–4.32. For Antenna 1, both responses show only a single pulse (Figures 4.29–4.30), because the loading has converted this dipole to a traveling-wave antenna. These results further substantiate the property of a long dipole antenna which states that its field radiated from the antenna is a replica of the input pulse. On the other hand, the form of the radiated field for $\sigma/\tau > 1$ (Antenna 2) in Figure 4.31 concurs with an earlier statement that radiation from the antenna is the second temporal derivative of the input pulse [7, 9–11]. The radiated and received (Figures 4.31–4.32) pulsed signals associated with Antenna 2 indicate the effectiveness of the resistive loading in reducing the reflections from the end of the structure.

The efficiency of the loaded dipole antenna with $\sigma/\tau < 1$ is shown in Figure 4.33, where a significant loss of the radiated energy from the loading is apparent at all frequencies (0.1–8.0 GHz). The efficiency monotonically increases with frequency, with a minimum of 16% at 0.1 GHz and a maximum of 50% at 8 GHz; whereas the efficiency of the unloaded antenna is expected to be 95% or better.

Finally, we discuss the excitation that needs to be applied to the unloaded dipole for $\sigma/\tau < 1$ so that it radiates a monocycle pulse. By using the

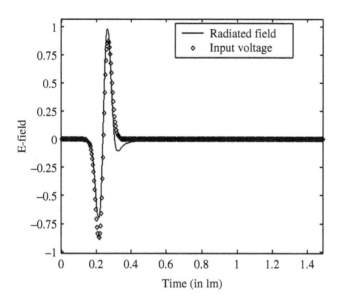

Figure 4.29 Radiation from Antenna 1 ($\sigma/\tau < 1$) with loading.

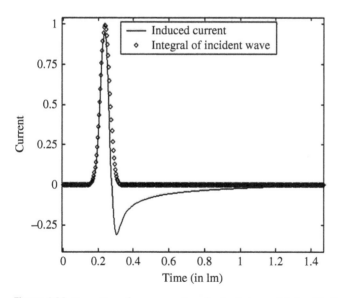

Figure 4.30 Reception of a monocycle pulse by Antenna 2 ($\sigma/\tau < 1$) with loading.

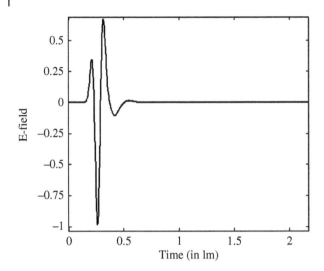

Figure 4.31 Radiation from Antenna 2 ($\sigma/\tau > 1$) with loading.

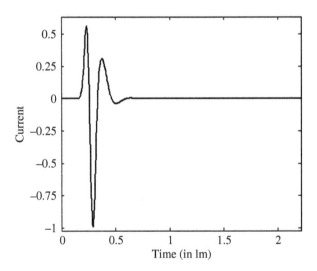

Figure 4.32 Reception of a monocycle pulse by Antenna 2 ($\sigma/\tau > 1$) with loading.

principle of reciprocity, the requisite exciting voltage is computed as follows. According to circuit theory for a two-port linear network, the reciprocity theorem states that,

$$V_1(\omega) I_1(\omega) = V_2(\omega) I_2(\omega) \tag{4.12}$$

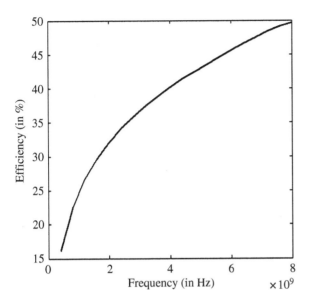

Figure 4.33 Radiation efficiency of the loaded dipole Antenna 1.

when a voltage V_1 applied at port 1 produces a current I_2 at port 2 and a voltage V_2 applied at port 2 will produce a current I_1 at port 1. This reciprocity theorem for circuits can be modified to make it applicable to the case of antennas [12]. Reciprocity states that the source and the measurement points can be interchanged without changing the system response. *It is related to the transfer impedance and not to the transfer function!* Consequently, the input voltage v_2 (t) required to produce the induced current $i_1(t)$ and $v_1(t)$ corresponding to i_2 (t), then are related by

$$v_2(t) = IFFT\left\{V_1(\omega)\, I_1(\omega)\, /\, I_2(\omega)\right\}$$ (4.13)

where *IFFT* denotes the inverse Fourier transform. Thus, by using two identical antennas, the input required for a specified induced current can be synthesized. Similarly, to synthesize the input for a specified radiation, one needs to have a single antenna and use the radiated fields instead of the induced currents in (4.13). If the radiation field due to input voltage V_1 is $\mathfrak{R}_1$ and if $\mathfrak{R}_2$ is the frequency domain characteristics of the required radiated pulse, the requisite time-domain input voltage is calculated from

$$v_2(t) = IFFT\left\{V_1(\omega)\mathfrak{R}_2(\omega)\, /\, \mathfrak{R}_1(\omega)\right\}$$ (4.14)

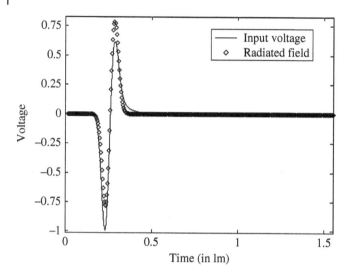

Figure 4.34 Input required for a monocycle transmission by the antenna.

using the reciprocity principle. In a similar manner, this principle can be modified for application to the reception problem [12]. The goal here is to illustrate that by proper waveform shaping any pulse shape can be transmitted or received. The principle of reciprocity is vector in nature whereas power is scalar.

The fields radiated by a straight current element $J(z')$ along the broadside direction is given by (4.1). Now if we consider a long straight wire irradiated by a plane wave of constant amplitude, then the open circuit voltage received by the wire will be given by (4.2). The term $j\omega$ in $\mathbf{E}_{far}$ indicates that it is proportional to V_{rec} when multiplied by a linear function of ω. This represents a derivative operation in the time domain. Therefore, the impulse response of an antenna in the transmit mode is the time derivative of the impulse response of the same antenna in the receive mode.

Now, the requisite exciting voltage is shown in Figure 4.34, along with the radiated field. In addition, Figure 4.35 provides the incident waveform that is needed to induce a monocycle of current at the feed point of the dipole.

4.5.2 Biconical Antennas

The radiation pattern of a biconical antenna is very similar to that of a long dipole antenna. As in the case of the dipole, the biconical antenna also suffers from the formation of standing waves due to the discontinuities on the structure. Reflections from the ends can be reduced by increasing the flare angle above 30° [13, 14].

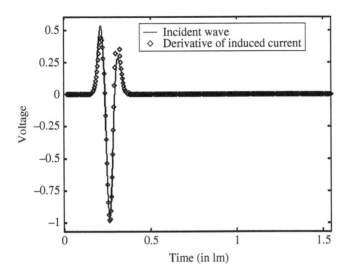

Figure 4.35 Incident pulse shape to generate a monocycle of current at reception.

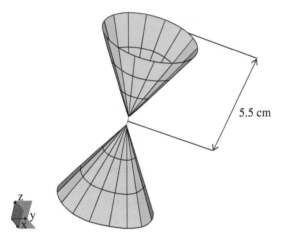

Figure 4.36 Structure of a biconical antenna.

First consider a truncated bicone with open ends, 0.11 m long along the lateral side, and a flare angle that varies from 24° to 90° (Figure 4.36). The antenna has a feed wire of length 4 mm and radius 0.1 mm connecting the vertices of the two cones. The performance of the antenna is simulated between 150 MHz and 30 GHz. The time support of the response determined by the lowest frequency of operation is 2 lm, and the duration of the input pulse is 0.038 lm. The antenna is polarized along the vertical axis (the axis of symmetry through the vertices), and following Jasik [15], the antenna height is taken to be at least $\lambda/4$.

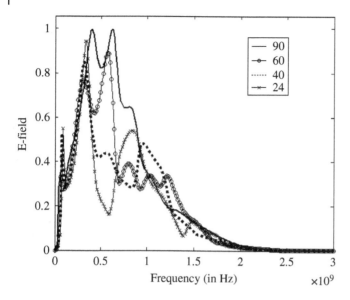

Figure 4.37 Frequency spectrum of the transmitted signal for different flare angles of the bicone.

The amplitudes of the frequency-domain radiated far fields for four flare angles (24°, 40°, 60°, 90°) illustrate that as the flare angle is increased, the responses have less oscillations, thereby resulting in less reflections for the time-domain fields (in Figures 4.37 and 4.38). When the biconical antenna is used as a receiver, the induced currents (in Figure 4.39) for the different flare angles are very similar for the first 0.1 s and are fairly closely banded for the remainder of the time. To reduce the end reflections, the open truncated bicone can be made continuous (smoother) by closing the open ends with identical hemispherical caps (Figure 4.40). Their inclusion prevents the sudden termination of the bicone. Each hemisphere has a radius of 0.055 m, and the origin of the coordinate system is placed at the midpoint of the feed wire that connects the apex of the two cones, with the z axis pointing upwards along the wire. Since this antenna is symmetric about the z axis, the computed radiated field at any angle in the xy plane is constant (Figure 4.41) and is polarized along the vertical length of the cone (z axis). Figure 4.42 plots the received current when a monocycle field is incident on the bicone perpendicular to the z axis. Both figures show that the use of end caps reduces the reflections to a certain level but cannot negate the reflections totally.

To obtain a reflection-free structure, we use a flare angle of 90° and apply a tapered resistive loading outward from each apex along the lateral surface of

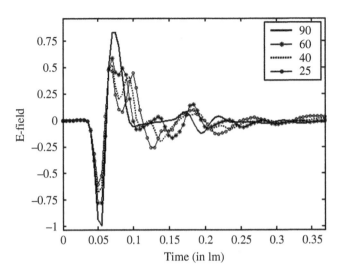

Figure 4.38 Radiated field from the bi- conical antenna for different flare angles.

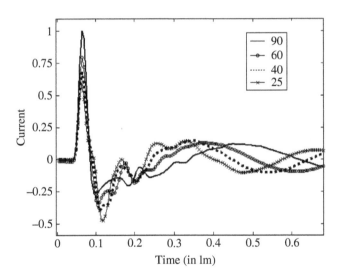

Figure 4.39 Reception of a monocycle by a biconical antenna for different flare angles.

the antenna. By using the principles stated in Section 4.3, we calculate the value of ψ at the frequency for which the bicone is a half-wavelength long by using (4.3). The effective radius for the load calculation in (4.3) is taken to be one-hundredth of the length of the bicone to meet the specification for a thin

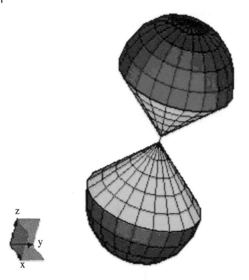

Figure 4.40 Structure of the biconical antenna with end caps.

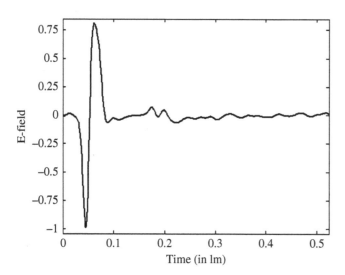

Figure 4.41 Radiation from a biconical antenna with end caps.

antenna. Thus one can obtain the resistive loading profile required to make the antenna a traveling wave structure. The initial model of the antenna has been divided into numerous plates along its length as shown in Figure 4.36. Distributed loading is applied on each layer of plates by calculating the

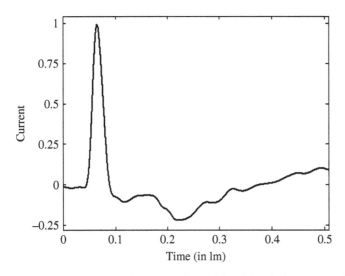

Figure 4.42 Reception of a monocycle pulse by a biconical antenna with end caps.

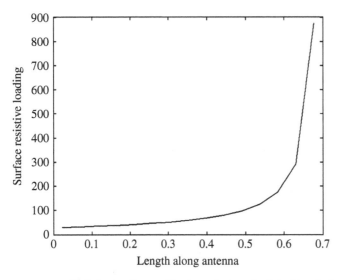

Figure 4.43 Resistive loading profile along the length of the biconical antenna.

resistivity at the midpoint of that layer. If the number of sections is sufficiently large, then the step-functional variation of the resistance will bear a close resemblance to the continuously varying resistive profile plotted in Figure 4.43.

The radiated field for this non-reflecting antenna nearly coincides with the input voltage (Figure 4.44), which verifies that the reflectionless bicone behaves

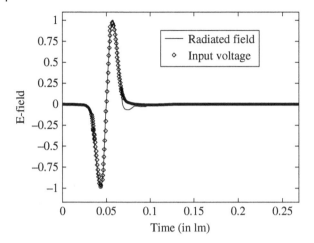

Figure 4.44 Radiation from the bicone antenna.

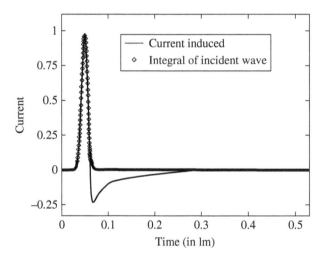

Figure 4.45 Reception of a monocycle by the bicone antenna.

like a long dipole. If this antenna is used as a receiver, the received current at the feed point of the antenna deviates slightly from the integral of the input voltage (Figure 4.45). This discrepancy is due to the absence of any dc current on the structure. As in the case of a dipole, we calculate the excitation that needs to be applied to the bicone so that it radiates a monocycle pulse (Figure 4.46) and the required incident field for inducing a monocycle of current at the feed point of the bicone (Figure 4.47).

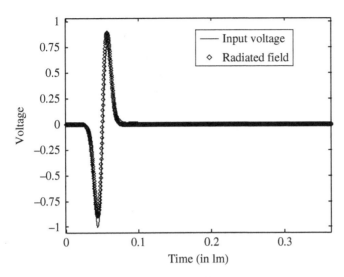

Figure 4.46 Input required for a monocycle transmission.

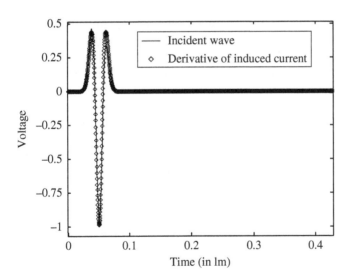

Figure 4.47 Incident field required for a monocycle of induced current.

4.5.3 TEM Horn Antenna

Theoretically, the TEM horn antenna gives a direct measurement of an incident wave when it is used as a receiving antenna. It is very popular in metrology. In our simulation, the overall length of the antenna is 0.15 m, the width at the mouth is 0.038 m, and the height at the mouth is 0.052 m. The feed wire of

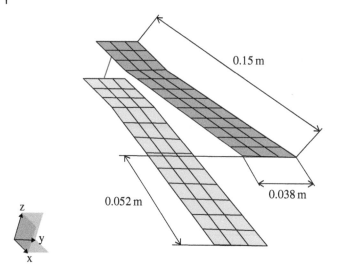

Figure 4.48 Structure of a TEM horn.

the antenna, which has a length of 4 mm and a radius of 0.05 mm, is situated at the narrower end of the antenna (Figure 4.48). In this example, the origin of the coordinate system shown is the midpoint of the feed wire, and the z-axis runs along the feed wire. The top and bottom plates, which stretch along the x-axis, are not tapered; that is, the height-to-width ratio of the antenna is not maintained constant. The antenna is simulated from 200 MHz to 40 GHz, the time support of the response determined by the lowest frequency of operation is 1.5 lm, and the width of the input pulse is 0.04 lm.

The radiated time-domain field is polarized along the z-axis and has significant ripple when evaluated along the x-axis (Figure 4.49), which represents a substantial departure from the monocycle. Similarly, if the antenna is receiving a z-polarized monocycle field traveling along the x-axis, the induced current has similar oscillatory behavior (Figure 4.50). Neither temporal signals generate a clean pulse, which indicates that the TEM horn has some reflections. On modifying the structure by tapering the plates of the horn antenna to maintain a constant height-to-width ratio of 0.68 (Figure 4.51) while keeping other parameters the same, the reflections are reduced (Figures 4.52 and 4.53) but are not completely eliminated. For example, the radiated field in Figure 4.52 is still not quite the derivative of the input pulse [14], because the structure resonates and still has some imperfections.

To convert this resonating structure to a traveling-wave structure, one must decrease the parallel-plate nature of the antenna near its feed point. Therefore, the lengths of the parallel plates are reduced to a minimum (Figure 4.54). To reduce the dispersion further a resistive loading is applied along the length of

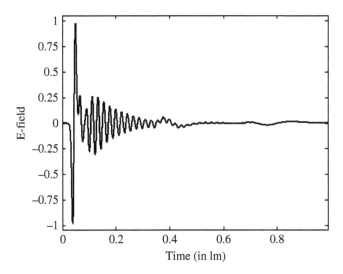

Figure 4.49 Radiation from a TEM horn antenna.

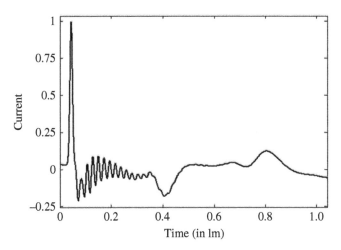

Figure 4.50 Reception of a monocycle by a TEM horn antenna.

the antenna, with the profile specified by (4.9) and is shown in Figure 4.55. Originally, (4.9) was developed for cylindrical antennas only, but it is now modified for use with horn antennas. Instead of using the radius of the antenna in (4.3), we use an effective radius which is calculated as half of the average thickness of the antenna. The effective radius in this case is 0.045 m. We divide the antenna into separate zones along the length and calculate the surface resistance at the midpoint of each zone.

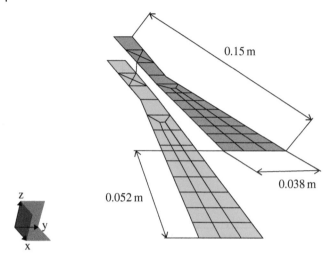

Figure 4.51 Structure of a tapered horn antenna.

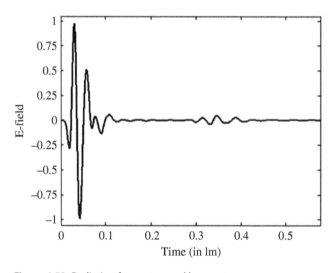

Figure 4.52 Radiation from a tapered horn antenna.

The radiated time-domain waveform and time-domain received current are displayed in Figures 4.56 and 4.57. Both the transmission and the reception properties show that by properly loading the antenna, the unwanted reflections are reduced and the energy is focused in a single pulse. Previously, it was experimentally demonstrated that the radiated field is the first derivative of the driving point voltage, whereas the current induced during reception is

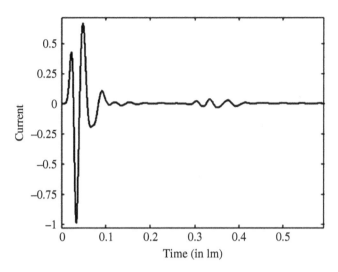

Figure 4.53 Reception of a monocycle by a tapered horn antenna.

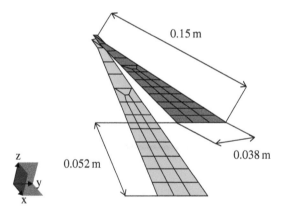

Figure 4.54 Structure of a non-resonating horn antenna.

identical to the incident wave [7, 11]. Not only do the simulations contained herein verify the experimental results of Ref. [16], they also indicate that the impulse response in the transmit mode is proportional to the time derivative of the impulse.

Similar to the unloaded dipole, the efficiency of the unloaded TEM horn is approximately 96–100%. Over the range of simulated frequencies (200 MHz to 40 GHz), the efficiency of the loaded horn monotonically increases, with a maximum of 65% at 40 GHz and nearly linear behavior between 7 GHz and

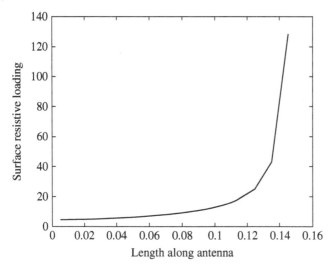

Figure 4.55 Resisting loading profile on the antenna.

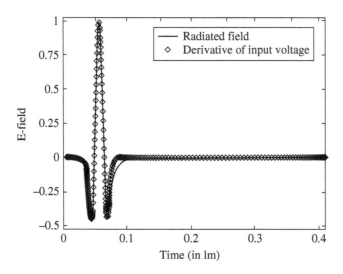

Figure 4.56 Radiation from non-resonating tapered horn antenna.

27 GHz (Figure 4.58). The loaded horn is inefficient (< 10%) for frequencies < 7 GHz.

Finally, we calculate the excitation that needs to be applied to the TEM horn so that it radiates a monocycle pulse (Figure 4.59) and the required incident field for inducing a monocycle of current at the feed point of the horn (Figure 4.60).

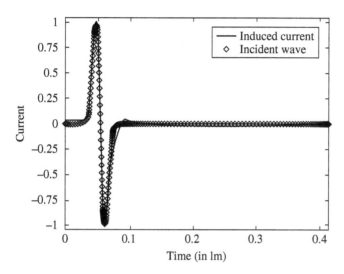

Figure 4.57 Reception of a monocycle by a non-resonating tapered horn antenna.

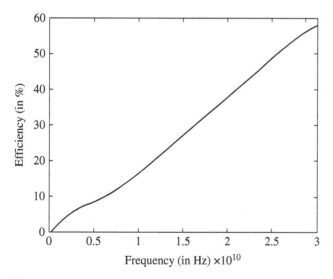

Figure 4.58 Radiation efficiency of the loaded tapered horn antenna.

4.6 Two Ultrawideband Antennas of Century Bandwidth

In this section we describe two special antennas both of which has a century bandwidth, i.e., 100:1.

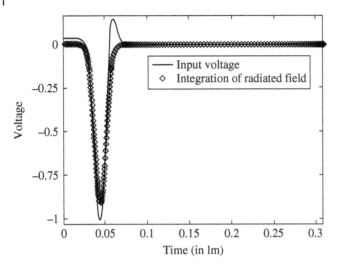

Figure 4.59 Input required for a monocycle transmission.

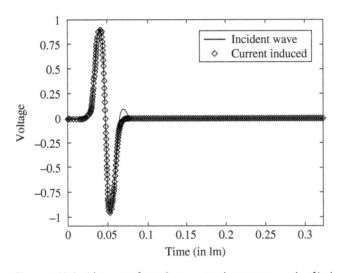

Figure 4.60 Incident waveform shape to produce a monocycle of induced current.

4.6.1 A Century Bandwidth Bi-Blade Antenna

The bi-blade antenna is designed for multi-octave transmission with a century bandwidth, i.e., a ratio of high to low frequency of 100:1. Due to its century bandwidth of operation, the bi-blade radiates and receives at UHF, L, C, S, and X bands [17, 18]. Each of the two blades has a throat (narrowest part), a mouth (widest part), and a tip. The throat serves as the feed point. The tip is an arc of constant radius, thereby giving rise to a low voltage standing wave ratio of about 1.19 to 1 [17]. The radius of the arc determines the slope of the antenna's surge impedance. The blades are designed such that the slot width between the two blades increases logarithmically from the throat to the mouth of the antenna as shown in Figure 4.61. The blade length is 0.56 m and the maximum slot width is 0.43 m. The blade is 0.11 m wide at its mouth. The antenna has a coplanar geometry and thus is easy to integrate into many systems. The antenna is fed by a gen-

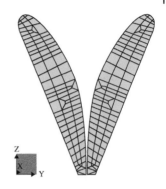

Figure 4.61 A bi-blade century band-width antenna.

erator placed between the two blades at the throat. The feed wire connecting the two blades of the antenna has a length of 4 mm and a radius of 0.01 mm. The origin of the coordinate system is the midpoint of the feed wire.

The antenna is simulated between 160 MHz and 16 GHz. The time support of the response determined by the lowest frequency of operation is 1.875 lm, and the width of the input pulse is 0.075 lm. The antenna radiates along the z-axis of Figure 4.61, and the polarization of the radiated field is along the y-axis on the yz-plane. The radiated field has a strong wave coming out and a weak reflected pulse corresponding to the reflection from the tip of the antenna (Figure 4.62), which implies that the antenna does not have significant phase dispersion. The first pulse radiated is a good approximation of the input

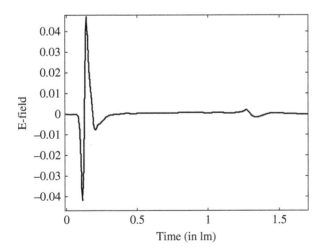

Figure 4.62 Radiation from a bi-blade antenna.

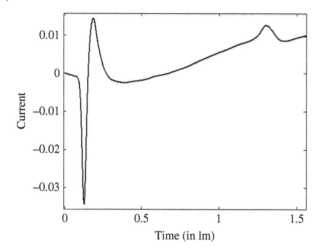

Figure 4.63 Reception of a monocycle by a bi-blade antenna.

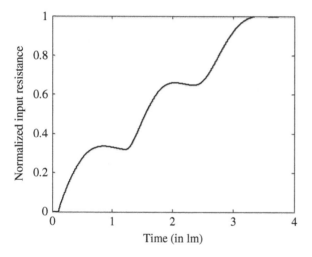

Figure 4.64 Transient input resistance of a bi-blade antenna.

monocycle. When the antenna is illuminated by a monocycle pulse arriving along the z-axis, the induced current roughly has the same general shape as the incident field but with significant differences (Figure 4.63). In particular, the current increases to a value that is commensurate with its maximum positive excursion. The transient input resistance of the antenna in response to a monocycle pulse is shown in Figure 4.64. This is an extremely wideband antenna and can also provide good temporal waveforms if the tips are properly shaped.

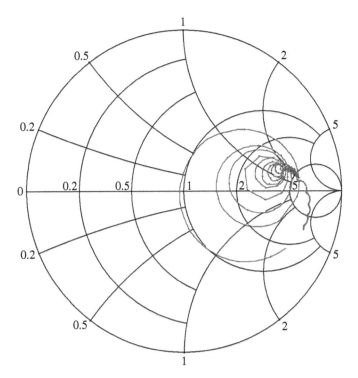

Figure 4.65 Plot of the input impedance of a bi-blade century bandwidth antenna with a 2.5 VSWR circle surrounding the input S-parameter data.

These types of bi-blade antennas indeed have a century bandwidth (i.e. 100:1), and the limitation in performance of circular polarization of using two of these antennas in space and time quadrature is not due to the antenna, but is dependent on the phase shifters used. The computed input impedance of the bi-blade antenna shown in Figure 4.61 is plotted in Figure 4.65. A 2.5 VSWR circle is also drawn to illustrate that the century bandwidth can operate from a theoretical point of view with a VSWR of 2.5 or less across the entire bandwidth of 300 MHz to 30 GHz for the antenna modeled in Figure 4.61. Figure 4.61 also provides the HOBBIES [2] discretization of the antenna structure and was used to numerically generate the input impedance data as shown in Figure 4.65. The important factor that limits the bandwidth of these antennas, are the feed structures. In a numerical electromagnetics code the wires that excite the feed structure can only be done by thin wires where the ratio of the length to radius has to be at least of the order of 10 due to the approximation of an axial current flow in a thin wire approximations. This leads to a higher input voltage standing wave ratio (VSWR) indicating the nature of the variation of the input impedance.

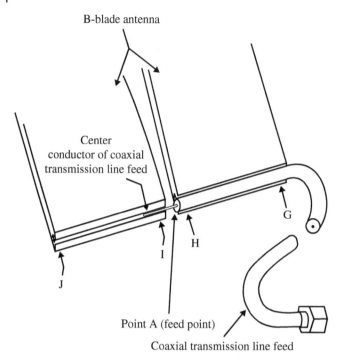

B-blade antenna

Center conductor of coaxial transmission line feed

G

H

I

J

Point A (feed point)

Coaxial transmission line feed

Figure 4.66 Details of feeding the bi-blade century bandwidth antenna with a coaxial transmission line.

Had the model was made for a thick wire as would be done in an actual deployment shown in Figure 4.66, then the input VSWR will approach that of the measured one of approximately 1.19 as stated in the patent [18].

4.6.2 Cone-Blade Antenna

The cone-blade [18] is a circularly polarized antenna structure based on the conical antenna concept and is a modification of the bi-blade century bandwidth antenna. The cone-blade structure is a truncated closed half cone that is surrounded by four equally spaced blades that are separated by 90° (Figure 4.67). The supporting base of the antenna is the flat truncated portion of the cone, and the blades are thin conducting plates that are placed near the base of the cone and extend upwards in the same direction as the vertex of the cone. The cone has a half angle of 11.53° and a height of 0.228 m. The antenna is fed via a quadriphase monocycle excitation by four generators, one at each of the four wires connecting the blade and the cone. Each generator has a 90° phase shift from the generator on the adjacent feed arm. Since the main beam of the radiation pattern at each frequency points along the axis of the cone (z-axis), the simulated far field

Figure 4.67 A cone-blade antenna structure for generating circular polarization.

is calculated at an observation point along the positive z-axis. Each of the feed wires has a length of 20 mm and a radius of 0.01 mm. The antenna is simulated between 200 MHz to 20 GHz, and the width of the input pulse is 0.06 lm.

The observed radiated field and the second derivative of the monocycle are shown in Figure 4.68. Since no reflected pulses are present in the field and the curves almost coincide, this antenna is a traveling-wave antenna. For reception, the current induced in the antenna from a circularly polarized monocycle field that is incident along the z-axis is shown in Figure 4.69. The induced

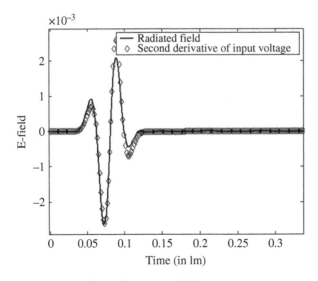

Figure 4.68 Radiation from a cone-blade antenna.

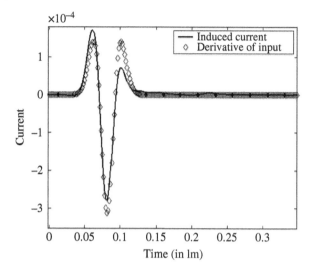

Figure 4.69 Reception of a monocycle by a cone-blade antenna.

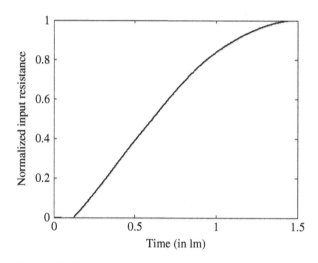

Figure 4.70 Transient input resistance of a cone-blade antenna.

current has no components associated with reflections along the antenna, thereby providing further verification that the antenna is a traveling-wave structure.

Consequently, applying resistive loading to broaden the cone blade's transfer function is unnecessary, which means that reflection less transmission and reception are achieved with no loss in efficiency. The dynamic transient input resistance of the antenna is shown in Figure 4.70.

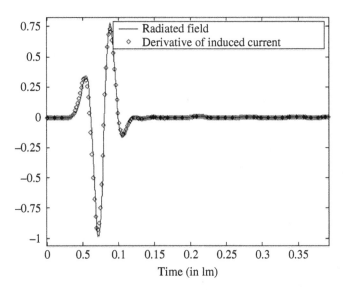

Figure 4.71 Comparison of the radiated field and the first derivative of the induced current when operating as a receiver.

Kanda has shown that the transient response of the antenna on transmit is proportional to the time derivative of the response of the same antenna on receive [9]. In order to illustrate this property of the antenna, the radiated field of the transmitting antenna is compared to the derivative of the induced current on the receiving antenna (Figure 4.71). By analyzing Figure 4.71, one can conclude that this property of antennas stated by Kanda is applicable to any antenna, irrespective of the structure of the antenna.

4.6.3 Impulse Radiating Antenna (IRA)

The structure is a 46-cm diameter IRA with 45° feed arms [19, 20]. The ratio of the focal length to the aperture diameter is 0.4, and the ratio of the focal length to the feed arm of the parabolic reflector is 0.4. This antenna is simulated between 20 MHz to 20 GHz with a 20 MHz frequency step using a numerical electromagnetic analysis code [2], and the time support of the response determined by the lowest frequency of operation is 0.15 lm. The structure of the IRA which was provided by E. Farr is his model IRA-2 [21, 22] and is shown in the Figures 4.72 along with the discretization of the structure used in the numerical simulation. This IRA is a class of antennas designed to radiate a short pulse in a narrow beam for applications in high-power pulse radiators and transient radars for mine detection. The IRA modeled here is a TEM-fed reflector. Baum et al. provide a basic design procedure for these antennas in Chapter 12 of Ref. [19]. This review summarizes the initial research, design,

(a)   (b)   (c)

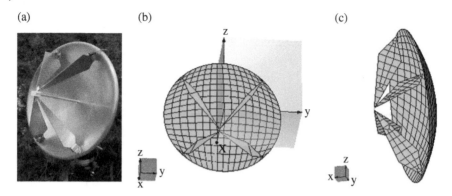

Figure 4.72 (a) An Impulse Radiating Antenna (IRA). (b) shows a model of the full IRA, with the reflector segmented using 16 subsections around the quarter of the circumference (n = 16) for carrying out an accurate electromagnetic simulation up to 20GHz. (c) side view of the IRA.

and development of the IRA and includes a comprehensive bibliography of references through 1999.

In this section, the radiation fields for the IRA are calculated in the frequency domain using HOBBIES [2] and then the results are inverse Fourier transformed to generate the time domain response. The time-domain radiation and reception of a monocycle pulse by the IRA are shown in Figures 4.73 and 4.74, respectively. The transient input resistance of the antenna is shown in Figure 4.75 and in a magnified form in Figure 4.76. Hence, this antenna is truly capable of generating impulses of extremely high voltages and they have been deployed in practice for such applications.

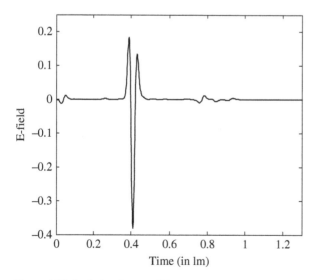

Figure 4.73 Radiation from an IRA.

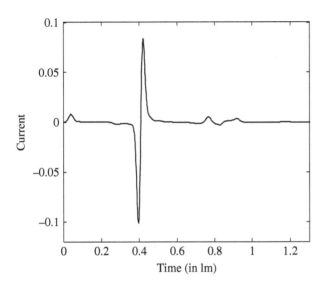

Figure 4.74 Reception of a monocycle pulse by the IRA.

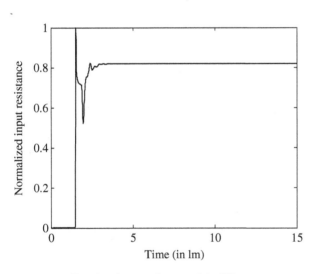

Figure 4.75 Transient input resistance of the IRA.

4.7 Experimental Verification of Distortionless Transmission of Ultrawideband Signals

Experimental results are provided to enforce the various properties related to the transmit and receive impulse responses of some antennas described so far. In general, it is experimentally verified that the impulse response in the transmit mode of any antenna is related to the time derivative of the impulse

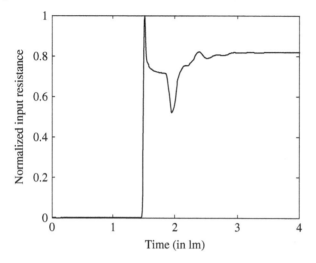

Figure 4.76 Magnified view of the transient input resistance of the IRA.

response of the same antenna when it is operating in the receive mode. In addition, a special transmit-receive system is discussed which is nondispersive over a very wide bandwidth of approximately tens of gigahertz.

To put it in the proper perspective we need to introduce the realistic properties of antennas and not the assumed ones usually used in the wireless literature. Often in the published literature an antenna is characterized by a point source generating an isotropic omni-directional radiation pattern. In reality, an isotropic omni-directional point radiator does not exist in electromagnetics. The smallest physically realizable radiating current element is a short (Hertzian) dipole, which can produce omni-directional patterns only along certain directions. The practical significance of this truth is that a propagation path loss is independent of frequency and by proper choice of transmitting and receiving antennas it is possible to transmit a wideband signal in free space, for example, without any distortion. This point will be illustrated using measured data and illustrate how a 40 GHz bandwidth waveform can be transmitted and received through free space without any distortion and what is the nature of the physics that make it possible. To begin with, it is absolutely essential to look at the Frii's transmission formula but in a form that is not very popular!

For illustration purposes, consider the following two identical forms of the Frii's transmission formula:

$$P_r = P_t G_r G_t \left(\frac{\lambda}{4\pi R}\right)^2 = \frac{P_t G_t A_R}{4\pi R^2}, \tag{4.15}$$

where P_r is the power obtained at the receiving antenna, P_t is the power radiated by the transmitting antenna, G_r and G_t are the gains of the receiving and transmitting antennas with respect to an isotropic radiator, λ is the wavelength of transmission, and R is the separation distance between the transmitting and receiving antennas. The first form (middle) is conventionally used in most applications. However, it is the second form that is of interest to us as it reveals a different dimension to the Frii's formula. The second one (right) is a special form that illustrates the principle of ultrawideband (UWB) transmission of waveforms when the transmitting and the receiving antennas are factored into the channel characterization. In this expression, A_r is the effective area of the receiving antenna and is related to the gain of the receiving antenna by $G_r = 4\pi A_r / \lambda^2$. Now if we focus our attention on the second form, the last part of (4.15) of Frii's transmission formula, we observe that a UWB transmission without any distortion can be achieved simply by transmitting a waveform using an antenna which has a nearly constant transmitting gain with frequency, instead of with a non-existing isotropic omni-directional point radiator. In addition, if the receive antenna has an effective receiving aperture whose effective size does not vary with frequency, we will achieve our goal of distortion less transmission and reception of UWB signals over a very high frequency bandwidth.

In order to experimentally verify these theoretical conjectures, we include here the various experimental results obtained by Dr. James R. Andrews which are presented here with his permission [16] and also presented earlier in [7]. A miniature UWB antenna range, Figure 4.77, was designed to demonstrate the principles of UWB transmission, reception and propagation. The UWB

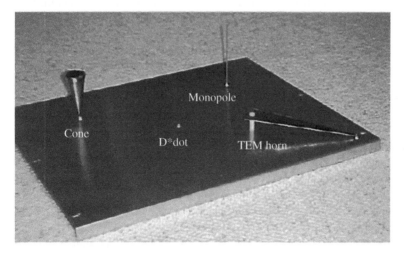

Figure 4.77 Miniature UWB Antenna Range. Antennas shown include a conical antenna, TEM Horn antenna, D*dot probe antenna, and a Vertical Monopole Antenna.

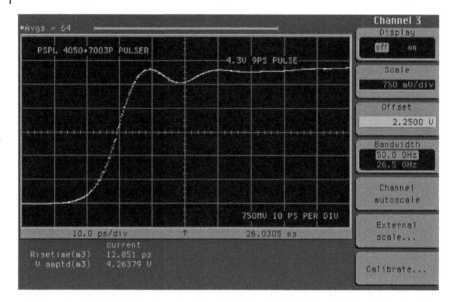

Figure 4.78 4 Volt, 9 ps risetime, step pulse used for UWB antenna testing. Measured by an HP 50 GHz, 9 ps risetime, sampling oscilloscope. Scales are 750 mV/div & 10 ps/div.

metrology antennas demonstrated on this range include the conical, TEM horn and a D*dot probe. Monopole antennas and wave-guide horn antennas were also studied. Both differentiation and integration effects in the time domain will be demonstrated by using these various antennas as transmitters and receivers. The transmitting and the receiving antennas were placed on a 36 cm × 44 cm aluminum plate.

The ultra-broadband test input signal used on this antenna range was a 4 V, 9 ps risetime step from a nonlinear transmission line (NLTL) pulse source of Picosecond Pulse Laboratory as shown in Figure 4.78. All of the waveforms shown in this section were measured using a HP-54752B, 50 GHz, 9 ps risetime oscilloscope. A 35 cm, Gore, SMA (Sub Miniature version A) coaxial cable was used to connect the antennas to the oscilloscope. The risetime of this cable was 9 ps. Thus the composite risetime of the pulse generator, coax cable and oscilloscope was 16 ps.

In the first example we use a conical antenna for transmission. The conical antenna suspended over a large metal ground plane is the preferred antenna for transmitting known transient electromagnetic waves. This type of antenna is used by National Institute of Standards and Technology (NIST) as their reference standard transient transmitting antenna. This antenna radiates an electromagnetic (EM) field that is a perfect replica of the driving point voltage waveform.

However, when the pulse from the feed point reaches the top of the cone it is reflected back and so the perfect replica property of the transmitted pulse no longer holds unless resistive loading on the far end of the antenna is used to help suppress multiple reflections. The upper bandwidth of a conical antenna is mainly determined by the fidelity of the coax connector to conical antenna transition region. If a conical antenna is used as a receiving antenna, its output is the integral of the incident E field. The conical antenna was made of brass sheet with a half solid angle of 12° and having a height of 7 cm. This is shown in Figure 4.77.

TEM (transverse electromagnetic mode) horns are the most preferred metrology receiving antenna for making a direct measurement of transient EM fields. The TEM horn antenna is basically an open-ended parallel plate transmission line. It is typically built using a taper from a large aperture at the receiving input down to a small aperture at the coax connector output. The height to width ratio of the parallel plate is maintained constant along the length of the antenna to maintain an uniform characteristic impedance. However, to optimize sensitivity, most TEM antennas are designed with a 100 Ω antenna impedance. Practical UWB TEM horn antennas are usually designed with resistive loading near the mouth of the antenna to help suppress multiple reflections. The upper bandwidth of a TEM antenna is mainly determined by the size of its aperture and secondarily by the parallel plate to coax connector transition. When the aperture is too large relative to the wavelength of incident fields, the parallel plate line becomes a waveguide with higher order TE and TM modes present which in turn limit its bandwidth. If a TEM antenna is used as a transmit antenna, the radiated E field is the first derivative of the input driving point voltage.

To demonstrate faithful UWB radiation and propagation the conical antenna and TEM horn shown on the test range, Figure 4.77 were used. The TEM horn was also fabricated from brass with a length of 15 cm. The height at the mouth was 2.6 cm and its width was 3.8 cm. At the mouth it was supported by a nylon spacer. The approximate incline angle was 9° as shown in Figure 4.77. The 4 V, 9 ps rise pulse, as shown in Figure 4.78, was connected directly to the conical antenna. The output of the TEM horn antenna was connected through the 35 cm cable to the 50 GHz oscilloscope. The conical antenna was 7 cm high and had an impedance of 132 Ω. The TEM horn was 15 cm long and had an impedance of 106 Ω. For most of the experimental data shown in this chapter, the separation between the antennas was 25 cm. Figure 4.79 is the output of the TEM horn antenna. This shows a 23 ps risetime step waveform that is in good agreement with the input waveform of Figure 4.78. Figure 4.80 shows the output from the TEM horn antenna when the two antennas were separated by 5 meters. This received signal is identical in wave shape to the 25 cm path signal of Figure 4.79, except that it is weaker in amplitude to Figure 4.79. It is clear that very little dispersion/distortion is introduced by the transmit/receive system as earlier seen through numerical simulations.

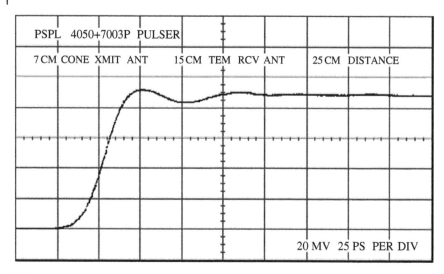

Figure 4.79 Transmission from a conical antenna to a TEM horn antenna over a 25 cm path.

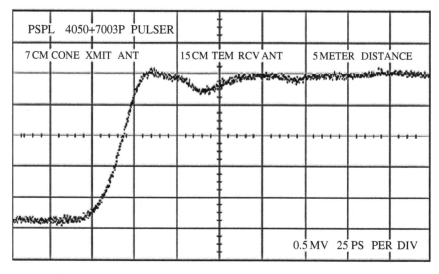

Figure 4.80 Transmission from a conical antenna to a TEM horn antenna over a 5 m path.

It is important to note that an antenna does not radiate DC. However, by observing Figures 4.78.–4.80 one would get the impression that DC is being radiated. This is not true. These are waveforms that are used in time domain reflectometry (TDR), where trapezoidal pulses are used with positive and negative polarities. On the figures we focus on only the rise time of a single trapezoidal pulse to illustrate the fidelity with which the waveshapes have been radiated and received by the different antennas.

The above figures have tremendous implications when dealing with broadband wireless systems. Many researchers are routinely performing channel modeling of free space for a broadband wireless system. In perspective of the Figures 4.78, 4.79, and 4.80 it appears that modeling of a broadband wireless system in free space is equivalent to channel modeling of free space which is dispersionless! Hence, channel modeling of free space sounds very surprising at the first look because in Maxwell's theory air is typically assumed to be dispersionless, and so, why model a dispersionless channel? However, many researchers are used to only looking at the channel and not considering the effect of the transmitting and the receiving antennas which are an integral part of a wireless system. Alternately antennas are often modeled as point sources thereby excluding the fundamental physics from the analysis. Now, if one looks at Friis's transmission formula one observes that any transmitted signal strength will decay as $1/\lambda^2$ in free space due to the propagation of a spherical wave from the source. If such were the case, it will really be a highly dispersive channel. However, particularly for a broadband wireless system, one cannot only think about the isolated channel without the associated transmitting and receiving antennas, as they form an integral part of a wireless system. With the transmitting and the receiving antennas included in the broadband model, things behave in quite a different way as illustrated by (4.15). Therefore, with the transmitting and the receiving antennas in a broadband wireless transmission system, Friis's transmission loss is compensated for by the two antennas making the channel completely dispersionless over a large band. The experimental data provided by Dr. James R. Andrews of the Picosecond Pulse Lab [16] verifies the physical principles that we have discussed before.

This is where the vector electromagnetic problems differ from the scalar acoustic problems. One can observe that an acoustic signal is not distorted when it is converted to electrical energy through any microphone or when the electrical energy is converted to acoustic energy through any loudspeaker. However, the problem is quite different for the vector electromagnetic problem. In the time domain, the impulse response of the antenna when it is operating in the transmit mode is the time derivative of the impulse response when the same antenna is operating in the receive mode. Hence, even a point source, which is often used for modeling the antennas, differentiates the input waveform on transmit [3]. Therefore, the modeling of any wireless system and moreover that of a broadband system must include the antenna effects to obtain a physically meaningful solution [16].

Figure 4.81 shows the received waveform when transmitting a step electromagnetic field between a pair of identical conical antennas. The resultant received output, rising ramp, waveform is the integral of the step pulse from the generator. This illustrates that a conical receive antenna integrates the

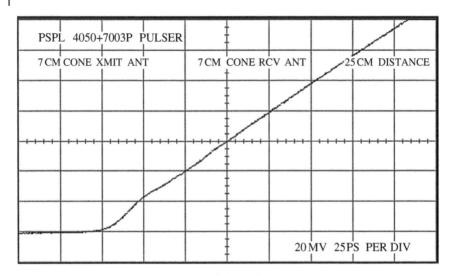

Figure 4.81 Transmission between a pair of identical conical antennas.

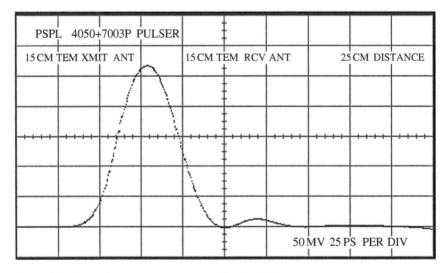

Figure 4.82 Transmission between a pair of identical TEM horn antennas.

incident electric field. Figure 4.82 shows the received waveform when using a pair of identical TEM horn antennas for transmit and receive. The resultant output waveform, an impulse, is the first derivative of the pulse generator's step pulse. This illustrates that a TEM horn transmit antenna radiates an electric field which is the first derivative of the signal generator's waveform.

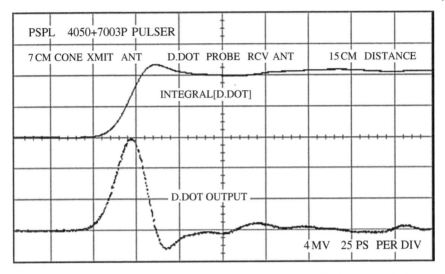

Figure 4.83 Transmission from conical antenna to D*dot antenna. The lower trace is the output from the receive antenna. The upper trace is the integral of the lower trace.

Next, we consider a D*dot antenna [16, 23] which is another popular UWB metrology antenna. The D*dot antenna is basically an extremely short, monopole antenna. The equivalent antenna circuit consists of a series capacitance and a voltage generator. For a very short monopole, the antenna capacitance is very small and the capacitor thus acts like a differentiator to transient electromagnetic fields. Therefore the received output from a D*dot probe antenna is the first derivative of the incident electric field and the transmitted field will be the double derivative of the waveform applied to the feed point. To determine the actual wave shape of the incident electric field, one must integrate the radiated output from the D*dot probe. When the frequencies become too high, the D*dot probe loses its derivative properties and it becomes a monopole antenna. This happens when the length of the D*dot probe approaches a quarter wavelength of the incident wave. Figure 4.83 is the output from a D*dot receive antenna when it is illuminated with a step electromagnetic field radiated from a conical transmit antenna. The lower trace is the actual receive antenna output. It is a 19 ps wide impulse which is consistent with the antenna's output being the first derivative of the incident field. The upper trace is the computed integral of the lower trace. It has a 19 ps risetime step and is a good representation of the incident electric field at the antenna. Figure 4.84 shows the radiated field from a D*dot transmit antenna. It was received by a TEM horn antenna. This shows that the transmitting transient response of the D*dot antenna is the second derivative of the driving generator voltage. Figure 4.85 is

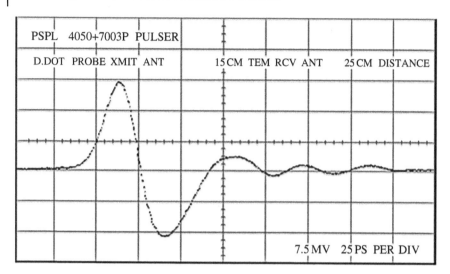

Figure 4.84 Transmission from a D*dot antenna to a TEM horn antenna.

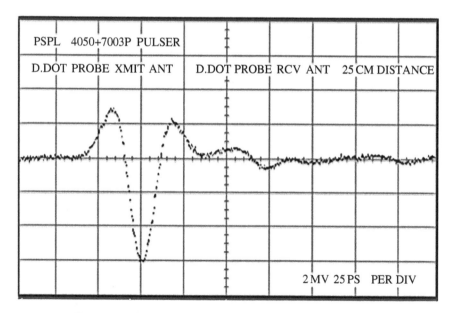

Figure 4.85 Transmission between a pair of identical D*dot antennas.

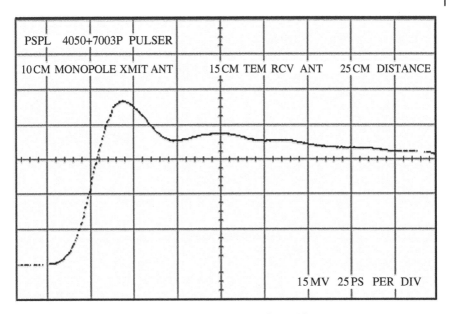

PSPL 4050+7003P PULSER								
10 CM MONOPOLE XMIT ANT			15 CM TEM RCV ANT		25 CM DISTANCE			

15 MV 25 PS PER DIV

Figure 4.86 Transmission from monopole antenna to the TEM horn.

the received output signal using a pair of D*dot antennas for both transmit and receive. This signal then is the third derivative of the step excitation.

Next, consider a monopole antenna which is a quarter wave whip antenna above an infinite ground plane. The monopole antenna is sometimes used as a cheap replacement for a conical antenna for transmitting UWB signals which are similar in wave shape to the driving point voltage. However, its radiated fields are not as uniform as those for the conical antenna. Its driving point impedance is not constant, but rises as a function of time. This leads to distortions in the radiated electromagnetic fields. Figure 4.86 shows the radiated electric field of a 10 cm monopole. It resembles the step electric field from the conical antenna. However its top line is not flat, but sags with increasing time. This is due to the non-uniform TDR impedance of this antenna. When a monopole is used for receiving transient electromagnetic fields, its output is the integral of the incident electric field. This is shown experimentally in Figure 4.87. The lower trace is the antenna output. It shows an almost monotonically rising ramp which is consistent with the integral of the step incident electric field. The upper trace is the calculated first derivative of the lower trace and is somewhat representative of the step incident electric field.

Figure 4.88 shows the output using a pair of 10 cm monopoles for both transmit and receive antennas. This is essentially the integral of the generator's step pulse. The integration effect can be explained simply. Assume an impulsive electric field is incident upon the monopole at a 90° angle to its axis. Thus a

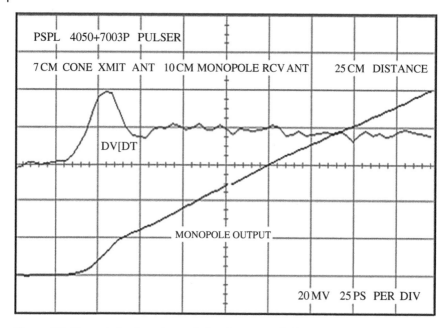

Figure 4.87 Transmission from the conical to the monopole. Lower trace receiving antenna output. Upper trace is dV/dt.

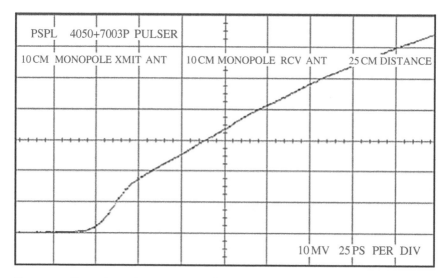

Figure 4.88 Transmission between a pair of identical 10 cm monopole antennas (25 ps/div).

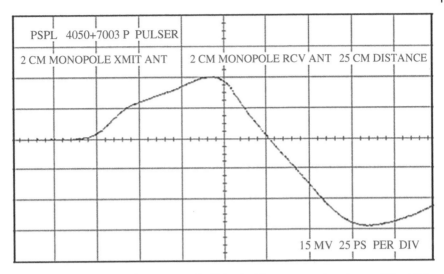

Figure 4.89 Transmission between a pair of identical 2 cm monopole antennas (25 ps/div).

current is induced simultaneously in each differential element, *dx*, of the antenna. These current elements, *dI*, thus start to flow towards the output connector of the antenna. They do not arrive at the output simultaneously, but in sequence. The output appears as a step function, which is the integral of the incident impulse. The integrating effect of this antenna only lasts for $t < \ell/c$, where ℓ is the length of the antenna. This effect is demonstrated in Figures 4.89 and 4.90 for a transmitting/receiving pair of shorter 2 cm monopole antennas.

Figure 4.89 shows the output starts off like a rising linear ramp, as in Figure 4.81, but due to the short length of the antenna, this stops after $\ell/c = 67$ ps and then the multiple reflections on the antennas predominate. This is shown more clearly in Figure 4.90 on a slower speed of 100 ps/div in which the resonant frequency of the 2 cm antenna is obvious as a damped, ringing sinusoid.

Next, we illustrate how to generate, transmit and receive an ultrawideband signal fitting the Federal Communications Commission (FCC) ultra-wideband (UWB) spectral mask.

4.8 Distortionless Transmission and Reception of Ultrawideband Signals Fitting the FCC Mask

A discrete finite time domain pulse is designed under the constraint of the Federal Communications Commission (FCC) ultra-wideband (UWB) spectral mask. This pulse also enjoys the advantage of having a linear phase over the

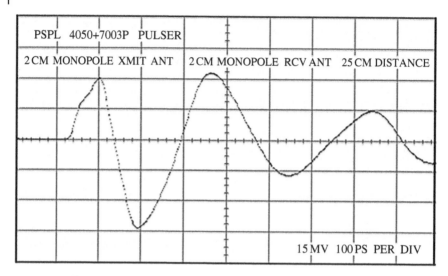

Figure 4.90 Transmission between a pair of identical 2 cm monopole antennas (100 ps/div).

frequency band of interest and is orthogonal to its shifted version of one or more baud times. [The term "baud" originates from the French engineer Emile Baudot, who invented the 5-bit teletype code. Baud rate refers to the number of signal or symbol changes that occur per second. A symbol is one of several voltage, frequency, or phase changes.] The finite time pulse is designed by an optimization method and concentrates its energy in the allowed bands specified by the FCC. Finally, an example is presented to illustrate how these types of wideband pulses can be transmitted and received with little distortion.

A method to design a T-pulse (a T-pulse stands for a strictly Time limited pulse) [24] that is of finite duration and its energy is concentrated primarily in a finite bandwidth whose envelope is an arbitrarily shaped frequency mask. So a T-pulse is not only a time limited pulse but for all practical purposes it is also practically bandlimited. For a practical design of a time domain pulse it has to be strictly limited in time otherwise it cannot be transmitted from a pragmatic point of view. However, from Fourier transform theory, a time limited pulse cannot simultaneously be band-limited, i.e., limited in frequency. Hence, we propose to design a strictly time limited pulse whose energy is primarily concentrated in a prescribed band so that for all practical purposes, the pulse can be considered to be also band-limited. Hence, the designed waveform can be realized in practice. In addition, an UWB signal that can totally fit in the FCC UWB spectrum mask is designed by this method.

T-pulse is proposed as a low-pass or band-pass signal which will concentrate its energy in the specified bands [24, 25]. Moreover, the T-pulse can extend beyond one baud time, and that necessitates that the waveform be orthogonal to its shifted versions and thereby reducing the inter symbol interference. In addition, if the pulse were to be transmitted then it also needs to have a zero dc value as an antenna cannot transmit DC. In order to meet these design criteria, an optimization method is utilized so that the T-pulse is orthogonal to its shifted versions of one or more baud times, and hence be a subset of orthogonal Nyquist signals [24, 25].

In this section, we describe the mathematical procedure to design a T-pulse that can fit within the FCC UWB spectral mask [26–28] and yet retain the various properties outlined previously including having a linear phase over the frequency band of interest. This will guarantee little dispersion and the group delay related to the derivative of the phase function will be almost constant over the design band of interest. We compute the various cost functions related to each criterion and evaluate their functional gradients, or equivalently the Frechet derivatives. Then we can minimize the composite cost function by various optimization methods. In the development we use, the method of steepest descent because it provides more robust results over other widely used methods.

Next, a T-pulse with a linear phase function that can completely fit in the FCC UWB spectral mask is designed and plotted to display its temporal and spectral structures. Finally, it is demonstrated that with the proper choice of transmitting and receiving antennas it is possible to transmit and receive such broadband pulses with little distortion.

4.8.1 Design of a T-pulse

The design of the T-pulse is carried out in the discrete time domain. Let us assume a discrete waveform of sequence $f(n)$, which is defined for $n = 0, 1, 2,...,$ N_n-1, and is identically zero outside these N_n values. Let us assume there are N_s samples in one-symbol duration. Then the total number of samples N_n is N_s times the number of symbols N_c, i.e.,

$$N_n = N_c \times N_s \tag{4.16}$$

The discrete Fourier transform (DFT) – both forward and backward – of the waveform $f(n)$, reflecting its spectral content is given by

$$F(p) = \frac{1}{\sqrt{N_p}} \sum_{n=0}^{N_n-1} f(n)\exp\left[-j\frac{2\pi np}{N_p}\right], \quad p = 0,1,...,N_p-1, \tag{4.17}$$

$$f(n) = \frac{1}{\sqrt{N_p}} \sum_{p=0}^{N_p-1} F(p)\exp\left[j\frac{2\pi np}{N_p}\right], \quad n = 0,1,...,N_n-1. \tag{4.18}$$

where N_p is the total number of samples in the frequency domain. We assume there are N_r samples per baud rate (the inverse of one-symbol duration), which satisfies $N_p = N_s \times N_r$. In addition, we have $N_r \geq N_c$, and increasing N_r increases the resolution of the waveform in the frequency domain. In the T-pulse construction, the objective is to maximize the in-band energy within the set $\Phi = \{-N_b \leq p \leq N_b\}$, which corresponds to a frequency band of N_s/N_r bauds. Or equivalently, we want to minimize the energy outside the N_b samples.

Thus we want to maximize the energy in a specific band $\Phi = \{-N_b \leq p \leq N_b\}$.

$$E_\Phi = \sum_{p=-N_b}^{N_b} |F(p)|^2 = \frac{1}{N_b} \sum_{n=0}^{N_n-1} \sum_{m=0}^{N_n-1} f(n)f(m) \frac{\sin\left[\frac{2\pi(n-m)}{N_p}\left(N_b + \frac{1}{2}\right)\right]}{\sin\left[\frac{\pi(n-m)}{N_p}\right]}$$

(4.19)

Therefore the energy and its gradient in the band $\Phi = \{N_{bl} \leq |p| \leq N_{bu}\}$ can be written as

$$E_\Phi = \frac{1}{N_b} \sum_{n=0}^{N_n-1} \sum_{m=0}^{N_n-1} f(n)f(m) \left\{\sin\left[\frac{2\pi(n-m)}{N_p}\left(N_{bu} + \frac{1}{2}\right)\right]\right.$$
$$\left. - \sin\left[\frac{2\pi(n-m)}{N_p}\left(N_{bl} - \frac{1}{2}\right)\right]\right\} \Big/ \sin\left[\frac{\pi(n-m)}{N_p}\right],$$

(4.20)

$$\frac{\partial E_\Phi}{\partial f(m)} = \frac{1}{N_b} \sum_{n=0}^{N_n-1} f(n) \left\{\sin\left[\frac{2\pi(n-m)}{N_p}\left(N_{bu} + \frac{1}{2}\right)\right]\right.$$
$$\left. - \sin\left[\frac{2\pi(n-m)}{N_p}\left(N_{bl} - \frac{1}{2}\right)\right]\right\} \Big/ \sin\left[\frac{\pi(n-m)}{N_p}\right].$$

(4.21)

where N_{bl} and N_{bu} represent the lower and the upper frequency bounds for the band, respectively.

Furthermore, we want to concentrate the energy of the waveform in some specific frequency bands, and we can simultaneously minimize the energy of this waveform out of these bands. In this section, our goal is to fit the spectrum delineated by the FCC UWB frequency mask. This is accomplished by applying different weights to different frequency bands as illustrated below

$$g_1 = w_{\Phi 1}E_{\Phi 1} + w_{\Phi 2}E_{\Phi 2} + w_{\Phi 3}E_{\Phi 3} + \ldots.$$

(4.22)

In the next step, the criterion to minimize the intersymbol interference is introduced. This is accomplished by making the waveform orthogonal to its shifted version as illustrated in Eq. (4.23).

$$\sum_{n=1}^{N_n-pN_s} f(n)f(n+pN_s) = \delta(p), \quad p = 0,1,\ldots,N_c - 1$$

(4.23)

The resulting cost function and its gradient can be written as follows:

$$g_2 = \sum_{p=0}^{N_c-1} w_p \left[\sum_{n=1}^{N_n-pN_s} f(n)f(n+pN_s)-\delta(p) \right]^2, \tag{4.24}$$

$$\frac{\partial g_2}{\partial f(n)} = \sum_{p=0}^{N_c-1} w_p \left[\left(\sum_{n=1}^{N_n-pN_s} f(n)f(n+pN_s)-\delta(p) \right) \right.$$
$$\left. \times \left(f(n-pN_s)+f(n+pN_s) \right) \right]. \tag{4.25}$$

The total cost function to be optimized for this problem then is obtained by adding g_1 and g_2 as

$$J = g_1 + g_2 \tag{4.26}$$

4.8.2 Synthesis of a T-pulse Fitting the FCC Mask

We apply the method of steepest descent to minimize the cost function of (4.26) in order to obtain a T-pulse that can fit in the given frequency mask. Here is a brief description of the procedure that has been implemented in designing the required waveform.

1) We start from an initial guess for the waveform $f_0(n)$ and choose the appropriate weights w_Φ and w_p to calculate the total energy.
2) Next, we compute the gradient of the cost function.
3) We start the process of optimization utilizing the method of steepest descent and update the waveform $f(n)$ as outlined in that recipe [25, 28].
4) We compare the norm of the gradient with the given threshold, if the gradient is not small enough, repeat the iteration and we go to step 2.

With this optimization method and with appropriate weights w_Φ and w_p, one can design a T-pulse that can fit into the FCC UWB mask possessing a linear phase. For the T-pulse designed in this work, $N_c = 10$, $N_s = 25$, $N_r = 250$, $w_p = 1$, and the values of w_Φ for the different bands are listed in Table 4.1. The weights chosen for this example have been decided by trial and error and the ones that provide the best results have been included. The duration of the T-pulse in the continuous time domain is 1.53 ns (nanosecond).

The designed T-pulse in the time and frequency domains is plotted in Figures 4.91(a) and (b), respectively. The waveform generated can actually fit in the FCC UWB mask displayed in Figure 4.91(b) while maintaining a linear phase as seen in Figure 4.91(c). The variation in the phase for Figure 4.91(c) occurs during the regions where the magnitude of the spectral component is small and hence these variations will not have any effect when deployed in a real system.

4.8.3 Distortionless Transmission and Reception of a UWB Pulse Fitting the FCC Mask

In this section, we illustrate the transmission and reception of this wideband T-pulse with little distortion. It has been shown by Jim Andrews [16] that transmitting any pulse through a biconical antenna does not distort the

Table 4.1 The different weights used over each frequency band.

Frequency band (GHz)	w_Φ
0.96~1.61	70000
1.61~1.99	10
1.99~3.1	5
3.1~10.6	−1
>10.6	1

(a)

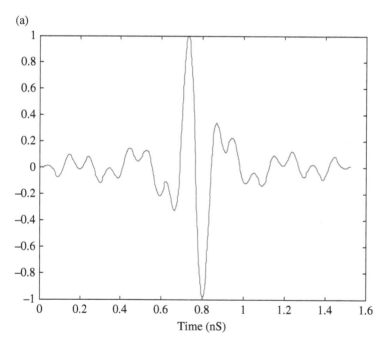

Figure 4.91 The designed T-pulse in the time and frequency domains. (a) The designed T-pulse in the time domain. (b) The spectral magnitude of the T-pulse in the frequency domain fitting the FCC Mask. The vertical axis is in dB and the red line represents the level of the FCC UWB mask. (c) The phase of the T-pulse in the frequency domain. The vertical axis represents the phase.

(b)

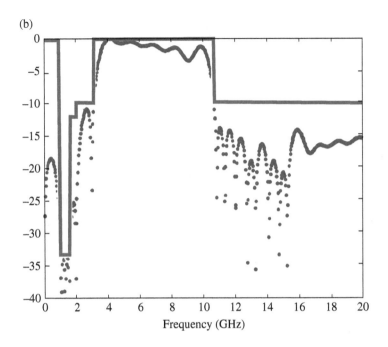

(c)

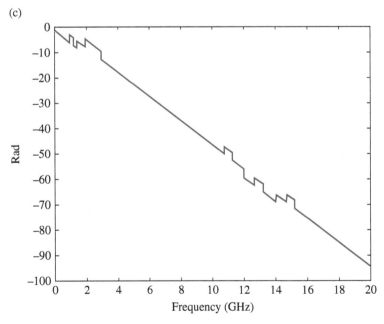

Figure 4.91 (Cont'd)

radiated waveforms. Similarly using a TEM horn as a receiving antenna does not distort the received waveform either [16]. Two different antennas must be chosen for transmission and reception because the impulse response of a transmitting antenna is the time derivative of the impulse response when the same antenna is used as a receiver [7]. Therefore, it is not possible to use similar antennas for transmit and receive without distorting the input pulse. Experimental results of Jim Andrews [16] reveal that when a bicone-TEM horn transmit/receive system is used for pulse propagation, this system will not distort any broadband waveforms. Experimental data has been provided in the previous section to support this claim. In this section we use the same combination of transmit and receive antennas to illustrate the transmission and reception of this wideband T-pulse without little distortion. The simulation is carried out using the electromagnetic software HOBBIES which can analyze radiation/scattering from arbitrary shaped complex material electromagnetic structures [2].

A biconical antenna is used as the transmitting antenna and a TEM horn is used as the receiving antenna in this setup as shown in Figures 4.92 and 4.93. A bicone radiates a field that is similar in nature to the driving voltage when $t < l/c$, where l is the length of the cone and c is velocity of light [27]. And the TEM horn can provide a direct measurement of the incident waveform [27, 28].

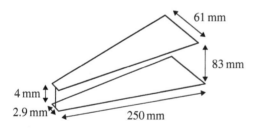

Figure 4.92 The schematic of a TEM horn.

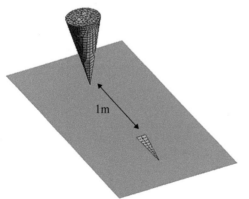

Figure 4.93 The bicone and the TEM horn.

In this example, the bicone and the TEM horn combination is set up as illustrated in [27]. In our analysis, the bicone is 0.5 m long and is situated over a ground plane. The solid half angle of the cone is 12°. The parameters of the full TEM horn are shown in Figure 4.92 [27]. In this simulation, the bicone and the TEM horn are placed 1 meter apart and both are located over an infinite perfectly electrically conducting ground plane as shown in Figure 4.93. This will be a typical near-field coupling scenario.

We use the T-pulse designed in the previous section as the driving point voltage to the bicone and observe the received pulse at the output of the TEM horn as shown in Figure 4.94. If we switch the function of these antennas, which means that we use the TEM horn as the transmitting antenna and use the bicone as the receiving antenna, then the functions of the transmit and the receive antennas will be completely different from the previous scenario. Since, the TEM horn does not distort the pulse on receive, on transmit it will differentiate the input waveform which is our T-pulse, as the transmit impulse response is the time derivative of the receive impulse response. On the other hand, since the bicone does not distort the waveform on transmit and therefore on receive it will perform an integration of the received waveform. Therefore in this system we have a differentiator in cascade with an integrator and therefore they will cancel each other and the received waveform will still be the same. Indeed, this is the case, as illustrated by Figure 4.95. The received current is then normalized and shifted by 3.3 ns and it overlaps with the driving

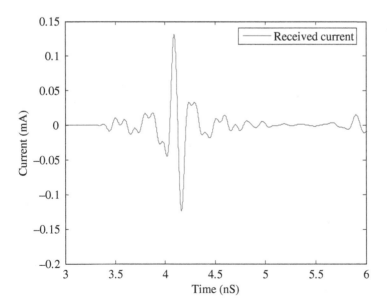

Figure 4.94 The output current at the TEM horn placed 1 m away.

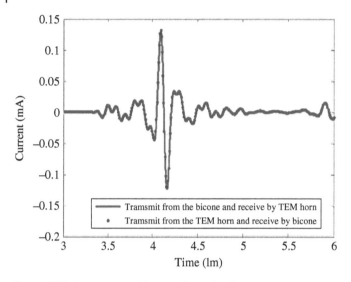

Figure 4.95 Comparison of the waveforms for the two scenarios.

voltage, which is the T-pulse and they are both displayed in Figure 4.96. It can be seen that the output induced current is similar to the driving voltage. There are some undesired reflection that arise from the finite length cone and the horn after the duration of the pulse. They can be eliminated by properly terminating the ends of the antennas.

In the next example, we change the distance between the antennas to ten meters. This will possibly be a far field scenario – as this signal is bandlimited and therefore can have a finite value for the limit of λ. The output current is plotted in Figure 4.97 and then normalized and shifted 33.3 ns resulting in Figure 4.98. It is seen that the received signal agrees well with the driving voltage.

It is rather interesting to observe that the received waveshapes are very similar irrespective of whether the separation between the bicone and the TEM horn is 1 m or 10 m. Does that imply as Jim Andrews [16] remarks: *that path loss is independent of frequency*! This is indeed quite true if the entire system, including the transmit and the receive antennas are considered. This has been illustrated in section 4.7. Typically, when one performs propagation modeling, it is only the channel that one considers. Under that circumstance indeed due to Huygens's principle the channel is quite dispersive as a function of frequency. Specifically the path loss will vary as the square of the frequency. However, the final result changes when the transmit and the receive antennas are considered as part of the complete channel as demonstrated in section 4.7. Hence, in summary, the gain of the transmit and receive antennas will compensate for the loss in the channel resulting in no apparent distortion.

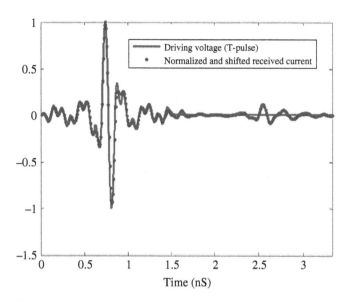

Figure 4.96 Comparison of waveforms between the T-pulse and the normalized and shifted output pulses when the antenna separation is 1 m.

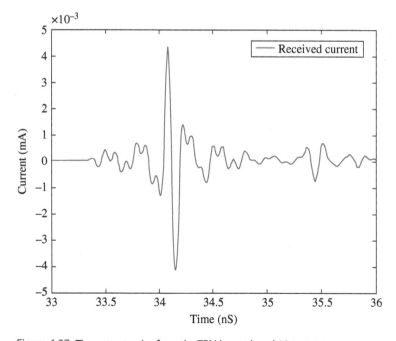

Figure 4.97 The output pulse from the TEM horn placed 10 m away.

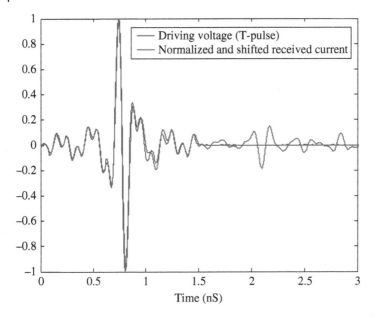

Figure 4.98 Comparison between the T-pulse and the normalized and shifted output current when the antenna distance is 10 m.

Indeed the second form of Frii's Transmission formula given in (4.15) illustrates this point.

In summary, we have presented an optimization procedure for the design of a time limited pulse that has a spectral mask satisfying the FCC UWB criterion. In addition, the waveform possesses a linear phase over the bandwidth of interest. The cost function in this optimization procedure is formulated in terms of the in-band and out-of-band energies, and additionally enforcing orthogonality between the waveform and its shifted versions. Numerical results also show that the method of steepest descent can be applied to generate the T-pulse of interest. Therefore through a suitable choice for the transmit and the receive antenna systems it is possible to transmit an ultrawideband T-pulse with little distortion.

4.9 Simultaneous Transmission of Information and Power in Wireless Antennas

4.9.1 Introduction

Antennas [29] are useful in information transfer [30, 31] as well as power transfer [32–34]. In some applications, the two functions should be applied simultaneously [35, 36]. Two practical applications are passive radio frequency identification (RFID) [37, 38] and wireless sensing system (WSN) [39, 40]. The role of information transfer in a RFID system is to detect objects, while its role in a WSN is to monitor the environment. The role of power transfer in both RFID system and WSN is powering or charging the tags used in the systems.

In this section, an analysis of wireless information versus power transfer over the same channel consisting of a transmit and receive antenna system is discussed. This frequency selective additive white Gaussian noise channel display a fundamental tradeoff between the rate at which energy and reliable information can be transmitted over the same channel as in a RFID system, a power-line communication system, or for an energy harvesting system. The optimal tradeoffs between power transferred and the channel capacity due to Shannon (which is additive white Gaussian noise limited), Gabor (which is interference limited), and Tuller (which is defined in terms of the signal and noise amplitudes and not power) are compared and the differences are discussed. The appropriate use of each of the channel capacity formulations for a frequency-selective transmitting/receiving antenna system in wireless communication is then computed as an illustrative example to describe the tradeoff between wireless power transfer and wireless information transfer over a transmitting/receiving antenna system.

Consider a practical wireless communication system where noise is present, and where the power available at the transmitting end is limited. All such systems are bound by a fundamental tradeoff between the power delivered to the receiver, and the rate of communication achievable. That is to say, for a fixed input power to the transmitter, the designer can choose to maximize the power transfer, maximize the capacity, or can obtain a tradeoff between the two, which satisfies the system's requirements. The relationship governing that tradeoff is extremely important for the proper design of wireless communication systems which in some cases can be an RFID system [41–43]. For example, maximum power can be delivered under a resonance condition between the transmitting and the receiving wireless systems. However, resonances are inherently narrow band, and therefore are not suitable for information transmission, as they require a finite bandwidth. On the other hand, the principle of information transmission is limited by the bandwidth, and not conducive to efficient wireless power transmission. In this section, we set up the optimization problem dealing with the tradeoffs between the simultaneous wireless power and information transfer and illustrate how they can be optimized. Our results are determined by what criteria one uses to define the channel capacity. There are three different ways to characterize the channel capacity. One due to Shannon, which is characterized in terms of an additive-white-Gaussian-noise channel, and is defined in terms of signal and noise power. A second is due to Gabor, which is primarily characterized in terms of interference, and not background noise. Finally, the third due to Tuller, where the signal-to-noise ratio is determined in terms of the signal and noise amplitudes, as practical receivers

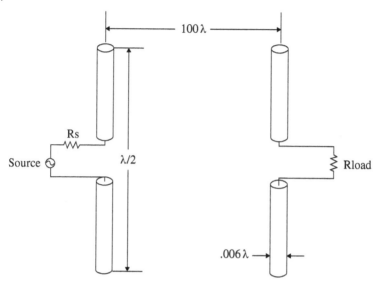

Figure 4.99 The electromagnetic simulation setup, showing the dimensions, spacing, and loading of the transmitting and receiving antennas.

are sensitive to a minimum value of the signal amplitude that must exist at the front end of a receiver, in contrast to the noise amplitude.

For this illustration, a simple setup was used consisting of two half-wave dipole antennas tuned in length to resonate at 1 GHz. The receiving antenna is placed in the far field of the transmitting antenna operating in free space, a distance of 100 λ where λ is the wavelength of operation. The antennas are arranged in free space without the influence of ground or other scatterers. Because the impedance of a thin (0.003λ radius), half wavelength dipole antenna is approximately 73 Ω at resonance, the source and the load resistances were also chosen to be 73 Ω. In this setup we consider the voltage measured across the load resistor to be the received signal. This setup is shown in Figure 4.99. The power transfer function and the voltage transfer function between the transmit and the receive systems is shown in Figure 4.100.

In the following analysis, the transmitting antenna delivers the maximum power to an identical receiving dipole placed in the far field when the frequency of excitation is 1 GHz. In this arrangement, the capacity for information transmission is zero, as at resonance we assume that the bandwidth is theoretically zero. In order to increase the capacity of the system for information transfer, the power input at the transmitting end may be removed from the frequency of maximum power transfer, and can be reallocated over a selected bandwidth, which in the literature is termed as the *water-filling algorithm* [44–46].

(a)   (b)

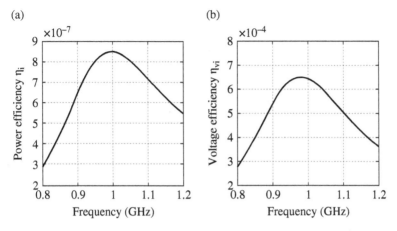

Figure 4.100 (a) The power transfer efficiency. (b) The voltage induced efficiency.

However, the present results of this section differ from Shannon's classical formulation for the water-filling algorithm, as we not only have a limitation on the input power, but are also trying to maximize the channel capacity, in addition to maximizing the received power.

There are several forms of channel-capacity theorems that may be useful for determining a suitable power-allocation strategy. These include the Shannon Channel Capacity Theorem [44], Gabor's formula for channel capacity [47, 48], and the channel-capacity theorem arrived at by Hartley, Nyquist, and Tuller (HNT) [49, 50]. The various concepts and their developments have been introduced in Chapter 1. It is important to note that the distinction between the approaches used to arrive at each of these formulations have been discussed in section 1.9. In order to make practical comparisons between each of the channel-capacity theorems, a simulation of a two dipole antenna setup was executed, using an electromagnetic field solver called *HOBBIES (Higher-Order Basis-Based Integral-Equation Solver)* [2]. This was done in order to show the results of applying each of the channel-capacity theorems to the vector electromagnetic problem [51, 52]. These comparisons were drawn from two simulations. In the first simulation, the power allocation was found to maximize the capacity while using variable bandwidths, according to the Shannon Channel Capacity Theorem with a fixed input power. The results were compared to the capacity curves found using Gabor's theorem and the Hartley, Nyquist, and Tuller theorem, when using the same power allocation. In the second simulation, all three of the channel-capacity theorems were optimized, in order to find the power allocations that maximized capacity and power delivered for the allotted bandwidth.

4.9.2 Formulation and Optimization of the Various Channel Capacities

The Shannon Channel Capacity theorem, described in Equation (1.117), is a function of the ratio of the power received to the power transmitted. It is based on an analysis of noise as a purely random quantity that is determined by statistical analysis, as was discussed in Chapter 1, section 1.9 [44]. This definition of capacity is based on power signal-to-noise ratio (SNR). The assumption for noise may not be adequate in all scenarios of antenna-based communications, such as receivers being located in the near field or the far field of the transmitting antenna, and also because the noise may not be completely random. In addition, the definition of power in the near field - where the Poynting vector is complex - as opposed to the far field is interesting, as that makes the channel capacity a complex number!

Gabor's channel capacity, discussed in Chapter 1, section 1.9 [48] uses the power definition for SNR, the proper definition of which is as ambiguous as it is in Shannon's definition. However, the difference is that Gabor's formulation takes into account the interaction between the thermal noise and the signal, which leads to actual interference (not just noise).

Finally, the Hartley, Nyquist, and Tuller case, discussed in Chapter 1, section 1.9 [49, 50] was derived using Maxwellian physics. It avoids the ambiguity of the definition of power used in the previous two definitions and does so by instead using the field quantities. Another important distinction is that the Hartley, Nyquist, and Tuller measure is defined as the capacity at which error may be made to be arbitrarily small [44–46]. However, the important distinction is that this formulation takes into account the physics of the entire system, when considered from the signal at the transmitting terminal, and also at the receiving terminal.

4.9.2.1 Optimization for the Shannon Channel Capacity

In order to determine the water-filling solution for an arbitrary bandwidth, we start with Shannon's Capacity per bandwidth segment ΔB over which power is allocated:

$$\frac{C_S}{\Delta B} = \log_2\left(1 + \frac{\eta_i P_i}{P_N}\right) \tag{4.27}$$

where P_i is the input power allocated to the band I of duration ΔB. η_i is the transfer power gain from the input to the output of the transmitting/receiving system. P_N is the input noise power, assumed to be constant over the bandwidth of interest. The bandwidth segments may be of variable size, depending on the desired accuracy of the simulation. By summing these capacities over M bandwidth segments over which we allocate power, we can find the total capacity. This analysis is analogous to that performed in [43], except that the analysis was performed on a simple inductive circuit.

Next, a functional $\mathfrak{I}$ is defined, which includes a term for the capacity that will be maximized against the constraint that the maximum power allotted over the bandwidth of the simulation be equal to the total power available. This term is weighted by the non-negative Lagrange multiplier, λ. The final term is intended to deliver a fixed amount of power scaled by another Lagrange multiplier, μ. The functional to be optimized is thus given by

$$\mathfrak{I} = \sum_{i=1}^{M} \log_2\left(1 + \frac{\eta_i P_i}{P_N}\right) - \lambda\left(\sum_{i=1}^{M} P_i - P_{av}\right) + \mu\left(\sum_{i=1}^{M} \eta_i P_i - P_{del}\right) \tag{4.28}$$

where, P_{av} is the total power available at the-transmitter, and P_{del} is the power to be delivered at the receiver. It is important to note that generally, the channel-capacity formula appears as an integral, but numerical computations can only be done in a discrete sense. The integral is hence replaced by a summation for computational reasons, as initially carried out in [41, 42]. When the power delivered to the receiver is dictated by the input power distribution, given by the water-filling algorithm, then the third term in the functional given by Equation (4.28) can be dropped [41]. However, if P_{del} is greater than the value given by the water-filling algorithm, then the third term using the Lagrange multiplier is required.

In the next step, we drop the summations and set the derivative to zero, in order to find the power allocation that maximizes the Shannon channel capacity. This results in [36, 41]

$$\frac{dF}{dP_i} = \frac{\log_2 e}{\left[\dfrac{P_N}{\eta_i} + P_i\right]} - \lambda + \mu\eta_i = 0 \tag{4.29}$$

Solving for P_i, one obtains

$$P_i = \left(\frac{\log_2 e}{\lambda - \mu\eta_i} - \frac{P_N}{\eta_i}\right)^{+} \tag{4.30}$$

The superscript + implies that only the positive values are selected. When the required power to be delivered to the receiver, P_{del} is less than the power to be delivered to the receiver through this water-filling algorithm, then the third term in Equation (4.28) can be neglected. Note that by dropping the last term in the functional equation of Equation (4.28) for the optimum power-distribution strategy, the power allocation over each band is given by

$$P_i = \left(\frac{\log_2 e}{\lambda} - \frac{P_N}{\eta_i}\right)^{+} \tag{4.31}$$

Note that the first term in Equation (4.31) is a constant, and is given by approximately $1.44/\lambda$. The power to be allocated over each individual band, P_i, thus related to a constant minus the noise power scaled by the input-output transfer function at that band. This is how the term *water-filling algorithm* originated, as it has to do with determining the total power allocated plus the scaled noise power to be a constant, independent of the bandwidth segment. The noise power is a constant, independent of the band [41–43]. The total signal and the noise power are thus constant over any bandwidth segment. The power allocated over the band therefore follows a water-filling algorithm, indicating that the level of signal and noise power over every bandwidth segment is a constant, resulting in the *water-filling* solution. The value for the Lagrange multiplier, λ, can be obtained by summing P_i over all the M bands, and equating it to the total power, P_{av}, available at the transmitter. However, if more power is desired at the receiver than the water-filling algorithm promises, then μ cannot be set equal to zero in Equation (4.28), and hence Equation (4.29) is introduced.

4.9.2.2 Optimization for the Gabor Channel Capacity

Gabor very thoroughly investigated the energy levels that can be used, in each cell of his *logon* concept, to represent distinguishable signals as described in section 1.9.2 [48]. In his analysis, Gabor introduced the phenomenon of beats between the signal and noise. These beats resulted in increased energy fluctuations. If we superimpose a signal and an interfering signal with a variable phase, we then obtain the different energy levels which can be used. Gabor then introduced his *logon* concept, to represent distinguishable signals [48]. Gabor's channel capacity formula then is given by (1.126).

Following the same technique used to optimize the Shannon Capacity, we start with Gabor's capacity per bandwidth segment as,

$$\frac{C_G}{\Delta B} = \log_2 \left(\frac{1 + \sqrt{1 + \frac{4\eta_i P_i}{P_N}}}{2} \right) \tag{4.32}$$

The functional in this case then becomes

$$\Im = \sum_{i=1}^{M} \log_2 \left(\frac{1 + \sqrt{1 + \frac{4\eta_i P_i}{P_N}}}{2} \right) - \lambda \left(\sum_{i=1}^{M} P_i - P_{av} \right) + \mu \left(\sum_{i=1}^{M} \eta_i P_i - P_{del} \right) \tag{4.33}$$

Taking the derivative and setting to zero in order to maximize capacity we find

$$\frac{dF}{dP_i} = \frac{2\log_2 e}{\left[1 + \sqrt{1 + \frac{4\eta_i P_i}{P_N}} \right]} \left(\frac{\eta_i}{P_N \sqrt{1 + \frac{4\eta_i P_i}{P_N}}} \right) - \lambda + \mu \eta_i = 0 \tag{4.34}$$

which reduces to

$$\left[1+\sqrt{1+\frac{4\eta_i P_i}{P_N}}\right]\left[\sqrt{1+\frac{4\eta_i P_i}{P_N}}\right] = \frac{2\eta_i \log_2 e}{(\lambda - \mu\eta_i)P_N}$$ (4.35)

Note that this equation can be expressed in quadratic form

$$(1+a)a = b$$ (4.36)

which results in the solution for a as

$$a = -\frac{1}{2} \pm \sqrt{\frac{1}{4}+b}$$ (4.37)

Using (4.35) and (4.37) one obtains,

$$a = \sqrt{1+\frac{4\eta_i P_i}{P_N}} \qquad b = \frac{2\eta_i \log_2 e}{(\lambda - \mu\eta_i)P_N}$$ (4.38)

Therefore, the power allocated over each band will be given by

$$P_i = \frac{P_N}{4\eta_i}\left[-\frac{1}{2}+\frac{2\eta_i \log_2 e}{(\lambda - \mu\eta_i)P_N} \mp \sqrt{\frac{1}{4}+\frac{2\eta_i \log_2 e}{(\lambda - \mu\eta_i)P_N}}\right]$$ (4.39)

Again, if the power delivered to the receiver is not greater than the one delivered by the water filling algorithm which maximizes the channel capacity then we can drop the second Lagrange multiplier μ from (4.39) resulting in

$$P_i = \frac{P_N}{4\eta_i}\left[-\frac{1}{2}+\frac{2\eta_i \log_2 e}{\lambda P_N} \mp \sqrt{\frac{1}{4}+\frac{2\eta_i \log_2 e}{\lambda P_N}}\right]$$ (4.40)

This tells us how the total power available at the transmitter needs to be distributed over each band to maximize the Gabor channel capacity under an input-power constraint.

4.9.2.3 Optimization for the Hartley-Nyquist-Tuller Channel Capacity
The Hartley, Nyquist, and Tuller formula for capacity per bandwidth segment [49, 50] is given by

$$\frac{C_{HNT}}{\Delta B} = 2\log_2\left(1+\frac{\eta_{vi}V_i}{V_N}\right)$$ (4.41)

where η_{vi} is the transfer voltage gain between the input and the output. The functional equation for maximization of the Tuller channel capacity is given by

$$\mathfrak{I} = \sum_{i=1}^{M} \log_2\left(1 + \frac{\eta_{vi}V_i}{V_N}\right) - \lambda\left(\sum_{i=1}^{M} \frac{V_i^2 \cos\theta}{|Z_i|} - P_{av}\right) + \mu\left(\sum_{i=1}^{M} \frac{(\eta_{vi}V_i)^2}{Z_L} - P_{del}\right) \qquad (4.42)$$

We now take the derivative of Equation (4.42) and set it to zero to find the voltage V_i to be applied at the source for each band to maximize the capacity for a given bandwidth. V_N is the amplitude of the noise voltage. Z_i and Z_L are the input impedance looking at the source V_i, and the load impedance (73 Ω), respectively. The input power has the power factor $\cos\theta$, which is taken into account in the equation. This result in

$$\frac{dF}{dP_i} = \frac{\log_2 e}{\left[\dfrac{V_N}{\eta_{vi}} + V_i\right]} - 2V_i\left(\frac{\lambda\cos\theta}{|Z_i|} - \frac{\mu\eta_{vi}^2}{Z_L}\right) = 0 \qquad (4.43)$$

Finally, by rearranging and solving for V_i we find

$$V_i = \frac{-\dfrac{V_N}{\eta_{vi}} \pm \sqrt{\left(\dfrac{V_N}{\eta_{vi}}\right)^2 + \dfrac{4\log_2 e}{\dfrac{\lambda\cos\theta}{|Z_i|} - \dfrac{\mu\eta_{vi}^2}{Z_L}}}}{2} \qquad (4.44)$$

Note that by dropping the last term in the functional equation of (4.42) when the required delivered power is given by the water filling algorithm, then we obtain

$$V_i = \frac{-\dfrac{V_N}{\eta_{vi}} \pm \sqrt{\left(\dfrac{V_N}{\eta_{vi}}\right)^2 + \dfrac{4|Z_i|\log_2 e}{\lambda\cos\theta}}}{2} \qquad (4.45)$$

This provides the optimum excitation that will maximize the Hartley, Nyquist, and Tuller channel capacity under the input power constraint.

Although the signal is transmitted using a channel bandwidth of ΔB, the voltages V_i and V_N here are accounted for as if the signal is transmitted with the center frequency for that band. Nevertheless, Equation (4.41) accounts for the ratio but not the magnitudes of the voltages, so the form of the representation is not too important.

4.9.3 Channel Capacity Simulation of a Frequency Selective Channel Using a Pair of Transmitting and Receiving Antennas

The first experiment comparing the channel-capacity formulations using the electromagnetic simulation described above was based on finding the optimal input-power allocation over a frequency band at the transmitting antenna, according to the optimization of Shannon's capacity theorem for capacity delivered. A finite input power distributed according to this allocation method is known as the water-filling algorithm.

To obtain the water-filling power allocation, we ran the electromagnetic simulation from 0.8 GHz to 1.2 GHz at 101 frequency steps. The plots for power-transfer efficiency, η_i, and voltage-induced efficiency, η_{Vi}, are shown in Figure 4.100. Note that the peak efficiency did not occur at the same frequency for the two cases. Looking at Equation (4.31) for the case of $\mu = 0$, we note that only the efficiency, η_i, the noise power at the receiver, P_N, and λ are necessary in order to determine the power allocation. By calculating the transfer-power efficiency for each bandwidth segment and always using the M most efficient bandwidths to transfer power in our algorithm, the results of Equation (4.31) for maximum Shannon channel capacity were obtained.

For example, when $M = 1$ a single bandwidth segment was used. The algorithm allocated power to the most efficient bandwidth segment. This power was numerically determined by solving for the value of λ in Equation (4.31) which satisfies the condition

$$\sum_{i=1}^{M} P_i = P_{av} \tag{4.46}$$

If $M = 5$ the algorithm allocates power to the 5 most efficient bandwidths for power transfer and then numerically solved for a new value of λ that satisfied (4.46). The process continued for M equals one to the number of bandwidth segments. The results are shown in Figures 4.101, 4.102, 4.103 for $P_{av} = 5.2$ W where the noise per bandwidth segment was constant over the entire bandwidth of interest. It is equal to the maximum value of η_i such that P_N/η_i is normalized to one at the most-efficient power-transfer frequency ($P_N = 8.3 \times 10^{-7}$ W). From the figures, when M increased, power was allocated to more bandwidth segments. However, with the fixed input power and the noise power, a solution for the maximum Shannon channel capacity when $M > 44$ was not possible. Therefore although there were 101 bandwidth segments, using 44 bandwidth segments gave the maximum Shannon channel capacity while for the other bandwidth segments zero power was allocated. When the number of bandwidth segments, M, increases, the power delivered to the receiver decreases, since more power is allocated to the bandwidth segments with smaller transfer-power gains. This is illustrated in Figure 4.104.

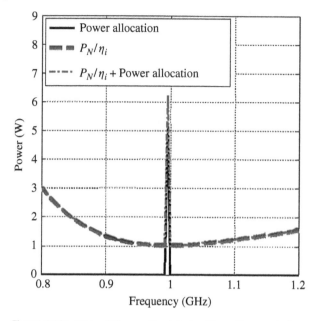

Figure 4.101 Water-filling method derived from Shannon's theorem showing the power allocated at one frequency segment.

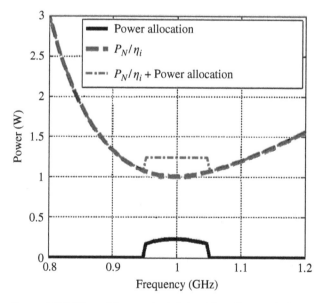

Figure 4.102 Water-filling method derived from Shannon's theorem showing the power allocated at 25 frequency segments.

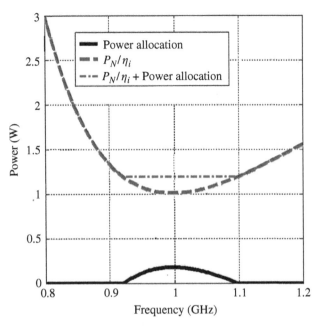

Figure 4.103 Water-filling method derived from Shannon's theorem showing the power allocated at 44 frequency segments.

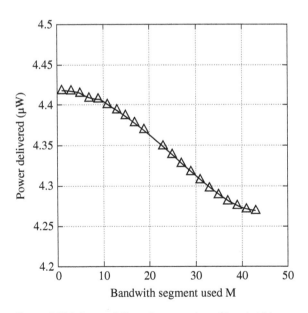

Figure 4.104 Power delivered per number of bandwidth segments used.

This illustrates that there is a tradeoff between wireless power transfer and information transfer. Wireless power transfer is optimum at a very narrow bandwidth, whereas the information transfer requires a sizable bandwidth, resulting in less power transfer. It is therefore important to point out that the meaning of the word *information* used in this section is quite different from the colloquial use of the word *information* [49].

The behavior of the water-filling algorithm for Shannon's theory was described above. We note that executing the algorithms for Gabor's and using the Hartley, Nyquist, and Tuller formulas, as derived, do not result in the same "water filling" behavior. An example of Gabor's distribution for $M = 101$ is shown in Figure 4.105. For $M = 101$, power was allocated over all bandwidth segments. However, the solution using Shannon's channel capacity restricted power to be allocated only for M < 45.

With the proper power allocation thus determined, we proceeded with the comparison of the capacity formulations. For a noise power per bandwidth segment of $P_N = 10^{-7}$ W, the capacity as a function of power delivered results are shown in Figure 4.106. This plot contains the solutions of the Shannon channel capacity, the Gabor channel capacity, and the Hartley, Nyquist, and Tuller channel capacity, i.e., Equations (4.31), (4.40), and (4.45), in the case of $\mu = 0$, with different numbers of M, from 1 to 101. As expected with Gabor's method, which takes into account the interaction between the signal and the

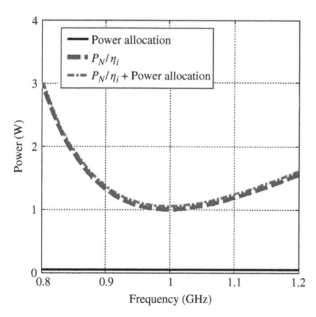

Figure 4.105 An example of Gabor's distribution for $M = 101$. As M increases, power is allocated over all bandwidth segments.

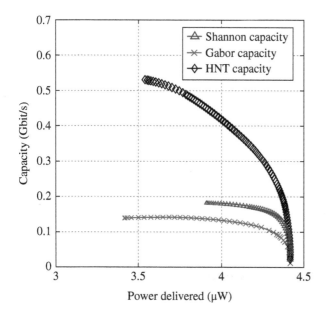

Figure 4.106 The capacity as a function of power delivered for $P_N = 10^{-7}$ W.

thermal noise (which Shannon's method does not), predicted a lower capacity per power delivered than Shannon's. The Hartley, Nyquist, and Tuller capacity related to the incident field strength received at the receiver displayed a much higher value for the capacity.

It is worth noting that if the signal-to-noise ratio for a bandwidth segment at the receiving end is less than or equal to 0 dB then the three formulations enter what is called the power limited regime of operation. If a restriction is imposed for the power delivered for Shannon's and Gabor's methods and the voltage received in the Hartley, Nyquist and Tuller method – such that if a segment has a SNR of less than 0 dB it is not counted towards the capacity – then we noticed that the plots for the capacity enter the power-limited regime of operation. This is shown in Figure 4.107 (note that the plots to the left of the maxima have little meaning).

Figure 4.108 displays similar plots but with different noise power levels. So for $P_N = 10^{-9}$ W and without the limitation on the signal-to-noise ratio (SNR) for capacity computation. It is interesting to note that the Shannon and Hartley, Nyquist, and Tuller capacities were in significantly better agreement as the SNR increased. This corresponded to Shannon's method leaving the power limited regime and entering the bandwidth limited regime. The relationship between the two methods was reduced to the ratio of the two capacity formulas plotted as a function of the SNR, shown in Figure 4.109. In this graph,

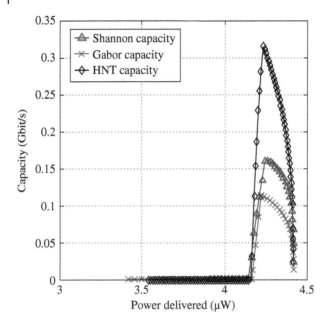

Figure 4.107 The capacity as a function of the power delivered for $P_N = 10^{-7}$ W with capacities based on a minimum SNR.

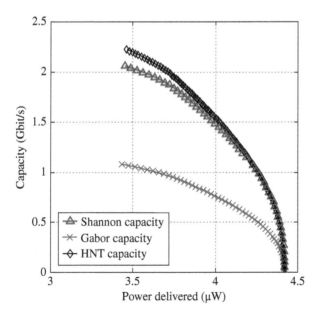

Figure 4.108 The capacity as a function of the power delivered for $P_N = 10^{-7}$ W. Note that the SNR always exceeded 0 dB in this case.

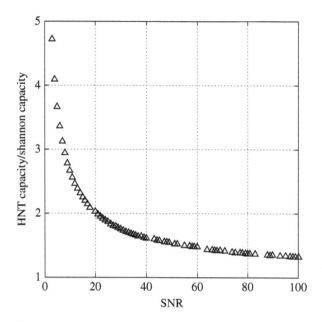

Figure 4.109 A comparison of the Hartley, Nyquist, and Tuller capacity with the Shannon Capacity as a function of SNR for M = 20.

η_{max}/P_N was normalized to one, where η_{max} is the maximum transfer-power gain. This resulted in $P_N = 8.5 \times 10^{-7}$ W. Here, SNR was defined as $\eta_{\text{max}}P_{av}/P_N$, which resulted in SNR = P_{av}. The number of bandwidth segments, M, was chosen to be 20 to generate this figure. When SNR increases, the ratio of the two capacity formulas significantly reduces, and converges to one.

4.9.4 Optimization of Each Channel Capacity Formulation

Instead of using the power allocation determined by the water-filling method of Shannon's theorem, we optimized each of the capacity formulations with respect to the capacity and power delivered. That is, instead of setting the weighting of power delivered to $\mu = 0$ as in the previous section, the optimization performed in this section used different values of μ, from 0 to 10^9, in optimizing Equations (4.28), (4.33), and (4.42), which resulted in the solutions of Equations (4.30), (4.39), and (4.44).

Figure 4.110 shows the optimized channel capacities for all three channel capacity theorems for $P_N = 10^{-8}$ W when all the bandwidth segments were used (M = 101). Note that these capacity plots were not restricted by a minimum SNR. When the value of μ is too large, there may not exist a possible solution so that all the channels have positive input power. This means that it is better

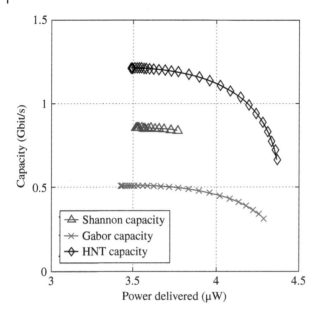

Figure 4.110 The capacity as a function of power delivered for $P_N = 10^{-8}$ W.

to give up using some of the channels of lower efficiency in order to increase the power delivery, and one then solves the equations with a smaller number of channels, M. This results in Figure 4.110, which consisted of only a few possible solutions for Shannon's channel capacity for $M = 101$.

Figure 4.111 shows the same plot for $P_N = 10^{-9}$ W. In this plot, the capacity was restricted by minimum SNR for all three methods, but it only affected the capacity of the Hartley, Nyquist, and Tuller method when the input signal was allocated over many bandwidth segments. This figure consisted of more possible solutions for the Shannon's channel capacity as compared to Figure 4.110, which covered a larger range of power delivered. As observed, a larger number of solutions could be found with a smaller noise power.

In summary, much work still needs to be carried out in order to further analyze the comparisons among the three methods, and to also validate the results. In particular, this is just the beginning of developing a basic understanding of the high capacity performance predicted by the Hartley, Nyquist, and Tuller method. The fact that Shannon's method and the method of Hartley, Nyquist, and Tuller agree for higher levels of SNR is in agreement with other literature. However, the high capacity predictions for the Hartley, Nyquist, and Tuller method are interesting. As expected, Gabor's interference limited method predicted less than Shannon's result for the capacity, which makes sense. The next steps for this work are to verify the results for the Hartley, Nyquist, and Tuller

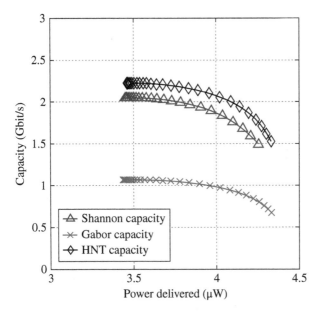

Figure 4.111 The capacity as a function of power delivered for $P_N = 10^{-9}$ W.

case, and to also run experiments again using the results from different electromagnetic environmental setups, such as near-field antenna positioning. Once these steps are complete, a more accurate representation of the differences between Shannon's method and the Hartley, Nyquist, and Tuller method, and the appropriate uses for each of the channel capacity theorems, will become clearer. Next the effect of matching the various impedances on the channel capacity is investigated.

4.10 Effect of Broadband Matching in Simultaneous Information and Power Transfer

The effect of broadband matching in simultaneous information and power transfer is considered in this section. The narrowband characteristic of antennas limits the applications of simultaneous information and power transfer. The performance improvement in terms of channel capacity and power delivery under broadband matching will be useful, where the matching techniques are the simplified real frequency technique (SRFT) and the non-foster matching. Electromagnetic (EM) simulation and multiobjective optimization are performed to analyze the tradeoff between the channel capacity and power delivery in different matching conditions. The performance gain using the matching networks has been demonstrated and analyzed.

In the previous section the tradeoff between the information transfer and power transfer have been analyzed using the Shannon channel capacity [44], Gabor channel capacity [48], and Hartley-Nyquist-Tuller (HNT) channel capacity [49, 50]. An optimization problem is now considered to allocate an average power P_{av} to M frequency channels for transmission such that the channel capacity and power delivery are maximized. In terms of the problem, the formulation is set up as:

$$\text{Maximize} \quad F_1 = \text{Channel capacity} \, (\text{bits/s}) \qquad (4.47)$$

$$\text{Maximize} \quad F_2 = \text{Power delivery} \, (\text{W}) \qquad (4.48)$$

Since the average power available is P_{av}, the power allocation should satisfy the following condition so that the total power allocated equal P_{av} is:

$$\sum_{m=1}^{M} P_i(f_m) = P_{av} \qquad (4.49)$$

where $P_i(f_m)$ is the power allocated to the frequency channel f_m, for $m = 1$, 2, ..., M. A set of solutions can be found showing the tradeoff between the channel capacity and power delivery. Some of the solutions have a larger channel capacity while the others have a larger power delivery. Those solutions are referred to as the non-dominated solution in multi-objective optimization [53].

In the power allocation of information and power transfer in [54], the power efficiency and voltage efficiency of different channels are considered in the power allocation process. However, the voltage standing wave ratios (VSWR) of the channels were not being considered in the previous section. It becomes a problem when VSWR of the antenna becomes large so that input power to the antenna would reflect back to the generator. The VSWR of the circuit related to the amount of reflected power is shown in Table 4.2.

When the VSWR is larger than 6.0, more than half of the power would be reflected to the generator. It would be dangerous in some transmitter design if it heats up and damages the signal generator and processor. Also, it is obviously a waste of energy. Therefore, it would be necessary to set a constraint such that the transmitter should not provide any input power to the channel with a large VSWR. However, using fewer channels in transmission would result in a reduction of channel capacity. Therefore, this section proposes applying a broadband matching circuit in designing the system. By reducing the VSWR at different frequencies, more channels can be used without worrying about the reflected power. The implementation of the matching network using a simplified real frequency technique (SRFT) [55–59] and a non-foster matching technique [60–63] and their effect on the simultaneous information transfer and power transfer would be analyzed in this section.

Table 4.2 The relationship between VSWR and the reflected power.

VSWR	Reflected power (%)	Reflected power (dB)
1.0	0.0	$-\infty$
1.5	4.0	−14.00
2.0	11.1	−9.55
3.0	25.0	−6.00
4.0	36.0	−4.44
5.0	44.0	−6.02
6.0	51.0	−2.92
7.0	56.3	−2.50
8.0	60.5	−2.18
9.0	64.0	−1.94
10.0	66.9	−1.74
20.0	81.9	−0.87
50.0	92.3	−0.35

4.10.1 Problem Description

Continuing from the previous section, a wireless system with transmitting and receiving dipole antennas is considered for simultaneous information and power transfer. The configuration is shown in Figure 4.112. The transmitter and receiver are separated by 100λ, where $\lambda = 0.3$ m is the wavelength at 1 GHz. The system impedance in this section is chosen as 50 Ω. The length of the dipole is 140 mm.

The power transfer efficiency $\eta_P(f)$ and the voltage transfer efficiency $\eta_V(f)$ can be calculated as:

$$\eta_P(f) = \frac{\left|I_L(f)^2 Z_L\right|}{\left|V_i(f)I_i(f)\right|\cos\left[\angle I_i(f) - \angle V_i(f)\right]} \tag{4.50}$$

$$\eta_V(f) = \frac{\left|V_L(f)\right|}{\left|V_i(f)\right|} \tag{4.51}$$

where the input current I_i, the input voltage V_i, the current I_L across the load, and the load impedance Z_L, are defined in Figure 4.112. V_L is the voltage across the load at the receiver side. The power transfer efficiency $\eta_P(f)$ is calculated as the ratio between the power dissipated at the load and the input power, and the effect of the input power factor is also in consideration. To calculate the above

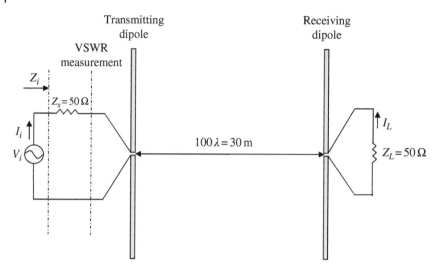

Figure 4.112 Antenna configuration considered in this section for simultaneous information and power transfer.

efficiency, an electromagnetic (EM) simulation is carried out. With a voltage excitation at the input, the currents in the circuit can then be found. Thus, the frequency response of the power transfer efficiency and voltage transfer efficiency can be obtained.

With the EM simulation [2], the VSWR of the dipole is shown in Figure 4.113 (blue dotted line). The impedance bandwidth defined by VSWR < 3 and VSWR < 2 are 0.92 – 1.09 GHz and 0.95 – 1.04 GHz, respectively. The power transfer efficiency $\eta_P(f)$ and the voltage transfer efficiency $\eta_V(f)$ of the transmitter are shown in Figure 4.114 and Figure 4.115 (blue dotted line), respectively.

4.10.1.1 Total Channel Capacity

The concept of channel capacity which focuses on the theoretical maximum data rate is applied to represent the performance of information transfer in this section. Three formulations of capacity have been considered in Chapter 1.9 and have been discussed in the previous section. However, the following provides a brief introduction.

Shannon capacity considers the maximum data rate in the presence of noise and is derived under the assumption that both the noises as well as the received signal are Gaussian distributed. This is not the case for Gabor and the HNT channel capacity to be presented. When the information is transferred through M channels each with a channel bandwidth ΔB, the Shannon capacity is given by

$$C_S = \sum_{m=1}^{M} \log_2\left[1 + \frac{\eta_P(f_m)P_i(f_m)}{P_N}\right]\Delta B \qquad (4.52)$$

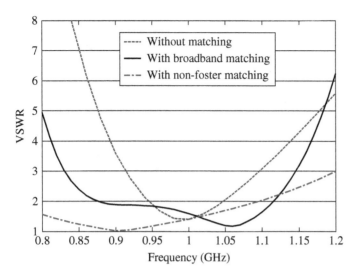

Figure 4.113 VSWR of the dipole antenna with different matching schemes.

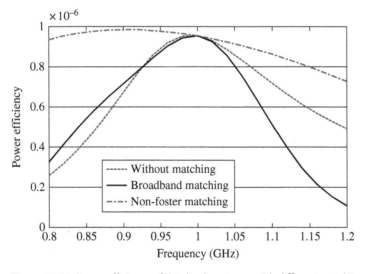

Figure 4.114 Power efficiency of the dipole antenna with different matching schemes.

where $P_i(f_m)$ and $\eta_P(f_m)$ are the input power and the power transfer efficiency at the m-th channel, respectively. P_N is the average noise power within the channel bandwidth ΔB.

Next, Gabor capacity considers the interaction between the signal and the thermal noise. With the step size chosen to be consistent to the Shannon capacity the Gabor capacity can be represented by

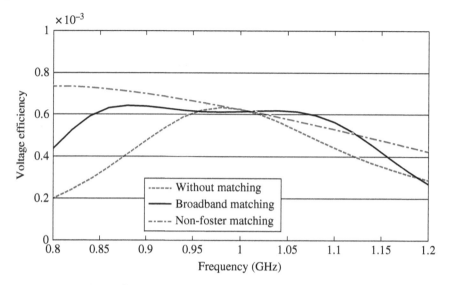

Figure 4.115 Voltage efficiency of the dipole antenna with different matching schemes.

$$C_G = \sum_{m=1}^{M} \log_2 \left[\frac{1}{2} + \frac{1}{2}\sqrt{1 + \frac{4\eta_P(f_m)P_i(f_m)}{P_N}} \right] \Delta B \tag{4.53}$$

Then, HNT capacity accounts for the signal and noise voltage but not the power, which is given by

$$C_{HNT} = \sum_{m=1}^{M} 2\log_2 \left[1 + \frac{\eta_V(f_m)V_i(f_m)}{V_N} \right] \Delta B \tag{4.54}$$

where V_i and V_N represent the signal voltage and noise voltage, respectively. Although the Shannon capacity and Gabor capacity are both calculated based on the noise power P_N, the HNT capacity is calculated based on the noise voltage V_N.

In the analysis in this paper, the three capacities will be compared on the same graph. When Shannon capacity and Gabor capacity are evaluated with P_N, the V_N used in HNT capacity would be assigned by

$$V_N = \sqrt{P_N Z_L} \tag{4.55}$$

Although the signal is transmitted using a channel bandwidth of ΔB, the voltages V_i and V_N here are accounted for as if the signal is transmitted with the center frequency. Nevertheless, (4.54) accounts for the ratio but not the magnitudes of the voltages, so the form of representation does not matter.

4.10.1.2 Power Delivery
The power that can be regenerated for powering or charging the receivers depends on the structure of the specific RF-DC rectifier circuit in use. The rectifier performance could be frequency dependent and could be nonlinear. This section would not look into the actual power that can be regenerated from the receiver. Instead, for the ease of analysis the term "power delivery" P_d is defined as the following:

$$P_d = \sum_{m=1}^{M} P_i(f_m) \eta_P(f_m) \tag{4.56}$$

or equivalently,

$$P_d = \sum_{m=1}^{M} \left| \frac{V_i(f_m)^2 \, n_V(f_m)^2}{Z_L} \right| \tag{4.57}$$

In the following analysis, an assumption has to be made that more the power delivery P_d, the more would be the power regenerated from the receiver. Therefore, the designers should consider the validity of Equations (4.56), (4.57), under the specific assumption for their rectifier circuits when they work out the same analysis described in this section. However, one can also derive an accurate representation of the average regenerated power for the specific RF-DC rectifier circuit, and use them to replace Equations (4.56) and (4.57) for a more accurate analysis.

4.10.1.3 Limitation on VSWR
To demonstrate the tradeoff between the total channel capacity and power delivery, for $M = 21$ channels with bandwidth $\Delta B = 20$ MHz, centered at $f_m = 0.8$ GHz, 0.82 GHz, ..., 1.20 GHz are used in this scenario. The transmitted power of $P_{av} = 5$ W, and the noise power of $P_N = 10^{-8}$ W or 10^{-7} W would be used throughout this section. Since the reflected power may heat up or damage the signal source, it would be preferable to set a limit on the VSWR, (voltage standing wave ratio) such that the transmitter will not provide any input power to the channel with a large VSWR. There is a common practice in microwave engineering to define the impedance bandwidth in terms of VSWR < 3 or VSWR < 2, which implies that the reflected power is limited to 25% and 11%, respectively. Multi-objective optimizations (by genetic algorithm [53, 57]) on the total channel capacity and power delivery (equations 4.47 and 4.48) are then performed with the constraint on the impedance bandwidth.

The non-dominated solutions showing the tradeoff between the total Shannon channel capacity and power delivery are shown in Figure 4.116. The achievable maximum capacity decreased from 1.67 Gbit/s to 1.01 Gbit/s and

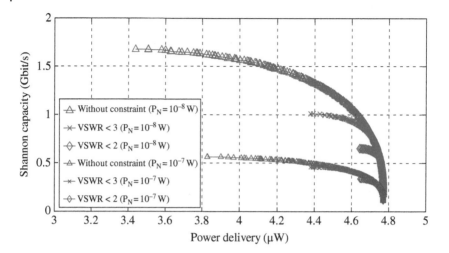

Figure 4.116 Shannon capacity versus power delivery in the case without any impedance matching.

0.66 Gbit/s when the constraints of VSWR < 3 and VSWR < 2 are applied, respectively, in the case of $P_N = 10^{-8}$ W.

The achievable maximum power delivery is the same in all cases, because the maximum power delivery is achieved by assigning all the power to one channel with the largest transmit-receive power efficiency. However, the solution is different when it is out of the maximum region of transmission. Considering the same capacity of 1 Gbit/s, the power delivery without VSWR constraint is larger than the power delivery with the constraint of VSWR < 3 (4.63 μW over 4.48 μW). The same decrement of maximum channel capacity applies when the noise power is increased by 10 times ($P_N = 10^{-7}$ W). In short, the sums of the capacity formulas are performed only for those frequencies where the VSWR is below the given mark.

4.10.2 Design of Matching Networks

From the last section, it is shown that the narrowband characteristic of the dipole antenna limited the channel capacity when the channels with large VSWR are not allowed in the system design. This section will introduce the use of a matching network to improve the VSWR of the circuits.

4.10.2.1 Simplified Real Frequency Technique (SRFT)

This section introduces the use of the SRFT [55–59] to build a broadband matching network. SRFT is a robust approach which can match antennas [56–58] as well as active amplifier circuits [59]. This method assumes the matching network is lossless and the scattering parameters are given by the Belevitch representation [60], i.e.,

$$S_{11}(s) = \frac{h(s)}{g(s)} = \frac{h_0 + h_1 s + \ldots + h_n s^n}{g_0 + g_1 s + \ldots + g_n s^n} \tag{4.58}$$

$$S_{21}(s) = S_{12}(s) = \pm \frac{s^2}{g(s)} \tag{4.59}$$

$$S_{22}(s) = -(-1)^k \frac{h(-s)}{g(s)} \tag{4.60}$$

where n is the number of reactive elements in the matching network and k is the integer specifying the order of the zero of transmission at the origin.

The flowchart for the SRFT methodology is shown in Figure 4.117 which includes an optimization loop characterized by the least square methods. At the start of the algorithm, it initializes the coefficients of $h(s)$, where typically initial values can be chosen in the range of $h_i = [-1, 1]$. After that, it goes into an optimization loop. In each iteration, first, the coefficients of $g(s)g(-s)$ are computed by:

$$g(s)g(-s) = h(s)h(-s) + (-1)^k s^{2k} = G_0 + G_1 s^2 + \ldots + G_n s^{2n} \tag{4.61}$$

where

$$G_0 = h_0^2$$
$$G_1 = -h_1^2 + 2h_2 h_0$$
$$\vdots$$
$$G_i = (-1)^i h_i^2 + 2 \left[2h_{2i} h_0 + \sum_{j=2}^{i} (-1)^{j-1} h_{j-1} h_{2i-j+1} \right]$$
$$\vdots$$
$$G_k = G_i \big|_{i=k} + (-1)^k$$
$$\vdots$$
$$G_n = (-1)^n h_n^2$$

Then, by computing the left half plane roots of $g(s)g(-s)$, the polynomial $g(s)$ can be formed. Next, the transducer power gain $T(\omega)$ of the network is calculated by:

$$T(\omega) = \frac{\left[1 - |\Gamma_S(\omega)|^2 \right] \left[1 - |\Gamma_L(\omega)|^2 \right] |\omega^{2k}|}{\left| g(j\omega) - \Gamma_S(\omega) h(j\omega) - (-1)^k \Gamma_S(\omega) \Gamma_L(\omega) g(-j\omega) + (-1)^k \Gamma_L(\omega) h(-j\omega) \right|^2} \tag{4.62}$$

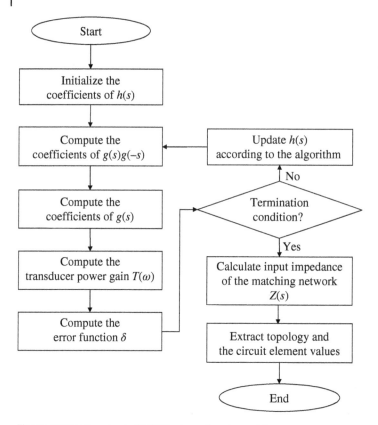

Figure 4.117 Flowchart of SRFT for broadband matching.

where

$$\Gamma_S(\omega) = \frac{Z_S(\omega) - R_0}{Z_S(\omega) + R_0} \quad \text{and} \quad \Gamma_L(\omega) = \frac{Z_L(\omega) - R_0}{Z_L(\omega) + R_0}$$

and R_0 is the normalizing resistance of the S-parameters ($R_0 = 50\ \Omega$ is considered in this section). $Z_S(\omega)$ and $Z_L(\omega)$ are the impedance of the source and the load connecting to the matching network, respectively.

Finally, at the end of the iteration, the error function δ is computed by:

$$\delta = \sum_{i=1}^{m} \left[\frac{T(\omega_i)}{T_0(\omega_i)} - 1 \right]^2 \tag{4.63}$$

where $\omega_1, \omega_2, \ldots, \omega_m$ are the m frequency points of interest, and $T(\omega_i)$ and $T_0(\omega_i)$ are the transducer power gain and the desired gain at ω_i, respectively. This error function δ is to be minimized in this least squares optimization loop.

Then, the optimization routine checks against the termination condition of the least squares algorithm. If it continues to the next iteration, the coefficients of $h(s)$ are updated according to the least squares algorithm to minimize δ. By continuing the iteration, the transducer power gain $T(\omega_i)$ converges to the desired gain $T_0(\omega_i)$.

After the least squares optimization, the input impedance $Z(s)$ of the matching network is calculated in the case when the load is perfectly matched $(Z_L = 50\ \Omega)$:

$$Z(s) = 50\frac{1 + S_{11}(s)}{1 - S_{11}(s)} \tag{4.64}$$

At the end, the input impedance of an equivalent matching network with lumped capacitors and inductors element is formulated. By comparing the coefficients of the Equation (4.64) and the real circuit, the values of the inductors and the capacitors in the real matching network can be found.

Now this method is applied to match the dipole which is previously considered in section 4.10.1 using the least squares optimization provided in MATLAB function lsqnonlin (Solves non-linear least squares problems).[1] The parameters are chosen as $n = 4$ (n is the number of reactive elements) and $k = 0$, so that the design is a 4 elements low pass ladder which gives good performance in antenna matching [56]. More complicated structures are not considered to make the implementation too difficult. A typical desired gain can be chosen as $T_0(\omega_i) \sim 0.95$ between 0.9 to 1.0 for good broadband performance.

According to [56], a shunt inductor L_p should be first connected to the load before designing the matching network to provide good performance. Therefore, the load impedance $Z_L(\omega)$ used in equation (4.62) is given by

$$Z_L(\omega) = \frac{j\omega L_p Z_A(\omega)}{j\omega L_p + Z_A(\omega)} \tag{4.65}$$

where $Z_A(\omega)$ is the input impedance of the dipole which can be obtained from the EM simulation [2]. The value of L_p is calculated by resonating this element to the input impedance of the dipole antenna at the lower edge of the passband.

After obtaining the optimized $Z(s)$, the coefficients of the input impedance a matching network consisting of two series connected inductors L_1 and L_2 and two shunt connected capacitors C_1 and C_2 are compared with $Z(s)$ to obtain the circuit values. The value of C_2 is too small, and thus, it is not taken

---

1 MATLAB is a Trademark of MathWorks, Inc.

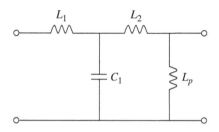

Figure 4.118 Matching network designed with SRFT.

into consideration in the circuit. Finally, the matching circuit is given in Figure 4.118, where L_P = 18.2 nH, L_1 = 14.3 nH, L_2 = 6.44 nH, and C_1 = 2.49 pF.

The VSWR of the dipole with the broadband matching network is given in Figure 4.113 (black solid line). The impedance bandwidth defined by VSWR < 3 and VSWR < 2 are 0.84 – 1.14 GHz and 0.88 – 1.11 GHz, respectively.

The impedance bandwidth is greatly improved compared with the case of not using the matching network. The power transfer efficiency $\eta_P(f)$ and the voltage transfer efficiency $\eta_V(f)$ of the transmitter are shown in Figures 4.114 and 4.115 (black solid line), respectively. Although the impedance bandwidth has been improved, the power efficiency is not better than the case of when there is no impedance matching. Nevertheless, the objective of reducing the reflected power and improving the impedance bandwidth has been achieved.

4.10.2.2 Use of Non-Foster Matching Networks

The idea of non-foster matching [61–65] is to use negative capacitance and negative inductance to cancel the internal capacitance and inductance of an antenna [61]. Hence, the input reactance can be minimized over a broadband. The negative capacitance and reactance can be realized by negative impedance converters (NICs) or negative impedance inverters (NIIs or NIVs) [61]. This approach is widely applied to extend the impedance bandwidth of dipole antennas [61], VHF (very high frequency) monopole antennas [62], and microstrip antennas [63, 64].

The non-foster matching circuit used in this section is shown as Figure 4.119, where L_n = 53.9 nH, C_n = 0.48 pF. Since there are only two elements, the circuit values can easily be obtained by optimization. The VSWR of the dipole with the non-foster matching network is given in Figure 4.113 (red dotted line).

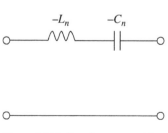

Figure 4.119 Non-Foster matching network.

The impedance bandwidth defined by VSWR < 3 and VSWR < 2 are 0.70 – 1.20 GHz and 0.76 – 1.11 GHz, respectively. The impedance bandwidth is significantly larger for the case when the matching network is not used as well as for the case when a broadband matching network designed by the SRFT technique is used.

The power transfer efficiency $\eta_P(f)$ and the voltage transfer efficiency $\eta_V(f)$ of the transmitter are shown in Figures 4.114 and

4.115 (red dotted line), respectively. The power efficiency and voltage efficiency are both better than the case when the matching network is not used and for the case when a broadband matching network deigned by the SRFT technique is used. However, the power required to feed an active device that is going to create the negative elements have not been taken into account. It is not clear what the conclusions will be when they are included.

4.10.3   Performance Gain When Using a Matching Network

Here, the performance of simultaneously information and power transfer is evaluated through the three circuit matching conditions: i) without matching; ii) broadband matching using the SRFT; and iii) non-Foster matching. The settings of the communication scenario and the constraint on the VSWR used here are the same as that described in Section 4.10.2. The performances concerning different constraints on VSWR are presented in the following subsections. The frequency range in use for different circuit matching conditions in the presence of VSWR constraint is tabulated in Table 4.3.

4.10.3.1   Constraints of VSWR < 2

When the constraint on VSWR is smaller than 2 in the power allocation, the non-dominated solutions showing the tradeoff between the total Shannon, Gabor, and HNT channel capacity versus power delivery are shown in Figures 4.120, 4.121 and 4.122, respectively.

The achievable maximum Shannon capacity increased from 0.66 Gbit/s to 1.22 Gbit/s and 1.57 Gbit/s when the SRFT broadband matching and the non-Foster matching are applied, respectively, in the case of $P_N = 10^{-8}$ W. The achievable maximum power delivery is slightly better than the other two matching conditions, because it has larger maximum power efficiency across the frequency band, as previously shown in Figure 4.114 (9.85×10^{-7} over 9.54×10^{-7}). It should be noticed that the maximum power delivery is achieved by the assigning all the power to the one channel with the largest power efficiency.

Table 4.3 Frequency range in use in the presence of VSWR constraint.

VSWR constraint	Frequency range (GHz)		
	Without matching	Broadband matching	Non-foster matching
Unconstrained	0.80 – 1.20	0.80 – 1.20	0.80 – 1.20
VSWR < 3	0.92 – 1.09	0.84 – 1.14	0.80 – 1.20
VSWR < 2	0.95 – 1.04	0.88 – 1.11	0.80 – 1.11

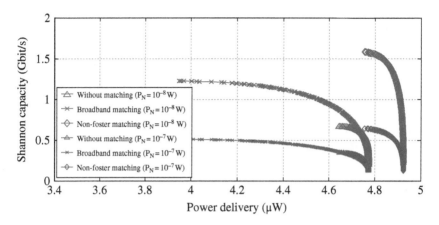

Figure 4.120 Shannon capacity versus power delivery with a constraint of VSWR < 2.

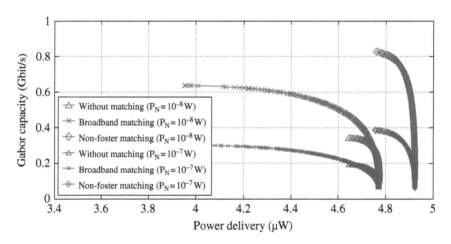

Figure 4.121 Gabor capacity versus power delivery with a constraint of VSWR < 2.

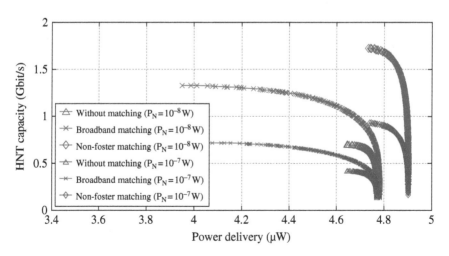

Figure 4.122 HNT capacity versus power delivery with a constraint of VSWR < 2.

Although the achievable maximum power delivery only differs by a little amount, the difference would be larger when the operation is out of the maximum power transfer region. Considering the same Shannon capacity of 1.21 Gbit/s, the power delivery with a non-Foster matching network is larger than the power delivery with SRFT broadband matching (4.89 μW over 4.00 μW).

The same conclusions from the aforementioned observations can be applied when the noise power is increased by 10 times ($P_N = 10^{-7}$ W). In addition, it is seen that the same conclusions apply also to Gabor capacity and HNT capacity.

It is interesting to note that the achievable maximum Shannon capacity with non-Foster matching when $P_N = 10^{-7}$ W is about the same as that of without any matching when the noise is ten times smaller in the case of $P_N = 10^{-8}$ W (0.63 Gbit/s and 0.65 Gbit/s).

4.10.3.2 Constraints of VSWR < 3

When the constraint on VSWR is smaller than 3 in the power allocation, the non-dominated solutions showing the tradeoff between the total Shannon, Gabor, and HNT channel capacity versus power delivery are shown in Figures 4.123, 4.124, and 4.125, respectively. Comparing this case with the constraint of VSWR < 2, it is seen that the difference between using the three matching conditions becomes smaller. The achievable maximum Shannon capacity increased from 1.01 Gbit/s to 1.42 Gbit/s and 1.87 Gbit/s when the SRFT broadband matching and the non-foster matching are applied, respectively, for the case of $P_N = 10^{-8}$ W. Again, the achievable maximum power delivery is slightly better than the other two matching condition (4.92 μW over 4.77 μW), and the difference becomes larger when operating out of the maximum region.

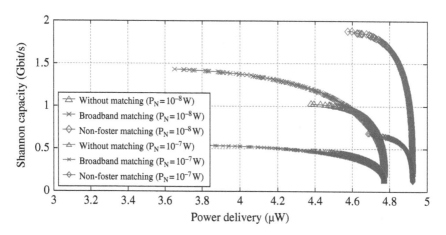

Figure 4.123 Shannon capacity versus power delivery with a constraint of VSWR < 3.

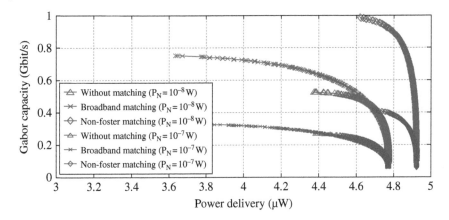

Figure 4.124 Gabor capacity versus power delivery with a constraint of VSWR < 3.

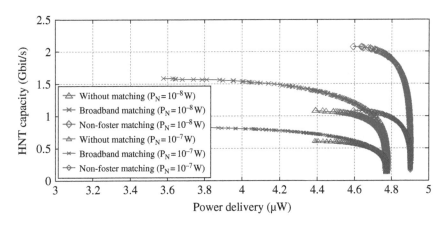

Figure 4.125 HNT capacity versus power delivery with a constraint of VSWR < 3.

Considering the same Shannon capacity of 1.42 Gbit/s, the power delivery with non-foster matching network is larger than the power delivery with SRFT broadband matching (4.86 µW over 3.65 µW). Again, the same conclusions from the aforementioned observations can be applied when the noise power increased by 10 times ($P_N = 10^{-7}$ W), or using the Gabor capacity or HNT capacity instead of Shannon capacity. It is interesting to note that the achievable maximum HNT capacity with non-Foster matching when $P_N = 10^{-7}$ W is slightly larger than that of without using any matching networks when the noise is ten times smaller in the case of $P_N = 10^{-8}$ W (1.09 Gbit/s over 1.08 Gbit/s).

4.10.3.3 Without VSWR Constraint

When there is no VSWR constraint on the power allocation, the non-dominated solutions showing the tradeoff between the total Shannon, Gabor, and HNT channel capacity versus power delivery are shown in Figures 4.126, 4.127, and 4.128, respectively.

By comparing with the case when there are constraints on the VSWR, it is seen that the differences between the three matching conditions become much smaller. It should be observed that the maximum Shannon capacity decreased from 1.68 Gbit/s to 1.56 Gbit/s when the SRFT broadband matching is applied in the case of $P_N = 10^{-8}$ W. This is because the power efficiency does not improve with the SRFT broadband matching as seen from the plots of Figure 4.114.

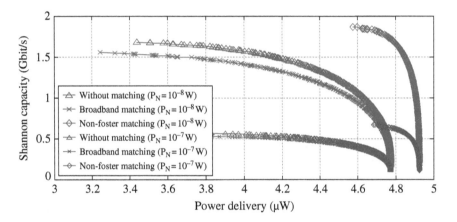

Figure 4.126 Shannon capacity versus power delivery without constraint on VSWR.

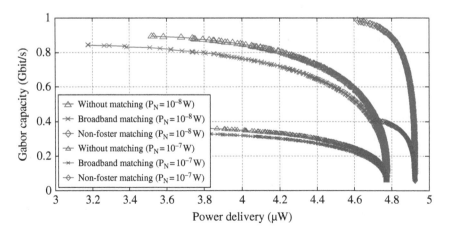

Figure 4.127 Gabor capacity versus power delivery without constraint on VSWR.

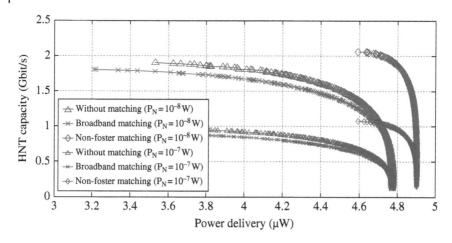

Figure 4.128 HNT capacity versus power delivery without constraint on VSWR.

However, there are improvements when using the non-Foster matching network. The achievable maximum Shannon capacity increased from 1.68 Gbit/s to 1.87 Gbit/s when the non-Foster matching is applied in the case of $P_N = 10^{-8}$ W. Again, the achievable maximum power delivery is slightly better than the other two matching condition (4.92 µW over 4.77 µW), and the difference becomes larger out of the region of the maximum response. Considering the same Shannon capacity of 1.68 Gbit/s, the power delivery with non-Foster matching network is larger than the power delivery without any matching (4.79 µW over 3.44 µW).

Again, the same conclusions from the aforementioned observations can be arrived at when the noise power increased by 10 times ($P_N = 10^{-7}$ W), or using the Gabor capacity or HNT capacity instead of Shannon capacity.

4.10.3.4 Discussions

From the results of the previous section, it is seen that the channel capacity increases significantly when the impedance bandwidth increases, particularly when the antenna is limited to operate in the regions of VSWR < 2 or VSWR < 3. This is because the channel capacity increases when there are more channels that are available for use, and this is embedded in the formula for the three channel capacities. Moreover, the impedance matching has improved the tradeoff in performance between the channel capacity and the power delivery as indicated in the various figures.

One important point to notice is that the current technology for non-Foster matching is based on the NICs or negative impedance inverters (NIIs or NIV s) [61]. These are active devices that consume energy. In this case, the energy consumption of these active circuits should be taken into account to perform a comparative comparison when designing a system. The presented results do not take into account the energy consumption that needs to be considered by these active devices. Although some rectifiers could be based on active devices (i.e., active rectifiers), it might not be true for every circuit or systems to produce a positive energy balance when active circuit elements are used. Therefore, the circuit and system structures should be carefully considered and designed. When the non-Foster matching method is not practical for some applications, the broadband matching method using positive capacitance and inductance should be considered instead. In any case, the theoretical results of non-Foster matching are included in this section for the completeness of the comparison.

Next, to familiarize the reader with the current technology using simultaneous information and power transfer, some possible applications are discussed. Passive RFID [37, 38] and WSN [39, 40] are two applications, which need to transfer data and convert EM waves to electrical power for functioning, preferably using the same antenna. RFID is used to detect an object, whereas WSN is used to sense and monitor the environment. By the way, the theories and the results described in this section not only apply for the design of a RFID or WSN application, but also can be applied to a general system.

4.10.4 PCB (Printed Circuit Board) Implementation of a Broadband-Matched Dipole

The PCB technology provides a convenient way to fabricate the antennas and microwave circuit for scientific verification and engineering purposes. It is well known that the transmission line structures can provide capacitance and inductance for impedance matching purposes. For instance, a short-circuited stub and an open-circuit stub act as a capacitor and an inductor, respectively, when the length of the stub is shorter than a quarter-wave. Moreover, a transmission line section connected within two circuit parts can provide impedance transformation between them. When the length of the transmission line is a quarter-wave, it is known as a quarter-wave transformer. Therefore, by optimizing a transmission line network, it is possible to simulate similar results when lumped capacitors and inductors are used. Here, the dipole and the matching network are implemented on a PCB. It is optimized to provide similar bandwidth as can be obtained using a lumped element matching network, seen in Figure 4.118. By implementing the matching network in a transmission line, the simulation errors caused by the package impedance of the lumped circuit elements and the effect of soldering can be totally eliminated. In addition, the error caused by soldering to connect the dipole to the matching network can also be eliminated. The transmission line matching network contains just a few parameters that can be optimized easily. The resultant circuit and the optimized dimensions for a system are given in Figure 4.129. The transmission line matching network is implemented by a parallel-strip line, where the strip

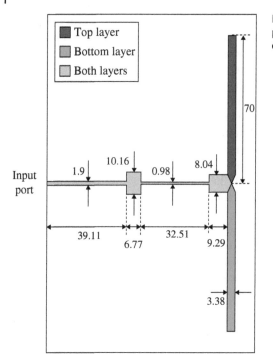

Figure 4.129 Configuration of a printed broadband matched dipole.

conductors on the top and on the bottom of the PCB are identical and broadside coupled. It is worthwhile to mention that the microstrip-line and the parallel-strip line are very similar in structure. The microstrip line under an infinite large ground plane can be modeled as a parallel-strip line using an image theory. Next, the dipole is also printed on the PCB, where the two line elements of the dipole are printed on the two sides of the PCB, respectively [63]. This implementation can replace the need for a ground plane in the implementation of the dipole, and the feeding structure of the dipole becomes symmetric at the location of the feed. The PCB is thin so that the two line elements of the dipole are approximately at the same height. A PCB of 1.524-mm thickness with a dielectric constant of 3.0 is used. The fabricated hardware is shown in Figure 4.130, indicating that it consists of only simple structures. The simulated and the measured VSWR are given in Figure 4.131. The simulated impedance bandwidth defined by VSWR < 2 is 0.87 GHz–1.12 GHz, which is very similar to the broadband-matched dipole using SRFT approach. The measured impedance bandwidth defined by VSWR < 2 is 0.91–1.19 GHz, where a small amount of frequency shift has occurred compared with the simulation result. The rationales, the implementation, and the results of broadband matching in simultaneous information and power transfer have been presented. Electromagnetic simulation has been performed to calculate the

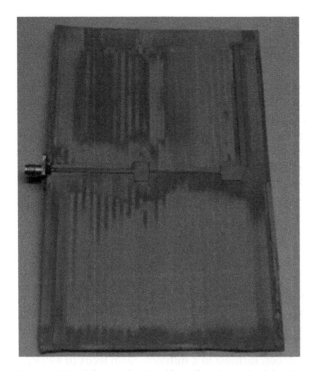

Figure 4.130 Fabricated printed broadband matched dipole.

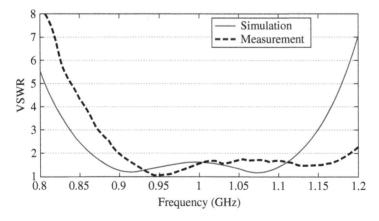

Figure 4.131 VSWR of the printed broadband matched dipole.

power efficiency and voltage efficiency for power allocation in utilizing a number of frequency channels.

It has been demonstrated that the narrow bandwidth of the antenna would limit the performance in information and power transfer. Two methods for broadband matching have been proposed and presented: SRFT and non-foster matching. From the analysis, it is demonstrated that the non-foster matching perform the best, which can improve the performance in terms of channel capacity and power delivery. However, the drawbacks of non-foster matching are the inclusion of negative impedance converters or negative impedance inverters which would increase the complexity of the circuit as well as power consumption. The broadband matching is a relatively simple approach and it only includes some lumped components which can also be implemented using a microstrip line.

4.11 Conclusion

This chapter presented a few of the subtleties in antenna theory in the time domain which is not pertinent in the frequency domain analysis. Specifically the impulse response of an antenna on transmit is approximately equal to the time derivative of the impulse response of the same antenna on receive. Also recasting the Friis's transmission formula in a different form leads to a design of an experimental system that can transmit/receive ultrawideband signals with very little distortion. In particular, the transmitting and receiving properties of a dipole, bicone and a TEM horn antenna are discussed particularly in the time domain as they illustrate some special properties. Although many antenna structures exhibit resonating properties when used in the ultrawideband range, in many cases, proper design and modification of such antennas can minimize the reflections due to structural discontinuities, whereby the antenna can be converted to a guiding-wave structure. Additionally, the application of a tapered resistive loading along the length of the antenna helps to reduce the outward traveling wave on the antenna so that reflections from the ends of the antenna are eliminated. This surface loading is designed according to the fundamental dimensions (length, average thickness, effective radius) of the antenna. Proper design of the loading profile minimizes the dispersion from the antenna, thus broadening its frequency response and focusing the energy into a single pulse. However, by loading the antenna, its efficiency and gain are reduced by a significant amount.

Also two specific design of a century bandwidth antenna are discussed. One of them can indeed radiate an impulse which is very useful for time domain measurements. This chapter also experimentally verifies that the impulse response of an antenna structure in the transmit mode is proportional to the time derivative of the impulse response of the same antenna when it is operating in the receive mode, irrespective of the antenna type. Moreover, observations of the output wave shapes (radiated field and received current) from the antennas provide important information about their transmitting and receiving properties, and certain relationships are obtained between the input and output wave shapes. Consequently, one can conclude, for example, that if a bicone transmits to a receiving TEM horn, then the induced current in the horn will be exactly identical to the driving point voltage of the bicone. Such observations can have very important ramifications in broadband high-speed information transmission.

Finally, a procedure is presented to design a time limited ultrawideband pulse fitting the FCC mask which has a good linear phase response. Also this pulse is practically bandlimited in addition to having a finite time domain support. Examples are given on how to transmit and receive such ultrawideband pulses without distortion using a special type transmit and receive antennas.

The implementation and the results of broadband matching in simultaneous information and power transfer have been described. Electromagnetic simulation has been performed to calculate the power efficiency and voltage efficiency for power allocation over a number of disjoint frequency channels. It has been demonstrated that the narrow bandwidth of the antenna would limit the performance of simultaneous information and power transfer. Two methods for broadband matching have been proposed and presented: SRFT and non-Foster matching. From the analysis, it is demonstrated that the non-Foster matching performs better, which can improve the design of a system in terms of channel capacity and power delivery. However, the drawbacks of non-Foster matching are the inclusion of NICS or NIIS, which would increase the complexity of the circuit and power consumption. The broadband matching is a relatively simple approach that only includes some lumped components that can also be implemented by using a microstrip line. A PCB implementation of a broadband-matched dipole using parallel-strip line has also been presented.

References

1 B. H. Jung, T. K. Sarkar, S. W. Ting, Y. Zhang, Z. Mei, Z. Ji, M. Yuan, A. De, M. Salazar-Palma, and S. M. Rao, *Time and Frequency Domain Solutions of EM Problems Using Integral Equations and a Hybrid Methodology*, IEEE Press/John Wiley & Sons, Inc., Hoboken, NJ, 2010.

2 Y. Zhang, T. K. Sarkar, X. Zhao, D. Garcia-Donoro, W. Zhao, M. Salazar-Palma, and S. W. Ting, *Higher Order Basis Based Integral Equation Solver (HOBBIES)*, John Wiley & Sons, Inc., Hoboken, NJ, 2012.

3 P. T. Montoya and G. S. Smith, "A Study of Pulse Radiation from Several Broad-Band Loaded Monopoles," *IEEE Transactions on Antennas and Propagation*, Vol. 44, No. 8, pp. 1172–1182, 1996.

4 T. T. Wu and R. W. P. King, "The Cylindrical Antenna with Non-Reflecting Resistive Loading," *IEEE Transactions on Antennas and Propagation*, Vol. AP-13, No. 3, pp. 369–373, 1965.

5 E. E. Altshuler, "The Traveling Wave Linear Antenna," *IEEE Transactions on Antennas and Propagation*, Vol. AP-9, No. 4, pp. 324–329, 1961.

6 J. Koh, W. Lee, T. K. Sarkar, and M. Salazar-Palma, "Calculation of Far-Field Radiation Pattern Using Nonuniformly Spaced Antennas by a Least Square Method," *IEEE Transactions on Antennas and Propagation*, Vol. 62, No. 4, pp. 1572–1578, 2014.

7 T. K. Sarkar, M. Salazar-Palma, and E. Mokole, *Physics of Multiantenna Systems and Broadband Processing*, John Wiley & Sons, Inc., Hoboken, NJ, 2008.

8 C. W. Harrison and R. W. P. King, "On the Transient Response of an Infinite Cylindrical Antenna," *IEEE Transactions on Antennas and Propagation*, Vol. 15, pp. 301–302, 1967.

9 M. Kanda, "Time Domain Sensors and Radiators," in *Time Domain Measurements in Electromagnetics*, edited by E. K. Miller, Ch. 5. Van Nostrand Reinhold, New York, 1986.

10 S. N. Samaddar and E. L. Mokole, "Some Basic Properties of Antennas Associated with Ultrawideband Radiation," in *Ultra-Wideband Short-Pulse Electromagnetics 3*, edited by C. E. Baum, L. Carin, and A. P. Stone, pp. 147–164, Plenum Press, New York, 1997.

11 D. Ghosh, A. De, M. C. Taylor, T. K. Sarkar, M. C. Wicks, and E. Mokole, "Transmission and Reception by Ultra-Wideband (UWB) Antennas," *IEEE Antennas and Propagation Magazine*, Vol. 48, No. 5, pp. 67–99, 2006.

12 Z. Ji, T. K. Sarkar, and B. H. Jung, "Transmitting and Receiving Wideband Signals Using Reciprocity," *Microwave & Optical Technology Letters*, Vol. 38, No. 5, pp. 359–362, Sept. 2003.

13 S. N. Samaddar, "Transient Radiation of a Single-Cycle Sinusoidal Pulse from a Thin Dipole," *Journal of the Franklin Institute*, Vol. 10, pp. 259–271, 1992.

14 C. P. Papas and R. W. P. King, "Radiation from Wide-Angle Conical Antennas Fed by a Coaxial Line," *Proceedings of the IRE*, Vol. 39, pp. 49–50, 1951.

15 H. Jasik, Ed., *Antenna Engineering Handbook*, Ch. 6, McGraw Hill, New York, 1961.

16 J. R. Andrews, "UWB Signal Sources, Antennas and Propagation," IEEE Tropical Conference on Wireless Communication Technology, Honolulu, HI, pp. 439–440, Oct. 2003 (also published as Application Note AN-14a, Picosecond pulse Lab, Aug. 2003, Boulder, CO. http://www.picosecond.com/objects/AN-14a.pdf/. (Excerpts presented here with permission from Dr. James R. Andrews.)

17 P. VanEtten and M. C. Wicks, "Bi-blade Century Bandwidth Antenna," US Statuary Invention Registration H1,913, Nov. 7, 2000.

18 M. C. Wicks and P. VanEtten, "Orthogonally Polarized Quadraphase Electromagnetic Radiator," US Patent 5,068,671, Nov. 26, 1991.

19 C. E. Baum, E. G. Farr, and D. V. Giri, "Review of Impulse-Radiating Antennas," in *Review of Radio Science 1996–1999*, edited by W. S. Stone, Ch. 12. Wiley-IEEE Press, New York, 1999.

20 L. H. Bowen, E. G. Farr, C. E. Baum, T. C. Tran, and W. D. Prather, "Results of Optimization Experiments on a Solid Reflector IRA," Sensor and Simulation Note 463, Jan. 2002.

21 E. Farr, "IRA-2 Dimensions", Farr Research, Inc., Albuquerque, NM, document dated March 15, 2000.

22 E. Farr, "Analysis of the Impulse Radiating Antenna," Sensor and Simulation Notes 329, Farr Research, Inc., Albuquerque, NM, July 1991.

23 R. Johnk and A. Ondrejka, "Time-Domain Calibrations of D-Dot Sensors," NIST Tech. Note 1392, NIST, Boulder, CO, Feb. 1998.

24 Y. Hua and T. K. Sarkar, "Design of Optimum Discrete Finite Duration Orthogonal Nyquist Signals," *IEEE Transactions on Acoustics, Speech & Signal Processing*, Vol. 36, No. 4, pp. 606–608, Apr. 1988.

25 T. K. Sarkar, M. Salazar-Palma, and M. C. Wicks, "*Wavelet Applications in Engineering Electromagnetics*," Artech House, Norwood, MA, 2002.

26 "Part 15.209, Radiated Emission Limits; General Requirements," Code of Federal Regulations—Title 47: Telecommunication, Dec. 2005.

27 Z. Mei, T. K. Sarkar, and M. Salazar-Palma, "The Design of an Ultrawideband T-Pulse with a Linear Phase Fitting the FCC Mask," *IEEE Transactions on Antennas and Propagation*, Vol. 59, No. 4, pp. 1432–11436, 2011.

28 S. H. Yeung, Z. Mei, T. K. Sarkar, and M. Salazar-Palma, "An Ultrawideband T-Pulse Fitting the FCC Mask Using a Multiobjective Genetic Algorithm," *IEEE Microwave and Wireless Components Letters*, Vol. 22, No. 12, pp. 615–617, 2012.

29 J. D. Kraus and R. J. Marhefka, *Antennas: For All Applications*, McGraw-Hill, New York, 2003.

30 A. Willig, "Recent and Emerging Topics in Wireless Industrial Communications: A Selection," *IEEE Transactions on Industrial Informatics*, Vol. 4, No. 2, pp. 102–124, May 2008.

31 M. Jonsson and K. Kunert, "Towards Reliable Wireless Industrial Communication with Real-Time Guarantees," *IEEE Transactions on Industrial Informatics*, Vol. 5, No. 4, pp. 429–442, Nov. 2009.

32 W. C. Brown, "The History of Power Transmission by Radio Waves," *IEEE Transactions on Microwave Theory and Techniques*, Vol. 32, No. 9, pp. 1230–1242, Sept. 1984.

33 J. A. G. Akkermans, M. C. van Beurden, G. J. N. Doodeeman, and H. J. Visser, "Analytical Models for Low-Power Rectenna Design," *IEEE Antennas and Wireless Propagation Letters*, Vol. 4, pp. 187–190, 2005.

34 C. C. Lo, Y. L. Yang, C. L. Tsai, C. S. Lee, and C. L. Yang, "Novel Wireless Impulsive Power Transmission to Enhance the Conversion Efficiency for Low Input Power," in Proceedings of the IEEE MTT-S International Microwave

Workshop Series on Innovative Wireless Power Transmission: Technologies, Systems, and Applications (IMWS), Kyoto, Japan, 2011, pp. 55–58.

35 P. Grover and A. Sahai, "Shannon Meets Tesla: Wireless Information and Power Transfer," in Proceedings of the IEEE International Symposium on Information Theory (ISIT), Austin, TX, June 13–18, 2010, pp. 2363–2367.

36 E. P. Caspers, S. H. Yeung, T. K. Sarkar, M. S. Palma, M. A. Lagunas, and A. Perez-Neira, "Analysis of Information and Power Transfer in Wireless Communications," *IEEE Antenna and Propagation Magazine*, Vol. 55, No. 3, pp. 82–95, 2013.

37 A. Ashry, K. Sharaf, and M. Ibrahim, "A Simple and Accurate Model for RFID Rectifier," *IEEE Systems Journal*, Vol. 2, No. 4, Dec. 2008.

38 Y. Yao, J. Wu, Y. Shi, and F. F. Dai, "A Fully Integrated 900-MHz Passive RFID Transponder Front End with Novel Zero-Threshold RF–DC Rectifier," *IEEE Transactions on Industrial Electronics*, Vol. 56, No. 7, pp. 2317–2325, July 2009.

39 K. M. Farinholt, G. Park, and C. R. Farrar, "RF Energy Transmission for a Low-Power Wireless Impedance Sensor Node," *IEEE Sensors Journal*, Vol. 9, No. 7, pp. 793–800, 2009.

40 F. Kocer and M. P. Flynn, "An RF-Powered, Wireless CMOS Temperature Sensor," *IEEE Sensors Journal*, Vol. 6, No. 3, pp. 557–564, June 2006.

41 M. Gastpar, "On Capacity under Receive and Spatial Spectrum Sharing Constraints," *IEEE Transactions on Information Theory*, Vol. 53, No. 2, pp. 471–487, Feb. 2007.

42 L. R. Varshney, "Transporting Information and Energy Simultaneously," Proceedings of the 2008 IEEE International Symposium on Information Theory (ISIT), Toronto, Ontario, Canada, pp. 1612–1616, 2008.

43 P. Grover and A. Sahai, "Shannon Meets Tesla: Wireless Information and Power Transfer," Proceedings of the 2010 IEEE International Symposium on Information Theory (ISIT), Austin, TX, June 13–18, 2010.

44 C. F. Shannon, "Communication in the Presence of Noise," *Proceedings of IRE*, Vol. 37, No. 1, pp. 10–21, 1949.

45 R. G. Gallagher, *Information Theory and Reliable Communication*, John Wiley & Sons, Inc., New York, 1971.

46 T. M. Cover and J. A. Thomas, *Elements of Information Theory*, John Wiley & Sons, Inc., New York, 1991; L. Brillouin, *Science and Information Theory*, Academic Press, New York, 1956, pp. 245–258.

47 D. Gabor, "Communication Theory and Physics," *Transactions on the IRE Professional Group on Information Theory*, Vol. 1, No. 1, pp. 48–59, 1953.

48 L. Brillouin, *Science and Information Theory*, 2nd edn., Academic Press, New York, 1962.

49 W. G. Tuller, "Theoretical Limitations on the Rate of Transmission of Information," *Proceedings of the IRE*, Vol. 37, No. 5, pp. 468–478, May 1949.

50 W. G. Tuller, "Information Theory Applied to System Design," *Transactions of AIEE*, Vol. 69, pp. 1612–1614, 1950.

51 T. K. Sarkar, S. Burintramart, N. Yilmazer, Y. Zhang, A. De, M. Salazar-Palma, M. A. Lagunas, E. L. Mokole, and M. C. Wicks, "A Look at the Concept of Channel Capacity from a Maxwellian Viewpoint," *IEEE Antennas and Propagation Magazine*, Vol. 50, No. 3, pp. 21–50, 2008.

52 T. K. Sarkar, S. Burintramart, N. Yilmazer, S. Hwang, Y. Zhang, A. De, and M. Salazar-Palma, "A Discussion About Some of the Principles/Practices of Wireless Communication under a Maxwellian Framework," *IEEE Transactions on Antennas and Propagation*, Vol. 54, No. 12, pp. 3727–3745, Dec. 2006.

53 K. Deb, *Multi-Objective Optimization Using Evolutionary Algorithms*, John Wiley & Sons, Inc., Chichester, U.K., 2001.

54 T. K. Sarkar, S. H. Yeung, M. Salazar-Palma, M. A. Lagunas, and A. I. Perez-Neira, "The Effect of Broadband Matching in Simultaneous Information and Power Transfer," *IEEE Antennas and Propagation Magazine*, Vol. 57, No. 1, pp. 192–203, 2015.

55 H. J. Carlin and B. S. Yarman, "The Double Matching Problem: Analytic and Real Frequency Solutions," *IEEE Transactions on Circuits and Systems*, Vol. 28, No. 1, pp. 15–28, Jan. 1983.

56 H. An, B. K. J. C. Nauwelaers, and A. R. Van de Capelle, "Broadband Microstrip Antenna Design with the Simplified Real Frequency Technique," *IEEE Transactions on Antennas and Propagation*, Vol. 42, No. 2, pp. 129–136, Feb. 1994.

57 H. An, B. Nauwelaers, and A. Van de Capelle, "Matching Network Design of Microstrip Antenna with the Simplified Real Frequency Technique," *Electronics Letters*, Vol. 27, No. 24, pp. 2295–2297, Nov. 1991.

58 K. Yegin and A. Q. Martin, "On the Design of Broad-Band Loaded Wire Antennas Using the Simplified Real Frequency Technique and a Genetic Algorithm," *IEEE Transactions on Antennas and Propagation*, Vol. 51, No. 2, pp. 220–228, Feb. 2003.

59 B. S. Yarman and H. J. Carlin, "A Simplified 'Real Frequency' Technique Applied to Broad-Band Multistage Microwave Amplifiers," *IEEE Transactions on Microwave Theory and Techniques*, Vol. 30, No. 12, pp. 2216–2222, Dec. 1982.

60 V. Belevitch, "Elementary Application of the Scattering Formalism to Network Design," *IRE Transactions on Circuit Theory*, Vol. 3, pp. 97–104, June 1956.

61 S. E. Sussman-Fort and R. M. Rudish, "Non-Foster Impedance Matching of Electrically-Small Antennas," *IEEE Transactions on Antennas and Propagation*, Vol. 57, No. 8, pp. 2230–2241, Aug. 2009.

62 C. R. White, J. S. Colburn, and R. G. Nagele, "A Non-Foster VHF Monopole Antenna," *IEEE Antennas and Wireless Propagation Letters*, Vol. 11, pp. 584–587, 2012.

63 S. Koulouridis, "Non-Foster Design for Antennas," in Proceedings of the IEEE International Symposium on Antennas and Propagation (APSURSI), Spokane, WA, 2011, pp. 1954–1956.

64 S. Koulouridis and S. Stefanopoulos, "A Novel Non-Foster Broadband Patch Antenna," in Proceedings of the 6th European Conference on Antennas and Propagation (EUCAP), Prague, Czech Republic, 2012, pp. 120–122.

65 J. T. Aberle, "Two-Port Representation of an Antenna with Application to Non-Foster Matching Networks," *IEEE Transactions on Antennas and Propagation*, Vol. 56, No. 5, pp. 1218–1222, May 2008.

Index

*The Physics and Mathematics of Electromagnetic Wave Propagation
in Cellular Wireless Communication*, First Edition. Tapan K. Sarkar,
Magdalena Salazar Palma, and Mohammad Najib Abdallah.
© 2018 John Wiley & Sons, Inc. Published 2018 by John Wiley & Sons, Inc.